SPREAD SPECTRUM COMMUNICATIONS

Volume II

Marvin K. Simon
Jet Propulsion Laboratory

Jim K. Omura
University of California

Robert A. Scholtz
University of Southern California

Barry K. Levitt
Jet Propulsion Laboratory

COMPUTER SCIENCE PRESS

Computer Science Press
11 Taft Court
Rockville, Maryland 20850

1 2 3 4 5 6   Printing   Year   89 88 87 86 85

Library of Congress Cataloging in Publication Data
Main entry under title:

Spread spectrum communications (V.II)
    Bibliography: p.
    Includes index.
    1. Spread spectrum communications—Collected works.
I. Simon, Marvin.
TK5102.5.S6662 1985   621.38′0413   84-4959
ISBN 0-88175-014-X (Vol. II)
ISBN 0-88175-017-4 (Set)

VOLUME II
CONTENTS

Contents to Other Volumes

VOLUME I
CONTENTS

Contents to Other Volumes

VOLUME III
CONTENTS

Preface

PART 5   SPECIAL TOPICS

PREFACE

Not more than a decade ago, the discipline of spread-spectrum (SS) communications was primarily cloaked in secrecy. Indeed, most of the information available on the subject at that time could be found only in documents of a classified nature.

Today the picture is noticeably changed. The open literature abounds with publications on SS communications, special issues of the *IEEE Transactions on Communications* have been devoted to the subject, and the formation of an annual conference on military communications, MILCOM, now offers a public forum for presentation of unclassified (as well as classified) papers dealing with SS applications in military systems. On a less formal note, many tutorial and survey papers have recently appeared in the open literature, in addition to which presentations on a similar level have taken place at major communications conferences. Finally, as further evidence we cite the publication of several books dealing either with SS communications directly or as part of the more general electronic countermeasures (ECM) and electronic counter-counter measures (ECCM) problem. References to all these forms of public documentation are given in Section 1.7 of Chapter 1, Volume I.

The reasons behind this proliferation can be traced to many sources. While it is undoubtedly true that the primary application of SS communications is still in the development of enemy jam-resistant communication systems for the military, a large part of which takes place within the confines of classified programs, the emergence of other applications in both the military and civilian sectors is playing a role of ever-increasing importance. For example, to minimize mutual interference, the flux density of transmissions from radio transmitters often must be maintained at acceptably low radiation levels. A convenient way of meeting these requirements is by spreading the power spectrum of the signal before transmission and despreading it after reception. This is the non-hostile equivalent of the military low-probability-of-intercept (LPI) signal design.

Another instance where SS techniques are particularly useful in a non-anti-jam application is in the area of multiple-access communications wherein many users desire to share a single communication channel. Here the assignment of a unique SS sequence to each user allows him or her to simultaneously transmit over the common channel with a minimum of mutual interference. This often simplifies the network control requirements to coordinate users of the available channel capacity.

Still another example is the requirement for extremely accurate position location using several satellites in synchronous and asynchronous orbits. Here, satellites transmitting pseudorandom noise sequences modulated onto the transmitted carrier signal provide the means for accomplishing the required range and distance determination at any point on the earth.

Finally, SS techniques offer the advantage of improved reliability of transmission in frequency-selective fading and multipath environments. Here the improvement stems from the fact that spreading the information bandwidth of the transmitted signal over a wide range of frequencies reduces its vulnerability to interference located in a narrow frequency band and often provides some diversity gain at the receiver.

At the heart of all these potential applications lies the increasing use of digital forms of modulation for transmitting information, which itself is driven by the tremendous advances that have been made over the last decade in microelectronics. No doubt this trend will continue, and thus it should not be surprising that more and more applications for spread-spectrum techniques will continue to surface. Indeed the state-of-the-art is advancing so rapidly (e.g., witness the recent improvements in frequency synthesizers boosting frequency hop rates from the Khops/sec to the Mhops/sec ranges over SS bandwidths in excess of a GHz) that today's primarily theoretical concepts in a particular situation will be realized in practice tomorrow.

Unclassified research and developments in spread-spectrum communications have reached a point of maturity necessary to justify a textbook on SS communications that goes far beyond the level of those available on today's market. Such is the purpose of *Spread Spectrum Communications*. Contained within the fourteen chapters of its three volumes is an in-depth treatment of SS communications that should appeal to the specialist already familiar with the subject as well as the neophyte with little or no background in the area. The book is organized into five parts within which the various chapters are for the most part self-contained. The exception to this is that Chapter 3, Volume I dealing with basic concepts and system models is a basis for many of the other chapters that follow it. As would be expected, the more traditional portions of the subject are treated in the first two parts, while the latter three parts deal with the more specialized aspects. Thus the authors envision that an introductory one-semester course in SS communications to be taught on a graduate level in a university might cover all or parts of Chapters 1, 3, 4, 5 of Volume I, Chapters 1 and 2 of Volume II, and Chapters 1 and 2 of Volume III.

In composing the technical material presented in *Spread Spectrum Communications*, the authors have intentionally avoided referring by name to specific modern SS systems that employ techniques such as those discussed in many of the chapters. Such a choice was motivated by the desire to offer a unified approach to the subject that stresses fundamental principles rather than specific applications. Nevertheless, the reader should feed confident that the broad experience of the four authors ensures that the material is practically significant as well as academically inspiring.

In writing a book of this magnitude, we acknowledge many whose efforts should not go unnoticed either by virtue of a direct or indirect contribution. Credit is due to Paul Green for originally suggesting the research that uncovered the material in Chapter 2, Volume I, and Bob Price for tireless sleuthing which led to much of the remarkable information presented there. Chapter 5, Volume I benefitted significantly from the comments of Lloyd Welch, whose innovative research is responsible for some of the elegant sequence designs presented there. Per Kullstam helped clarify the material on DS/BPSK analysis in Chapter 1, Volume II. Paul Crepeau contributed substantially to the work on list detectors. Last, but by no means least, the authors would like to thank James Springett, Gaylord Huth, and Richard Iwasaki for their contribution to much of the material presented in Chapter 4, Volume III.

Several colleagues of the authors have aided in the production of a useful book by virtue of critical reading and/or proofing. In this regard, the efforts of Paul Crepeau, Larry Hatch, Vijay Kumar, Sang Moon, Wei-Chung Peng, and Reginaldo Polazzo, Jr. are greatly appreciated.

It is often said that a book cannot be judged by its cover. The authors of *Spread Spectrum Communications* are proud to take exception to this commonly quoted cliche. For the permission to use the historically significant noise-wheel cover design (see Chapter 2, Volume I, Section 2.2.5), we gratefully acknowledge the International Telephone and Telegraph Corp.

Marvin K. Simon
Jim K. Omura
Robert A. Scholtz
Barry K. Levitt

To

Sidney, Belle, Anita, Brette, and Jeffrey Simon
Shomatsu and Shizuko Omura
Lolly, Michael, and Paul Scholtz
Beverly Kaye

for a variety of reasons known only to the authors

Part 2

CLASSICAL SPREAD-SPECTRUM COMMUNICATIONS

Chapter 1

COHERENT DIRECT-SEQUENCE SYSTEMS

In its most general form a direct-sequence spread anti-jam communication system takes a binary data sequence and multiplies it by a higher rate pseudorandom (PN) binary sequence. The result is a binary sequence at the PN binary sequence rate which is then modulated. Compared to the usual modulation of the data, the data multiplied by the PN sequence causes the modulated signal spectrum to spread by a factor of N, the ratio of the PN sequence bit rate to the data bit rate. Such systems have been discussed in recent books and proceedings on spread-spectrum communications [1]–[11] and in tutorial papers [12]–[19]. Clark and Cain [9] provide the most complete discussion.

Figure 1.1 illustrates the general direct-sequence spread modulation. The data waveform is given by

$$d(t) = d_n, \qquad nT_b \leq t < (n+1)T_b$$

$$d_n \in \{-1, 1\}$$

$$n = \text{integer} \tag{1.1}$$

where $\{d_n\}$ is the data sequence. The PN binary waveform is

$$c(t) = c_k, \qquad kT_c \leq t < (k+1)T_c$$

$$c_k \in \{-1, 1\}$$

$$k = \text{integer} \tag{1.2}$$

where $\{c_k\}$ is the PN sequence. Here,

$$N = \frac{T_b}{T_c} \tag{1.3}$$

is the signal spectrum-spreading factor. Typically, N is on the order of 1000 or more. T_b is the bit time and T_c is referred to as the "chip" time interval.

In the above multiplication of the data and the PN binary sequence, it is important that the data clock and the PN sequence clock are coincident [12]. That is, the data transition times must be at the transition time of a PN

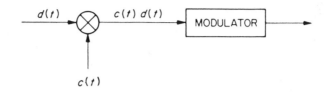

Figure 1.1.   Direct-sequence modulation.

sequence binary symbol. Figure 1.2 shows examples of data multiplied by a PN sequence for $N = 10$ where in (a) we do not have coincidence of the two sequence clocks. The problem here is that it may be possible for anyone to read the data directly from a clean copy of the multiplied binary sequence even when the PN sequence is unknown. One can first estimate the PN sequence clock (see (c)) and then determine unscheduled transitions (solid arrows in Figure 1.2(a)) in the sequence which must be due to the data. In (b) we show coincident data and PN sequence clocks where it is impossible to read the data without knowledge of the PN sequence. In (c) we show the clock times for the PN sequence.

In this book it is assumed that the data clock is divided down from the PN sequence clock so that possible transition times in the data line up with transition times of the PN sequence and no unscheduled transitions occur. Systems which have coincident data and PN sequence clocks are often said to have a data "privacy" feature since the data is hidden by the PN sequence.

The most common data modulation is coherent binary phase-shift-keying (BPSK) for direct-sequence spread systems. This coherent direct-sequence

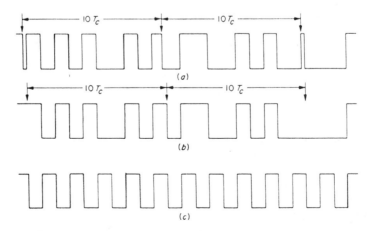

Figure 1.2.   Data times PN sequence with (a) no coincident, (b) coincident, and (c) PN clocks.

system is the one considered in this chapter. At the receiver ideal chip and phase synchronization is assumed. Acquisition and synchronization techniques are considered in Volume III.

1.1 DIRECT-SEQUENCE SPREAD COHERENT BINARY PHASE-SHIFT-KEYING

The simplest form of direct-sequence spread-spectrum communication systems uses coherent binary phase-shift-keying (BPSK) data modulation and binary PN modulation. However, the most common form employs BPSK data modulation and quaternary phase-shift-keying (QPSK) PN modulation. This has some potential advantage against a single tone jammer at the transmitted signal carrier frequency. It forces the jammer power to be evenly distributed over the cosine and sine signal coordinates.

Section 3.4 of Chapter 3, Volume I, describes the uncoded direct-sequence spread binary phase-shift-keying (DS/BPSK) signal which is given by

$$x(t) = c(t)\,d(t)\sqrt{2S}\,\cos\omega_0 t$$
$$= c(t)s(t) \tag{1.4}$$

where

$$s(t) = d(t)\sqrt{2S}\,\cos\omega_0 t \tag{1.5}$$

is the unspread BPSK signal. Defining T_b as the data bit time interval and T_c as the PN sequence bit time interval, $s(t)$ has a $\sin^2 x/x^2$ spectrum of bandwidth roughly $1/T_b$, while the spread-spectrum signal $x(t)$ has a similar-shaped spectrum but of bandwidth roughly

$$W_{ss} = 1/T_c. \tag{1.6}$$

The processing gain is

$$\mathrm{PG} = \frac{W_{ss}}{R_b}$$
$$= \frac{T_b}{T_c}$$
$$= N. \tag{1.7}$$

The jamming signal is represented by $J(t)$ and in the absence of noise (assume the jammer limits performance) the signal at the receiver is

$$x(t) + J(t). \tag{1.8}$$

The receiver multiplies this by the PN waveform to obtain the signal[1]

$$r(t) = c(t)[x(t) + J(t)]$$
$$= s(t) + c(t)J(t) \tag{1.9}$$

---

[1] Perfect chip synchronization is assumed for the intended receiver. Chip sequence acquisition and synchronization techniques are presented in Chapters 1 and 2, Volume III.

since

$$c^2(t) = 1. \tag{1.10}$$

Here $c(t)J(t)$ is the effective noise waveform due to jamming.

The conventional BPSK detector (see Figure 3.6 of Chapter 3, Volume I) output is thus

$$r = d\sqrt{E_b} + n \tag{1.11}$$

where d is the data bit for the T_b second interval, $E_b = ST_b$ is the bit energy, and n is the equivalent noise component given by

$$n = \sqrt{\frac{2}{T_b}} \int_0^{T_b} c(t)J(t)\cos\omega_0 t\, dt. \tag{1.12}$$

For QPSK modulation the inphase and quadrature data waveforms are denoted $d_c(t)$ and $d_s(t)$, respectively, and corresponding PN binary waveforms are $c_c(t)$ and $c_s(t)$. These form a QPSK signal

$$x(t) = c_c(t)d_c(t)\sqrt{S}\,\cos\omega_0 t + c_s(t)d_s(t)\sqrt{S}\,\sin\omega_0 t \tag{1.13}$$

where each QPSK pulse is of duration $T_s = 2T_b$. The channel output waveform given by (1.8) is multiplied by $\sqrt{2/T_s}\,c_c(t)\cos\omega_0 t$ and integrated over T_s seconds to obtain the inphase component

$$r_c = d_c\sqrt{E_b} + n_c \tag{1.14}$$

where

$$n_c = \sqrt{\frac{2}{T_s}} \int_0^{T_s} c_c(t)J(t)\cos\omega_0 t\, dt. \tag{1.15}$$

Here we assume that[2]

$$\int_\tau^{\tau + T_c} \cos\omega_0 t \sin\omega_0 t\, dt = 0 \tag{1.16}$$

for all τ where T_c is the "chip" time interval for both PN sequences. Similar results hold for the quadrature component

$$r_s = d_s\sqrt{E_b} + n_s \tag{1.17}$$

where

$$n_s = \sqrt{\frac{2}{T_s}} \int_0^{T_s} c_s(t)J(t)\sin\omega_0 t\, dt. \tag{1.18}$$

Thus, as in the conventional case, QPSK modulation can be viewed as two independent BPSK modulations each at half the data rate.

---

[2] This is the usual assumption that inphase and quadrature signal components are orthogonal. Perfect phase synchronization is assumed for the receiver. Orthogonality is approximately true for the weaker condition $W_{ss} < f_0 = \omega_0/2\pi$.

1.2 UNCODED BIT ERROR PROBABILITY FOR ARBITRARY JAMMER WAVEFORMS

Without coding, the BPSK detector decision rule is to decide $\hat{d}$ is the data bit where

$$\hat{d} = \begin{cases} 1, & \text{if } r \geq 0 \\ -1, & \text{if } r < 0 \end{cases} \qquad (1.19)$$

The bit error probability is thus

$$P_b = \Pr\{r \geq 0 | d = -1\}$$
$$= \Pr\left\{n \geq \sqrt{E_b}\right\}. \qquad (1.20)$$

Naturally, this bit error probability depends on the random variable n given by (1.12). For QPSK the same decision rule (1.19) can be applied to r_c and r_s to obtain $\hat{d}_c$ and $\hat{d}_s$. This results in the same bit error probability expression (1.20) with n_c and n_s replacing n.

For BPSK modulation the noise component given by (1.12) has the form

$$n = \sqrt{\frac{2}{T_b}} \sum_{k=0}^{N-1} \int_{kT_c}^{(k+1)T_c} c(t) J(t) \cos \omega_0 t \, dt$$
$$= \sqrt{\frac{2}{T_b}} \sum_{k=0}^{N-1} c_k \int_{kT_c}^{(k+1)T_c} J(t) \cos \omega_0 t \, dt \qquad (1.21)$$

where $c_0, c_1, \ldots, c_{N-1}$ are the N PN bits occurring during the data bit time interval. Defining the jamming component

$$J_k = \sqrt{\frac{2}{T_c}} \int_{kT_c}^{(k+1)T_c} J(t) \cos \omega_0 t \, dt \qquad (1.22)$$

we have

$$n = \sqrt{\frac{1}{N}} \sum_{k=0}^{N-1} c_k J_k \qquad (1.23)$$

as the final form for the noise component.

The PN sequence is approximated as an independent identically distributed binary sequence where

$$\Pr\{c_k = 1\} = \Pr\{c_k = -1\} = \tfrac{1}{2}. \qquad (1.24)$$

Then for any fixed jammer sequence

$$J = (J_0, J_1, \ldots, J_{N-1}) \qquad (1.25)$$

the noise component given by (1.23) is a sum of independent random variables. We next examine ways of evaluating the conditional bit error probability

$$P_b(J) = \Pr\left\{n \geq \sqrt{E_b} \middle| J\right\} \qquad (1.26)$$

for given jammer components J. The final bit error probability may be in terms of a parameter set characterizing a deterministic jammer model or a statistical characterization of the jammer with evaluation of the overall average bit error probability by

$$P_b = E\{P_b(J)\} \tag{1.27}$$

where the expectation is over the jammer statistics.

1.2.1 Chernoff Bound

For a given jammer coordinate sequence J fixed consider the Chernoff bound to the bit error probability of (1.26) as follows:

$$
\begin{aligned}
P_b(J) &= \Pr\left\{ n \geq \sqrt{E_b} \,\middle|\, J \right\} \\
&= \Pr\left\{ n - \sqrt{E_b} \geq 0 \,\middle|\, J \right\} \\
&\leq E\left\{ e^{\lambda[n - \sqrt{E_b}]} \,\middle|\, J \right\} \\
&= e^{-\lambda\sqrt{E_b}} E\left\{ \exp\left[\lambda\sqrt{\frac{1}{N}} \sum_{k=0}^{N-1} c_k J_k \right] \,\middle|\, J \right\} \\
&= e^{-\lambda\sqrt{E_b}} \prod_{k=0}^{N-1} E\left\{ \exp\left[\lambda\sqrt{\frac{1}{N}} c_k J_k \right] \,\middle|\, J_k \right\} \\
&= e^{-\lambda\sqrt{E_b}} \prod_{k=0}^{N-1} \left\{ \tfrac{1}{2} e^{\lambda\sqrt{1/N} J_k} + \tfrac{1}{2} e^{-\lambda\sqrt{1/N} J_k} \right\} \\
&= e^{-\lambda\sqrt{E_b}} \prod_{k=0}^{N-1} \cosh\left(\frac{\lambda J_k}{\sqrt{N}} \right)
\end{aligned}
\tag{1.28}
$$

for any $\lambda \geq 0$. Next, following the approach of Kullstam [20], [21], use the inequality

$$\cosh x \leq e^{x^2/2} \tag{1.29}$$

to obtain the form

$$P_b(J) \leq e^{-\lambda\sqrt{E_b}} e^{(\lambda^2/2N)\sum_{k=0}^{N-1} J_k^2}. \tag{1.30}$$

The $\lambda \geq 0$ that minimizes this Chernoff bound is

$$\lambda^* = \sqrt{E_b} \middle/ \left(\frac{1}{N} \sum_{k=0}^{N-1} J_k^2 \right) \tag{1.31}$$

giving the result

$$P_b(J) \leq \exp\left\{ -E_b \middle/ \left(\frac{2}{N} \sum_{k=0}^{N-1} J_k^2 \right) \right\}. \tag{1.32}$$

In addition, note that a factor of $1/2$ can be applied[3] to the above Chernoff bound (see Appendix 4B, Chapter 4, Volume I). The final Chernoff bound is then

$$P_b(J) \le \frac{1}{2} \exp\left\{ -E_b \middle/ \left(\frac{2}{N} \sum_{k=0}^{N-1} J_k^2 \right) \right\}. \tag{1.33}$$

This bound applies for all N and J and only assumes the PN sequence $\{c_k\}$ is an i.i.d sequence of binary symbols equally likely to be 1 or -1.

1.2.2 Gaussian Assumption

The Central Limit Theorem [22] states that the normalized sum of independent random variables approaches (in distribution) a Gaussian random variable as the number of terms increases. This basically applies to cases where the variances of the terms are more or less evenly distributed over some bounded range. For fixed J the Gaussian assumption means n is a Gaussian random variable with zero mean since for each k

$$E\{c_k J_k | J_k\} = \tfrac{1}{2} J_k - \tfrac{1}{2} J_k$$
$$= 0, \tag{1.34}$$

and variance given by

$$\text{Var}\{n|J\} = \frac{1}{N} \sum_{k=0}^{N-1} J_k^2. \tag{1.35}$$

The conditional bit error probability is then

$$P_b(J) = Q\left(\sqrt{E_b \middle/ \left(\frac{1}{N} \sum_{k=0}^{N-1} J_k^2 \right)} \right). \tag{1.36}$$

This approximation to the conditional bit error probability applies for large values of N, which is typically the case with direct-sequence spread systems.
Using the inequality

$$Q(x) \le \frac{1}{2} e^{-x^2/2}, \qquad x \ge 0 \tag{1.37}$$

we have

$$P_b(J) \le \frac{1}{2} \exp\left\{ -E_b \middle/ \left(\frac{2}{N} \sum_{k=0}^{N-1} J_k^2 \right) \right\}. \tag{1.38}$$

This inequality shows that in applying the Chernoff bound *assuming n is a Gaussian random variable*, we obtain exactly the same result as the general Chernoff bound derived in the previous section where no Gaussian assump-

---

[3]Although this factor of $1/2$ does not always apply in theory, it is a realistic approximation for all practical cases of interest.

tions were made. *Thus, the Gaussian assumption results in a bit error probability that is appropriate for direct-sequence spread systems for any spread factor N.*

1.3 UNCODED BIT ERROR PROBABILITY FOR SPECIFIC JAMMER WAVEFORMS

Consider the set of orthonormal basis

$$\phi_k(t) = \begin{cases} \sqrt{\dfrac{2}{T_c}} \cos \omega_0 t, & kT_c \leq t < (k+1)T_c \\ 0, & \text{elsewhere} \end{cases}$$

$$\tilde{\phi}_k(t) = \begin{cases} \sqrt{\dfrac{2}{T_c}} \sin \omega_0 t, & kT_c \leq t < (k+1)T_c \\ 0, & \text{elsewhere} \end{cases}$$

$$k = 0, 1, \ldots, N-1. \tag{1.39}$$

The transmitted direct-sequence spread BPSK signal during the data bit time interval $[0, T_b]$ has the form

$$x(t) = c(t)\, d(t)\sqrt{2S} \cos \omega_0 t$$

$$= d\sqrt{ST_c} \sum_{k=0}^{N-1} c_k \phi_k(t)$$

$$0 \leq t \leq T_b. \tag{1.40}$$

The jammer would like to place all of its signal energy into the same signal space as the transmitted signal, otherwise its power would be wasted. It, however, can only know the signal bandwidth and not the signal phase. Thus, in general it has the form

$$J(t) = \sum_{k=0}^{N-1} J_k \phi_k(t) + \sum_{k=0}^{N-1} \tilde{J}_k \tilde{\phi}_k(t) \tag{1.41}$$

where recall $\boldsymbol{J} = (J_0, J_1, \ldots, J_{N-1})$ consists of the cosine components. Note that only these cosine components enter into the bit error probability bound. The total jammer signal energy during the data bit time is

$$JT_b = \int_0^{T_b} J^2(t)\, dt$$

$$= \sum_{k=0}^{N-1} J_k^2 + \sum_{k=0}^{N-1} \tilde{J}_k^2 \tag{1.42}$$

or

$$N_J = \frac{1}{N} \sum_{k=0}^{N-1} J_k^2 + \frac{1}{N} \sum_{k=0}^{N-1} \tilde{J}_k^2. \tag{1.43}$$

The worst case jammer is one that places all its energy in the cosine coordinates so that

$$\sum_{k=0}^{N-1} J_k^2 = JT_b.$$ (1.44)

Normally, however, the jammer can only place equal energy in the cosine and sine coordinates with the result that

$$\sum_{k=0}^{N-1} J_k^2 = \frac{JT_b}{2}.$$ (1.45)

Figure 1.3 shows the bit error probability for the Gaussian assumption and

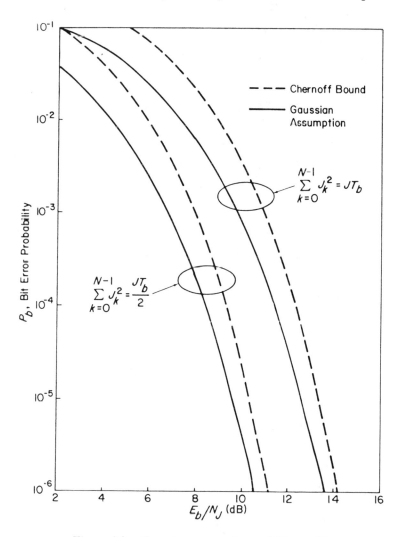

Figure 1.3. Gaussian assumption and Chernoff bound.

the general Chernoff bound for these two cases. Note that there is only about a one dB difference for fixed bit error rates. This suggests that the Gaussian assumption is reasonable for all N.

The impact of jamming signals by characterizing two types of jammers will now be illustrated. The first model assumes a deterministic jammer waveform that is characterized by a set of parameters θ. The second model assumes the jammer is a stationary random process with statistical characterizations. In both cases the jammer is assumed to be transmitting continuously with constant power J. In the next section pulse jamming will be examined along with its generalization where the jammer power can be varied in time while maintaining a time-averaged power J.

1.3.1 CW Jammer

The most harmful jammer waveform is one that maximizes J_k given by (1.22) for each value of k. Since the jammer does not know the PN sequence $\{c_k\}$ this means the jammer should place as much energy as possible in the cosine coordinate which is achieved with a CW signal. Generally, the jammer may not know the transmitted signal carrier phase so consider the deterministic jammer waveform model

$$J(t) = \sqrt{2J}\,\cos[\omega_0 t + \theta]\qquad(1.46)$$

which is characterized by the phase parameter θ. Thus

$$J_k = \sqrt{JT_c}\,\cos\theta,\quad\text{all }k\qquad(1.47)$$

which is maximized when $\theta = 0$.

For this CW jammer the conditional variance of n given in (1.35) is

$$\text{Var}\{n|J\} = JT_c\cos^2\theta.\qquad(1.48)$$

For the Gaussian approximation (Central Limit Theorem applied for large $N = T_b/T_c$) the conditional bit error probability is

$$P_b(\theta) = P_b(J)$$

$$= Q\left(\sqrt{\frac{E_b}{JT_c\cos^2\theta}}\right).\qquad(1.49)$$

Using $JT_c = N_J$ this becomes

$$P_b(\theta) = Q\left(\sqrt{\frac{E_b}{N_J\cos^2\theta}}\right)\qquad(1.50)$$

for the CW jammer. The choice of $\theta = 0$ maximizes $P_b(\theta)$ yielding the bound

$$P_b(\theta) \le Q\left(\sqrt{\frac{E_b}{N_J}}\right).\qquad(1.51)$$

This upper bound is the same as that for binary orthogonal signals in an additive white Gaussian noise channel of single-sided spectral density N_J. For a jammer with constant power J this is the worst performance of the direct-sequence spread BPSK system.

Against a CW jammer an effective technique is to use BPSK data modulation with QPSK PN spreading. This is a DS/BPSK signal of the form

$$x(t) = d(t)\sqrt{S}\left[c_c(t)\cos \omega_0 t + c_s(t)\sin \omega_0 t\right] \qquad (1.52)$$

where $c_c(t)$ and $c_s(t)$ are PN waveforms. This is the special case of $d_c = d_s = d$ in the QPSK modulation given in (1.13) where the QPSK symbol energy is also the bit energy (one bit per QPSK signal). For this case (1.14) and (1.17) have the form

$$r_c = d\sqrt{E_b/2} + n_c \qquad (1.53)$$

and

$$r_s = d\sqrt{E_b/2} + n_s \qquad (1.54)$$

where n_c and n_s are zero mean independent with conditional variances

$$\mathrm{Var}\{n_c|\theta\} = JT_c\cos^2\theta \qquad (1.55)$$

and

$$\mathrm{Var}\{n_s|\theta\} = JT_c\sin^2\theta. \qquad (1.56)$$

Next use

$$r = \frac{r_c + r_s}{2} - d\sqrt{E_b/2} + \frac{n_c + n_s}{2} \qquad (1.57)$$

as the statistic for the decision rule given in (1.19). Since

$$\mathrm{Var}\left\{\frac{n_c + n_s}{2}\Big|\theta\right\} = \frac{1}{4}\left\{JT_c\cos^2\theta + JT_c\sin^2\theta\right\}$$

$$= \frac{JT_c}{4}, \qquad (1.58)$$

the bit error probability for the Gaussian approximation is

$$P_b(\theta) = Q\left(\sqrt{\frac{2E_b}{N_J}}\right). \qquad (1.59)$$

This is independent of θ and is a 3 dB improvement over the worst case ($\theta = 0$) BPSK PN spreading system. Thus, to minimize the maximum possible degradation due to a CW jammer, one should use QPSK modulation with the same data in both inphase and quadrature coordinates. (For further discussion on this see Kullstam [20].)

1.3.2 Random Jammer

Consider next characterizing the jammer as a stationary random process with autocorrelation

$$R_J(\tau) = E\{ J(t + \tau) J(t) \} \tag{1.60}$$

and power spectral density

$$S_J(f) = \int_{-\infty}^{\infty} R_J(\tau) e^{-j2\pi f\tau} \, d\tau \tag{1.61}$$

where

$$\int_{-\infty}^{\infty} S_J(f) \, df = J, \tag{1.62}$$

the constant power of the jammer.

The PN waveform $c(t)$ is also stationary (introduce a uniformly distributed time shift) with autocorrelation

$$R_c(\tau) = E\{ c(t + \tau) c(t) \} = \begin{cases} 1 - \dfrac{|\tau|}{T_c}, & |\tau| \le T_c \\ 0, & |\tau| > T_c \end{cases} \tag{1.63}$$

and power spectral density

$$S_c(f) = \int_{-\infty}^{\infty} R_c(\tau) e^{-j2\pi f\tau} \, d\tau. \tag{1.64}$$

Naturally, $J(t)$ and $c(t)$ are independent of each other.

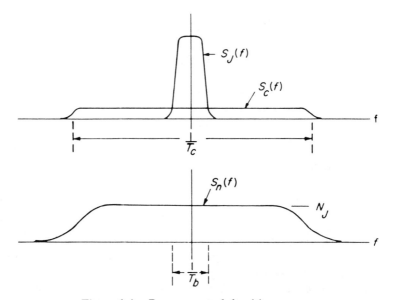

Figure 1.4. Power spectral densities.

Recall from (1.9) that when received waveform $x(t) + J(t)$ is multiplied by the PN waveform $c(t)$, the resulting noise term is

$$n(t) = c(t)J(t). \tag{1.65}$$

Since $J(t)$ and $c(t)$ are independent, the autocorrelation of $n(t)$ is

$$\begin{aligned}
R_n(\tau) &= E\{n(t + \tau)n(t)\} \\
&= E\{c(t + \tau)c(t)\}E\{J(t + \tau)J(t)\} \\
&= R_c(\tau)R_J(\tau),
\end{aligned} \tag{1.66}$$

which has the power spectral density

$$S_n(f) = S_c(f) * S_J(f) \tag{1.67}$$

where $*$ indicates the convolution operation.

Figure 1.4 illustrates the power spectra of $c(t)$, $J(t)$, and the product $n(t) = c(t)J(t)$. Here $S_J(f)$ is arbitrary and $S_c(f)$ is a broad $\sin^2 x/x^2$ spectrum of bandwidth roughly $1/T_c$. The resulting noise spectrum $S_n(f)$ has value at $f = 0$ given by

$$\begin{aligned}
S_n(0) &= \int_{-\infty}^{\infty} S_c(f)S_J(f)\,df \\
&\leq S_c(0)\int_{-\infty}^{\infty} S_J(f)\,df \\
&= S_c(0)J
\end{aligned} \tag{1.68}$$

since

$$S_c(f) \leq S_c(0) \tag{1.69}$$

and J is the total jammer power given by (1.62). There is equality in (1.68) when $J(t)$ has a narrow power density spectrum compared to the PN waveform $c(t)$. In addition

$$\begin{aligned}
S_c(0) &= \int_{-\infty}^{\infty} R_c(\tau)\,d\tau \\
&= \int_{-T_c}^{T_c}\left(1 - \frac{|\tau|}{T_c}\right)d\tau \\
&= T_c.
\end{aligned} \tag{1.70}$$

Thus

$$\begin{aligned}
S_n(0) &\leq JT_c \\
&= N_J
\end{aligned} \tag{1.71}$$

where W_{ss} is given by (1.6) and N_J is our usual definition given as J/W_{ss}.

Note that the equivalent noise of power spectral density bounded by N_J represents the total interference power. If we were to divide this power equally between the sine and cosine coordinates of the narrowband (unspread) signal, then each coordinate would have a noise component of variance less than or equal to $N_J/2$. With BPSK data modulation and QPSK PN spreading, equal distribution of the jammer power is guaranteed. In this case the jammer appears as white noise of power spectral density bounded by $N_J/2$.

Multiplying the channel output signal by the PN waveform results in the narrowband signal $s(t)$ in broadband noise $n(t)$. Here $s(t)$ has bandwidth $1/T_b$ while $n(t)$ has bandwidth greater than $W_{ss} = 1/T_c = N/T_b$ which is N times wider than the signal bandwidth. Since $S_n(f)$ is essentially flat over the narrowband signal bandwidth, the detection problem reduces to demodulation of a BPSK signal in white noise of double-sided power spectral density less than or equal to $N_J/2$. Thus, making the Gaussian approximation, the uncoded bit error probability is bounded by

$$P_b \leq Q\left(\sqrt{\frac{2E_b}{N_J}} \right). \tag{1.72}$$

Here equality is achieved for narrowband jamming signals. This gives the same result as with a CW jammer and QPSK modulation where the same data is entered in both the inphase and quadrature components (BPSK data modulation with QPSK PN spreading). It also is the baseline performance where the jammer simply transmits broadband Gaussian noise.

Based on these results for CW and random jammers, we would expect the direct-sequence BPSK anti-jam systems to be robust and insensitive to all jammers that produce waveforms which are independent of the transmission. They appear to always give as good a performance as the baseline jammer case and that is all one can expect from a good anti-jam communication system. This, however, is true only for constant power jammers.

1.4 PULSE JAMMING

Now consider the impact of pulse jamming and its generalization where the jammer can arbitrarily distribute its energy in time under a long time average power constraint. The worst time distribution of energy by the jammer is the two-level jammer corresponding to pulse jamming. This distribution results in considerable degradation to the uncoded DS/BPSK system.

1.4.1 Arbitrary Time Distribution

Consider an arbitrary continuous time distribution of the jammer power which is approximated as having L discrete levels $\hat{J}_1, \hat{J}_2, \ldots, \hat{J}_L$ with corresponding fraction of time occurrences given by $\rho_1, \rho_2, \ldots, \rho_L$. Thus, assume that at any given time the jammer power is $\hat{J}_l$ with probability ρ_l for

$l = 1, 2, \ldots, L$ where

$$J = \sum_{l=1}^{L} \rho_l \hat{J}_l \tag{1.73}$$

is the average power constraint. Also during the transmission time of a data bit, T_b, the jammer power is assumed to be constant with value equal to one of the L levels. Thus compared to T_b, the jammer power is slowly varying. Even with faster jammer power time variations this assumption is reasonable, since the receiver correlates or integrates the received channel output signal over every T_b seconds.

It is convenient to define

$$\gamma_l = \rho_l \frac{\hat{J}_l}{J}; \quad l = 1, 2, \ldots, L \tag{1.74}$$

where the average power constraint (1.73) becomes

$$\sum_{l=1}^{L} \gamma_l = 1. \tag{1.75}$$

For a DS/BPSK system when the jammer is transmitting with power $\hat{J}_l$ the uncoded bit error bound is analogous to (1.36)

$$P_b(\hat{J}_l) \leq Q\left(\sqrt{\frac{2E_b W_{ss}}{\hat{J}_l}}\right)$$

$$= Q\left(\sqrt{\frac{2\rho_l E_b}{\gamma_l N_J}}\right). \tag{1.76}$$

For CW jamming assume the same data is sent in both the inphase and quadrature components of a direct-sequence spread QPSK system. Next, using the inequality (1.37) we obtain

$$P_b(\hat{J}_l) \leq \tfrac{1}{2} e^{-\rho_l E_b/(\gamma_l N_J)} \tag{1.77}$$

and the average uncoded bit error rate bound

$$P_b = \sum_{l=1}^{L} \rho_l P_b(\hat{J}_l)$$

$$\leq \tfrac{1}{2} \sum_{l=1}^{L} \rho_l e^{-\rho_l E_b/(\gamma_l N_J)}. \tag{1.78}$$

Using the inequality

$$\rho e^{-\rho A} \leq \max_{\rho'} \rho' e^{-\rho' A} = \frac{e^{-1}}{A} \tag{1.79}$$

(1.78) is further bounded by

$$P_b \leq \frac{1}{2} \sum_{l=1}^{L} \gamma_l \frac{e^{-1}}{(E_b/N_J)} \tag{1.80}$$

or

$$P_b \le \frac{e^{-1}}{2(E_b/N_J)} \tag{1.81}$$

when constraint (1.75) is imposed.

The upper bound (1.81) can be achieved with two jammer power levels ($L = 2$) given by

$$\hat{J}_1 = 0$$
$$\hat{J}_2 = J/\rho \tag{1.82}$$

where

$$\rho_1 = 1 - \rho$$
$$\rho_2 = \rho \tag{1.83}$$

and

$$\rho = \frac{1}{(E_b/N_J)} \tag{1.84}$$

provided

$$E_b/N_J > 1. \tag{1.85}$$

When (1.85) is not satisfied we have the special case of constant jammer power, i.e., $\rho = 1$.

1.4.2 Worst Case Jammer

The jammer time distribution of power that maximizes the bound on the uncoded bit error probability is the pulse jammer with peak power

$$J_{\text{peak}} = J(E_b/N_J) \tag{1.86}$$

for a fraction ρ of the time where ρ is given by (1.84). This is not necessarily the jammer that maximizes the true bit error probability. Assuming a narrowband pulse jammer (or CW pulse jammer at the carrier frequency) which is on a fraction ρ of the time, the exact uncoded bit error probability is

$$P_b(\rho) = \rho Q\left(\sqrt{2\rho(E_b/N_J)}\right). \tag{1.87}$$

The choice of ρ that maximizes (1.87) is

$$\rho^* = \begin{cases} \dfrac{0.709}{E_b/N_J}, & E_b/N_J > 0.709 \\ 1, & E_b/N_J \le 0.709 \end{cases} \tag{1.88}$$

resulting in the bit error probability

$$P_b(\rho^*) = \begin{cases} \dfrac{0.083}{E_b/N_J}, & E_b/N_J > 0.709 \\ Q\left(\sqrt{\dfrac{2E_b}{N_J}}\right), & E_b/N_J \le 0.709. \end{cases} \tag{1.89}$$

The worst jammer time distribution of power will result in a bit error probability P_b^* which satisfies

$$P_b(\rho^*) \leq P_b^* \leq \begin{cases} \dfrac{e^{-1}}{2(E_b/N_J)}, & E_b/N_J > 1 \\[2mm] \tfrac{1}{2}e^{-E_b/N_J}, & E_b/N_J \leq 1. \end{cases} \tag{1.90}$$

Figure 1.5 shows the constant power bit error probability $Q(\sqrt{2E_b/N_J})$, its bound $(1/2)e^{-(E_b/N_J)}$, $P_b(\rho^*)$, and the bound given in (1.90). The cross-hatched region indicates the unknown bit error probability P_b^* of the worst case jammer with average power J. It is clear that pulse jamming can degrade the performance of uncoded DS/BPSK systems to a considerable

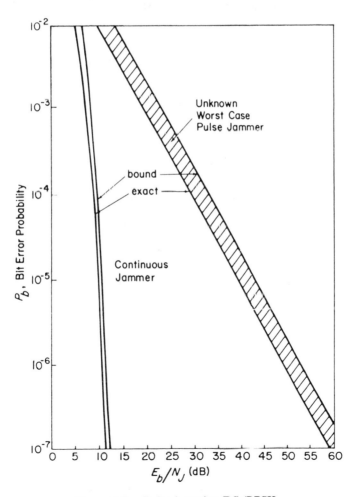

Figure 1.5. Pulse jamming DS/BPSK.

degree compared to the baseline jammer. At bit error probability of 10^{-6} there is approximately a 40 dB degradation due to pulse jamming. These jammers, however, can be effectively neutralized with coding and interleaving.

1.5 STANDARD CODES AND CUTOFF RATES

In Chapter 3, Volume I, the point was made that the use of coding with spread-spectrum systems does not require any reduction in data rate or increase in the spread-spectrum signal bandwidth. Even with simple repeat codes, the large degradation due to pulse jamming in DS/BPSK systems can be neutralized. With more powerful codes the pulse jammer neutralization plus additional coding gain can be achieved.

Recall that in the general coding analysis in Chapter 4, Volume I, the coded bit error bound was given in the form

$$P_b \le G(D) = B(R_0) \qquad (1.91)$$

where $G(\cdot)$ and $B(\cdot)$ are functions specified by the particular code, while D and R_0 are parameters of the channel. For DS/BPSK systems, D and R_0 have the relationship

$$R_0 = 1 - \log_2(1 + D) \quad \text{bits/symbol} \qquad (1.92)$$

where Figure 1.6 shows the general system block diagram. The system enclosed in dotted lines represents an equivalent binary input channel for which D and R_0 can be evaluated. In this case, a symbol entering the equivalent channel is a coded bit and the energy per coded bit (channel symbol) E_s is related to the energy per bit E_b by

$$E_s = RE_b \qquad (1.93)$$

where R is the code rate in bits per channel symbol. Parameter D, and hence parameter R_0, depend on E_s/N_J, the symbol energy to equivalent jammer noise ratio.

1.5.1 The Additive White Gaussian Noise Channel

This section summarizes well-known results for standard coherent BPSK and QPSK modulations with some binary codes and the additive white Gaussian noise channel model. These results will serve as a reference for the DS/BPSK system with jamming.

For the additive white Gaussian noise channel R_0 is given by (1.92) where

$$D = e^{-E_s/N_0} \qquad (1.94)$$

and N_0 is the single-sided noise spectral density. Coded bit error bounds of the form given by (1.91) are shown in Figure 1.7 for several binary codes.

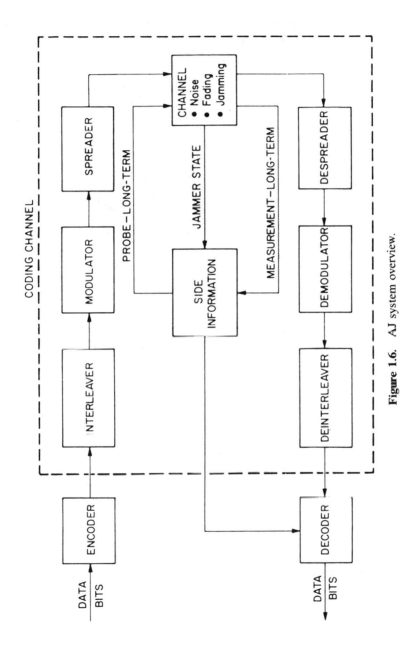

Figure 1.6.   AJ system overview.

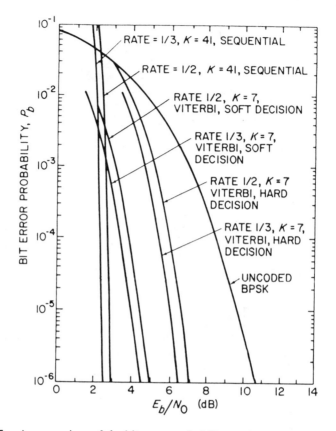

Figure 1.7. A comparison of the bit error probability performances of rate $1/2$ and $1/3$, hard and soft decision Viterbi and sequential decoders.

These curves serve as basic reference coded bit error probabilities from which we can derive coded performance for all other jamming channels. The only difference in the various jamming cases will be in the value of D and R_0 for each coded symbol to equivalent jammer noise ratio, E_b/N_J.

1.5.2 Jamming Channels

For the constant power jammer, regardless of the waveform used by the jammer, the bit error probability was bounded by the bit error probability of coherent BPSK modulation with the additive white Gaussian noise channel where $N_0 = N_J$, the equivalent jammer noise density. This assumes the Central Limit Theorem applies and in the case of CW jammers the system uses the same data in the inphase and quadrature coordinates of a direct-sequence spread QPSK modulation. In this case, the results for the additive white Gaussian noise channel shown in Figure 1.7 apply where E_b/N_0 is replaced by E_b/N_J.

The only remaining case of interest is when there is a pulse jammer that transmits pulses with power J/ρ for ρ fraction of the time. When the pulse is on, assume the jammer waveform is equivalent to additive white Gaussian noise with single-sided spectral density $N_0 = N_J/\rho$.

For pulse jamming, values of D are shown in Table 4.1 of Chapter 4, Volume I, for various cases of hard and soft decision metrics, which may or may not include jammer state information. These are all shown as a function of ρ. If for each of the four possible cases in Table 4.1, Volume I, the worst value of ρ that maximizes D is chosen, then the resulting values of the cutoff rates as a function of E_s/N_J are shown in Figure 1.8. Also shown here is the additive white Gaussian noise channel reference. For each case the worst case value of ρ is given in Figure 1.9.

To obtain the bit error bounds from standard error rate curves such as those shown in Figure 1.7 when there are worst case pulse jammers consider the constraint length $K = 7$, rate $R = 1/2$ convolutional code shown in Figure 1.7 for the additive white Gaussian noise channel with soft decision decoding. Suppose we wanted the bit error bound for this code with a hard decision decoder operating with no jammer state information and where the jammer is a worst case pulse jammer for this coded DS/BPSK system.

At 10^{-5} bit error probability, the $K = 7$, $R = 1/2$ convolutional code shown in Figure 1.7 requires

$$E_b/N_0 = 4.5 \text{ dB} \tag{1.95}$$

or since there is a rate $R = 1/2$ code, it requires

$$E_s/N_0 = 1.5 \text{ dB}. \tag{1.96}$$

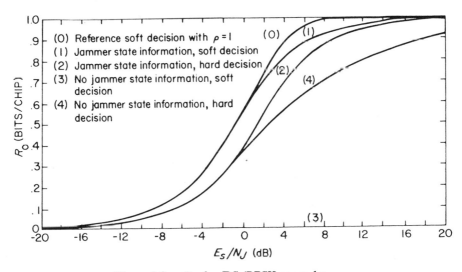

Figure 1.8. R_0 for DS/BPSK examples.

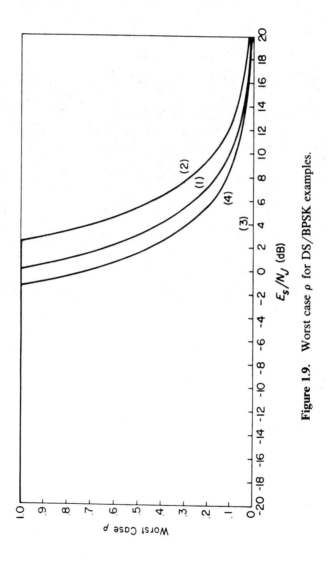

Figure 1.9. Worst case ρ for DS/BPSK examples.

From Figure 1.8 this choice of E_s/N_0 gives the cutoff rate

$$R_0 = 0.7 \qquad (1.97)$$

on the reference cutoff rate corresponding to the additive white Gaussian noise channel with soft decision decoding. This gives the value of R_0 required to achieve 10^{-5} bit error probabilities regardless of the channel. For this choice of R_0, the worst case pulse jamming channel with hard decision and no jamming state information requires (see curve labelled (4))

$$E_s/N_J = 8.0 \text{ dB} \qquad (1.98)$$

or

$$E_b/N_J = 11.0 \text{ dB}. \qquad (1.99)$$

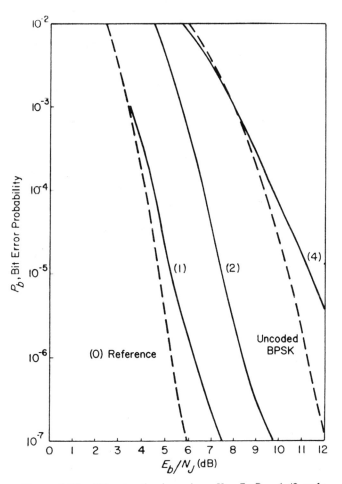

Figure 1.10. Worst pulse jamming: $K = 7$, $R = 1/2$ code.

This is the energy per bit required for this case to achieve 10^{-5} bit error probability with this code.

The above procedure repeated for various choices of the bit error probability results in the bit error bounds shown in Figure 1.10 for the reference case and three worst case pulse jammer examples. Note that the bit error probability curves shown here are shifted by exactly the same shift shown in Figure 1.8 for the cutoff rates. This is simply because for any given code, the coded bit error bound depends only on the value of R_0. Similar bit error bounds can be found for other codes by shifting the standard bit error probability curves by the same amount.

Figure 1.10 also illustrates the effectiveness of coding and interleaving in combatting pulse jamming. Comparing the coded bit error probabilities of Figure 1.10 with the uncoded case of Figure 1.5, we see that most of the large degradation due to pulse jamming has been neutralized and additional coding gain achieved. At 10^{-6} bit error probability, for example, there is about a 45 dB coding gain for the soft decision known jammer state decoder for this $K = 7$ convolutional code compared to no coding. These are against the worst case pulse jammer for each point on the bit error probability curves.

1.6 SLOW FREQUENCY NON-SELECTIVE FADING CHANNELS

Here we examine the performance of DS/BPSK systems in a fading channel [23]–[32]. The fading is assumed to be constant across the total spread spectrum bandwidth, W_{ss}, and varies slowly in time such that during any data bit time interval, T_b, the fading is assumed to be constant. This slowly varying frequency non-selective fading is sometimes referred to as flat-flat fading [23]. Although our analysis allows for arbitrary fade statistics, the examples will assume the commonly used Rayleigh fading amplitude statistics which models the case where there are many similar scatterers causing the fading [33].

1.6.1 Continuous Jammer with No Coding

Suppose the received signal amplitude is A. Then the energy per bit is

$$E_b = \frac{A^2 T_b}{2}.$$

$$(1.100)$$

Assuming the maximum-likelihood demodulator for the AWGN channel, then the uncoded bit error probability is bounded by (see (1.37) and (1.72))

$$P_b(A) = Q\left(\sqrt{\frac{A^2 T_b}{N_J}}\right) \le \frac{1}{2} e^{-A^2 T_b/(2N_J)}$$

$$(1.101)$$

where as before N_J is given by J/W_{ss} for the continuous jammer of average power J.

For a fading channel A is a random variable with some probability density function $p_A(\cdot)$ and the average uncoded bit error probability is bounded by

$$P_b = \int_0^\infty Q\left(\sqrt{\frac{a^2 T_b}{N_J}}\right) p_A(a)\, da$$

$$\leq \tfrac{1}{2} \int_0^\infty e^{-a^2 T_b/(2N_J)} p_A(a)\, da. \tag{1.102}$$

For Rayleigh statistics where

$$p_A(a) = \frac{a}{\sigma^2} e^{-a^2/(2\sigma^2)}, \qquad a \geq 0 \tag{1.103}$$

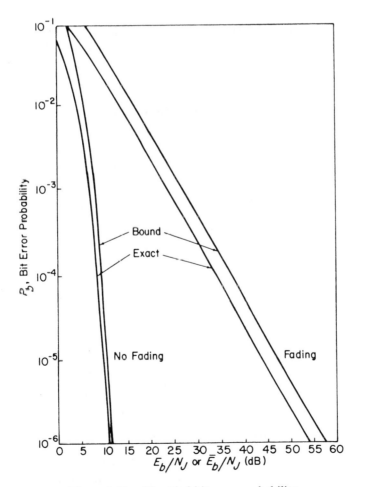

Figure 1.11. Uncoded bit error probability.

and the received signal energy has average value

$$\bar{E}_b \triangleq \frac{T_b}{2} \int_0^\infty a^2 p_A(a) \, da = \sigma^2 T_b,$$ (1.104)

the exact uncoded bit error probability is

$$P_b = \int_0^\infty Q\left(\sqrt{\frac{a^2 T_b}{N_J}} \right) \frac{a}{\sigma^2} e^{-a^2/(2\sigma^2)} \, da$$

$$= \frac{1}{2} \left\{ 1 - \sqrt{\frac{\bar{E}_b/N_J}{1 + \bar{E}_b/N_J}} \right\}.$$ (1.105)

The bound in (1.102) is

$$P_b \leq \frac{1}{2} \left\{ \frac{1}{1 + \bar{E}_b/N_J} \right\}.$$ (1.106)

Figure 1.11 shows the uncoded bit error probabilities and their bounds with no fading versus E_b/N_J and with Rayleigh fading versus $\bar{E}_b/N_J$. Note that the effects of Rayleigh fading are like the worst case pulse jammer shown in Figure 1.5.

1.6.2 Continuous Jammer with Coding—No Fading Estimate

With coding, assume ideal interleaving where each coded symbol is assumed to have an independent fading term. Consider two binary sequences of length N denoted

$$x = (x_1, x_2, \ldots, x_N)$$ (1.107)

and

$$\hat{x} = (\hat{x}_1, \hat{x}_2, \ldots, \hat{x}_N)$$

where

$$x, \hat{x} \in \{-1, 1\}.$$ (1.108)

Let the Hamming weight be $H(x, \hat{x}) = d$ and assume the fading amplitude sequence is

$$A = (A_1, A_2, \ldots, A_N)$$ (1.109)

where $\{A_n\}$ are independent identically distributed random variables with common probability density function $p_A(\cdot)$.

For the equivalent white Gaussian noise channel due to the constant power jammer, the unquantized channel outputs are

$$r_k = x_k A_k \sqrt{T_s/2} + n_k; \qquad k = 1, 2, \ldots, N$$ (1.110)

where T_s is the coded symbol time and n_k is zero mean with variance $N_J/2$. Assuming the soft decision metric $m(r, x) = rx$, the pairwise error probability that the receiver chooses $\hat{x}$ when x is sent is Chernoff bounded by

$$
\Pr(x \to \hat{x}) = \Pr\left\{ \sum_{k=1}^{N} r_k \hat{x}_k \geq \sum_{k=1}^{N} r_k x_k \middle| x \right\}
$$

$$
= \Pr\left\{ \sum_{k=1}^{N} r_k(\hat{x}_k - x_k) \geq 0 \middle| x \right\}
$$

$$
\leq E\left\{ \exp\left[\lambda \sum_{k=1}^{N} r_k(\hat{x}_k - x_k) \right] \middle| x \right\}
$$

$$
= \prod_{k=1}^{N} E\left\{ e^{\lambda r_k(\hat{x}_k - x_k)} \middle| x_k \right\} \tag{1.111}
$$

for all $\lambda \geq 0$. Note that here

$$
E\left\{ e^{\lambda r_k(\hat{x}_k - x_k)} \middle| A_k, x_k \right\} = \begin{cases} 1, & \hat{x}_k = x_k \\ e^{\lambda^2 N_J} e^{-\lambda A_k \sqrt{2T_s}}, & \hat{x}_k \neq x_k \end{cases} \tag{1.112}
$$

and

$$
E\left\{ e^{\lambda r_k(\hat{x}_k - x_k)} \middle| x_k \right\} = \begin{cases} 1, & \hat{x}_k = x_k \\ e^{\lambda^2 N_J} \int_0^\infty e^{-\lambda a \sqrt{2T_s}} p_A(a)\, da, & \hat{x}_k \neq x_k. \end{cases}
$$

$$
\tag{1.113}
$$

Defining

$$
D = \min_{\lambda \geq 0} e^{\lambda^2 N_J} \int_0^\infty e^{-\lambda a \sqrt{2T_s}} p_A(a)\, da \tag{1.114}
$$

the pairwise error bound has the form

$$
\Pr\{x \to \hat{x}\} \leq D^d \tag{1.115}
$$

and a general coded error bound of the form

$$
P_b \leq \sum_k a_k D^k. \tag{1.116}
$$

For the simple repeat m code described in Chapters 3 and 4, Volume I, the bit error bound is

$$
P_b \leq \tfrac{1}{2} D^m \tag{1.117}
$$

where D is given by (1.114). For the constraint length $K = 7$, rate $R = 1/2$ convolutional code the bit error bound is given by

$$
P_b \leq \tfrac{1}{2}[36D^{10} + 211d^{12} + 1404D^{14} + 11633D^{16} + \cdots] \tag{1.118}
$$

for the same D where T_s is the coded symbol time.

For the case of Rayleigh fading D is given by

$$D = \min_{\lambda \geq 0} e^{(1/2)\lambda^2(\bar{E}_s/N_J)}\left[1 - \sqrt{2\pi}\,\lambda(\bar{E}_s/N_J)e^{(1/2)\lambda^2(\bar{E}_s/N_J)}Q(\lambda(\bar{E}_s/N_J))\right]$$

$$(1.119)$$

where $\bar{E}_s = \sigma^2 T_s$ is the average coded symbol energy. Figure 1.12 shows coded bit error bounds for several repeat codes and the $K = 7$, $R = 1/2$ convolutional code when there is Rayleigh fading. Here $\bar{E}_s = R\bar{E}_b$, where R is the code rate in bits per coded symbol. Note that coding with interleaving

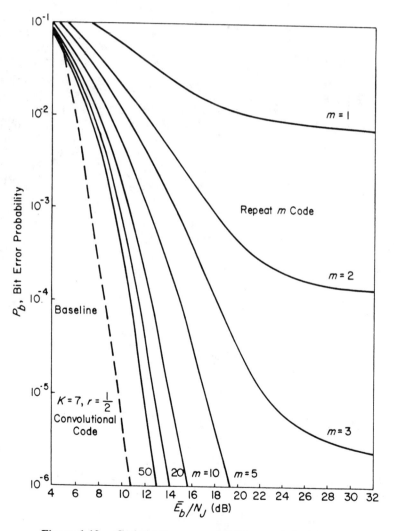

Figure 1.12. Coded bit error probability—soft decision.

combats fading much like it did for the worst case pulse jammer. Indeed, pulse jamming is similar to a fading channel where the signal-to-noise ratio can change in time.

The results illustrated above assume the soft decision metric

$$m(r, x) = rx \tag{1.120}$$

as implied in (1.111). A hard decision channel with ideal interleaving results in a binary symmetric channel where the crossover probability would depend on the specific amplitude term A. Here the crossover probability is

$$\varepsilon(A) = Q\left(\sqrt{\frac{A^2 T_s}{N_J}}\right) \tag{1.121}$$

which is the same as (1.101) with T_b replaced by T_s, the coded symbol time.

Now consider the fading channel with ideal interleaving and the usual hard decision metric when there are only two binary sequences of length N as shown in (1.107). Assuming x is transmitted and the channel output binary sequence is $y = (y_1, y_2, \ldots, y_N)$, the probability that the decoder incorrectly chooses $\hat{x}$ is the probability that

$$w(y, x) \geq w(y, \hat{x}). \tag{1.122}$$

The hard decision pairwise probability of error is thus Chernoff bounded by

$$
\begin{aligned}
P(x \to \hat{x}) &= \Pr\{w(y, x) \geq w(y, \hat{x})|x\} \\
&= \Pr\{w(y, x) - w(y, \hat{x}) \geq 0|x\} \\
&= \Pr\left\{\sum_{n=1}^{N} \left[w(y_n, x_n) - w(y_n, \hat{x}_n)\right] \geq 0\Big|x\right\} \\
&\leq E\left\{\exp \lambda \sum_{n=1}^{N} \left[w(y_n, x_n) - w(y_n, \hat{x}_n)\right]\Big|x\right\} \\
&= \prod_{n=1}^{N} E\left\{e^{\lambda[w(y_n, x_n) - w(y_n, \hat{x}_n)]}\Big|x_n\right\}
\end{aligned} \tag{1.123}
$$

where $\lambda \geq 0$ and the Hamming distance between two elements is a special case of the Hamming distance between two vectors,

$$w(y, x) = \begin{cases} 1, & y \neq x \\ 0, & y = x. \end{cases} \tag{1.124}$$

Thus

$$w(y, x) = w(y_1, x_1) + w(y_2, x_2) + \cdots + w(y_N, x_N). \tag{1.125}$$

Note that

$$y_n = \begin{cases} x_n, & \text{with probability } 1 - \varepsilon(A_n) \\ \hat{x}_n, & \text{with probability } \varepsilon(A_n) \end{cases} \tag{1.126}$$

for those d components where $\hat{x}_n \neq x_n$. Thus,

$$E\left\{ e^{\lambda[w(y_n, x_n) - w(y_n, \hat{x}_n)]} \big| x_n, A_n \right\}$$

$$= \begin{cases} 1, & \hat{x}_n = x_n \\ (1 - \varepsilon(A_n))e^{-\lambda} + \varepsilon(A_n)e^{\lambda}, & \hat{x}_n \neq x_n \end{cases} \tag{1.127}$$

and

$$E\left\{ e^{\lambda[w(y_n, x_n) - w(y_n, \hat{x}_n)]} \big| x_n \right\}$$

$$= \begin{cases} 1, & \hat{x}_n = x_n \\ (1 - \bar{\varepsilon})e^{-\lambda} + \bar{\varepsilon}e^{\lambda}, & \hat{x}_n \neq x_n \end{cases} \tag{1.128}$$

where

$$\bar{\varepsilon} = \int_0^\infty \varepsilon(a) p_A(a) \, da \tag{1.129}$$

is the averaged crossover probability. Defining

$$D(\lambda) = (1 - \bar{\varepsilon})e^{-\lambda} + \bar{\varepsilon}e^{\lambda} \tag{1.130}$$

we have

$$P(x \to \hat{x}) \leq [D(\lambda)]^{w(x, \hat{x})}, \qquad \text{all } \lambda \geq 0. \tag{1.131}$$

The choice of λ that minimizes $D(\lambda)$ satisfies

$$e^{\lambda} = \sqrt{\frac{1 - \bar{\varepsilon}}{\bar{\varepsilon}}} \tag{1.132}$$

giving the parameter

$$D = \min_{\lambda \geq 0} D(\lambda) = \sqrt{4\bar{\varepsilon}(1 - \bar{\varepsilon})} \tag{1.133}$$

The Rayleigh fading amplitude statistics has

$$\bar{\varepsilon} = \frac{1}{2}\left\{ 1 - \sqrt{\frac{\bar{E}_s/N_J}{1 + \bar{E}_s/N_J}} \right\}. \tag{1.134}$$

For the repeat m code with the bit error bound given by (1.117) and the constraint length $K = 7$, rate $R = 1/2$ convolutional code with the bit error bound (1.118) where D is given above by (1.133), the coded bit error probabilities are shown in Figure 1.13. Again note that coding can overcome the degradation due to fading.

In practice quantized amplitude histograms of a fading channel may be available through measurements. Suppose, for example, the measured quantized amplitude values are $\alpha_1, \alpha_2, \ldots, \alpha_L$ with measured probabilities

$$p_l = \Pr\{ A = \alpha_l \}; \qquad l = 1, 2, \ldots, L. \tag{1.135}$$

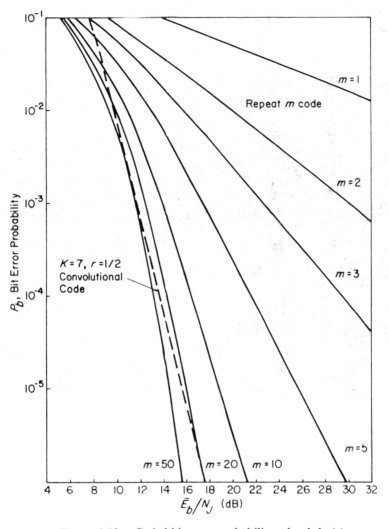

Figure 1.13. Coded bit error probability—hard decision.

For this channel the uncoded bit error probability is

$$P_b = \int_0^\infty Q\left(\sqrt{\frac{a^2 T_b}{N_J}}\right) p_A(a)\, da$$

$$= \sum_{l=1}^{L} p_l Q\left(\sqrt{\frac{\alpha_l^2 T_b}{N_J}}\right). \tag{1.136}$$

For soft decision decoding (1.114) becomes

$$D = \min_{\lambda \geq 0} e^{\lambda^2 N_J} \sum_{l=1}^{L} p_l e^{-\lambda \alpha_l \sqrt{2T_s}} \tag{1.137}$$

while for hard decision decoding D is given by (1.133) where

$$\bar{\varepsilon} = \sum_{l=1}^{L} p_l Q\left(\sqrt{\frac{\alpha_l^2 T_s}{N_J}}\right).$$
(1.138)

With this empirically measured fading, the impact of coding with ideal interleaving can be computed.

1.6.3 Continuous Jammer with Coding—Fading Estimate

With no coding, the optimum demodulator with a fading channel is the same as that for the non-fading channel. Thus, for the uncoded case in Section 1.6.1 knowledge of the fade term A does not change the performance. This is not true, however, for coded DS/BPSK systems with fading (see Proakis [33]).

Consider the same coding case described in Section 1.6.2 except that we now assume ideal estimates[4] for the fade values $A_1, A_2, \ldots, A_N$. That is, during any coded symbol transmission, complete knowledge of the random fade amplitude is assumed. In this case the maximum-likelihood decision rule obtained by comparing $p_N(r|x, A)$ with $p_N(r|\hat{x}, A)$ results in metric

$$m(r_k, x_k|A_k) = A_k r_k x_k.$$
(1.139)

Without loss of generality take the first d places of x and $\hat{x}$ to be different and obtain the conditional pairwise error bound

$$
\begin{aligned}
P(x \to \hat{x}|A) &= \Pr\left\{ \sum_{k=1}^{N} A_k r_k \hat{x}_k \geq \sum_{k=1}^{N} A_k r_k x_k \,\middle|\, x, A\right\} \\
&= \Pr\left\{ \sum_{k=1}^{N} A_k r_k (\hat{x}_k - x_k) \geq 0 \,\middle|\, x, A\right\} \\
&= \Pr\left\{ \sum_{k=1}^{N} A_k \left(x_k A_k \sqrt{T_s/2} + n_k\right)(\hat{x}_k - x_k) \geq 0 \,\middle|\, x, A\right\} \\
&= \Pr\left\{ \sum_{k=1}^{d} A_k n_k \geq \sum_{k=1}^{d} A_k^2 \sqrt{T_s/2} \,\middle|\, x, A\right\} \\
&= Q\left(\sqrt{\left(\sum_{k=1}^{d} A_k^2\right)(T_s/N_J)}\right) \\
&\leq e^{-(\sum_{k=1}^{d} A_k^2) T_s/(2N_J)} \\
&= \prod_{k=1}^{d} e^{-A_k^2 T_s/(2N_J)}.
\end{aligned}
$$
(1.140)

---

[4] This is reasonable for slowly varying fading channels where these estimates must also be deinterleaved.

Averaging this bound over the common fading density function $p_A(\cdot)$ gives

$$P(x \rightarrow \hat{x}) \leq \left\{ \int_0^\infty e^{-a^2 T_s/(2N_J)} p_A(a) \, da \right\}^d$$

$$= D^d \tag{1.141}$$

where

$$D = \int_0^\infty e^{-a^2 T_s/(2N_J)} p_A(a) \, da. \tag{1.142}$$

This can be compared with (1.114) for the case with no fading estimate.
When the fades are i.i.d. Rayleigh random variables

$$D = \frac{1}{1 + \overline{E}_s/N_J}. \tag{1.143}$$

This result can be compared with (1.119) where there is no fading estimate.
In Figure 1.14, the cutoff rates of these two cases using R_0 given by (1.92)
are shown. Note that for small values of $\overline{E}_s/N_J$ there is about a 1.25 dB
difference for the same value of R_0.

An exact expression can be obtained for the pairwise error probability for
the Rayleigh fading case where

$$Z = \sum_{k=1}^d A_k^2(T_s/N_J) \tag{1.144}$$

is a chi-square random variable with $2d$ degrees of freedom having the
probability density function

$$p_z(z) = \frac{z^{d-1} e^{-z/(\overline{E}_s/N_J)}}{(d-1)! (\overline{E}_s/N_J)^d}, \qquad z \geq 0. \tag{1.145}$$

Averaging the exact expression in (1.140) over this probability density
function results in the closed form solution

$$P(x \rightarrow \ddot{x}) = \int_0^\infty Q(\sqrt{z}) \, p_z(z) \, dz$$

$$= \left(\frac{1-\gamma}{2} \right)^d \sum_{k=0}^{d-1} \binom{d-1+k}{k} \left(\frac{1+\gamma}{2} \right)^k \tag{1.146}$$

where

$$\gamma = \sqrt{\frac{\overline{E}_s/N_J}{1 + \overline{E}_s/N_J}}. \tag{1.147}$$

In Figure 1.15, the exact bit error probability is compared with the Chernoff
bound for repeat codes having diversity $m = 1, 2,$ and 4. There is about a
1.5 dB to 3 dB difference between the exact bit error probability and the
Chernoff bound. As in earlier examples, the Chernoff bound is looser for

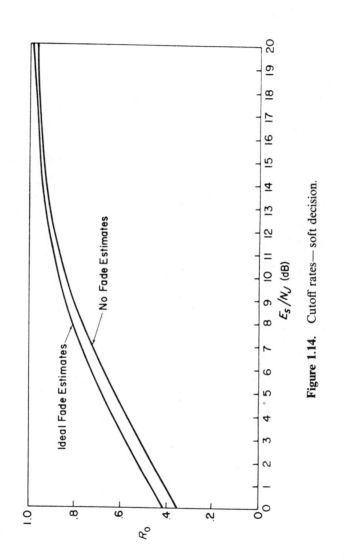

Figure 1.14. Cutoff rates—soft decision.

cases when the error probabilities fall off more slowly with signal-to-noise ratio.

In Figure 1.15 the impact of having ideal fade estimates and using the optimum metric (1.139) compared to the suboptimum metric (1.120) where no fade estimates are used can be seen. Using the Chernoff bounds for these two cases and assuming repeat codes with diversity m, we have the results shown in Figure 1.16. For smaller values of signal-to-noise ratio, there is approximately a 1.25 dB difference between these two metrics. At higher signal-to-jammer noise ratios, however, there is a much greater difference.

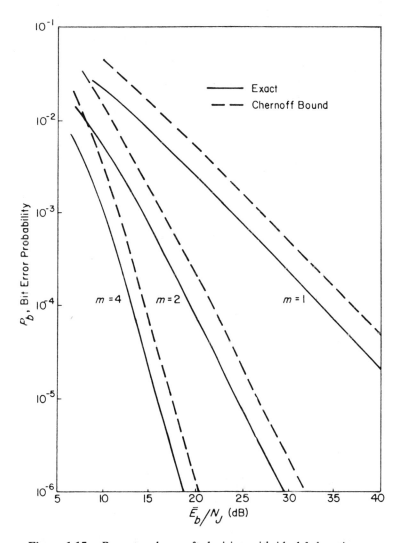

Figure 1.15. Repeat code— soft decision with ideal fade estimate.

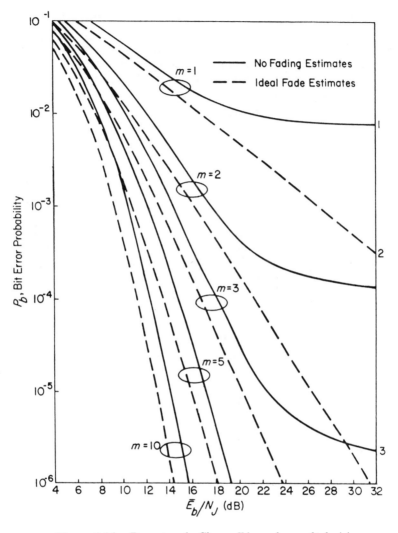

Figure 1.16. Repeat code Chernoff bounds— soft decision.

For the hard decision channel with ideal fade estimates the conditional probability of y given x and A is

$$p_N(y|x, A) = \prod_{k=1}^{N} (1 - \varepsilon(A_k))^{1-w(y_k, x_k)} \varepsilon(A_k)^{w(y_k, x_k)}$$

$$= \prod_{k=1}^{N} \left(\frac{\varepsilon(A_k)}{1 - \varepsilon(A_k)} \right)^{w(y_k, x_k)} \cdot \prod_{n=1}^{N} (1 - \varepsilon(A_n))$$

$$(1.148)$$

where $\varepsilon(A)$ is given by (1.121). Here the maximum-likelihood metric is

$$m(y_k, x_k|A_k) = w(y_k, x_k)\log\left[\frac{\varepsilon(A_k)}{1 - \varepsilon(A_k)}\right]. \qquad (1.149)$$

Assuming this optimum metric, the Chernoff bound for the conditional pairwise error probability is the Bhattacharyya bound

$$P(x \rightarrow \hat{x}|A) \le \sum_y \sqrt{p_N(y|x, A)p_N(y|\hat{x}, A)}$$

$$= \prod_{k=1}^{N}\left\{\sum_y \sqrt{p(y|x_k, A_k)p(y|\hat{x}_k, A_k)}\right\}$$

$$= \prod_{k=1}^{d} \sqrt{4\varepsilon(A_k)(1 - \varepsilon(A_k))} \qquad (1.150)$$

where, without loss of generality, only the first d places in x and $\hat{x}$ are assumed to differ. The pairwise error probability is thus

$$P(x \rightarrow \hat{x}) \le \left[\int_0^\infty \sqrt{4\varepsilon(a)(1 - \varepsilon(a))}\, p_A(a)\, da\right]^d. \qquad (1.151)$$

Here

$$D = \int_0^\infty \sqrt{4\varepsilon(a)(1 - \varepsilon(a))}\, p_A(a)\, da. \qquad (1.152)$$

For the suboptimum hard decision metric with no fade estimates, D is given by (1.133) where $\bar{\varepsilon}$ is given by (1.134). Since the function

$$f(x) = \sqrt{4x(1 - x)} \qquad (1.153)$$

is concave for $0 \le x \le 1$, using Jensen's inequality [22],

$$E\{f(X)\} \le f(E\{X\}), \qquad (1.154)$$

for the random variable X,

$$\int_0^\infty \sqrt{4\varepsilon(a)(1 - \varepsilon(a))}\, p_A(a)\, da \le \sqrt{4\bar{\varepsilon}(1 - \bar{\varepsilon})} \qquad (1.155)$$

as expected.

1.6.4 Pulse Jammer with No Coding

Now consider the case of a fading channel together with a pulse jammer with average power J and jamming with power J/ρ for a fraction ρ of the time. Again assuming that when there is no jammer pulse the bit error probability is negligible, then the average uncoded bit error probability is

$$P_b = \rho \int_0^\infty Q\left(\sqrt{\frac{a^2 T_b}{N_J}\rho}\right) p_A(a)\, da. \qquad (1.156)$$

For the case of Rayleigh fading, this has the form

$$P_b = \tfrac{1}{2}\rho\left\{1 - \sqrt{\frac{\rho(\overline{E}_b/N_J)}{1 + \rho(\overline{E}_b/N_J)}}\right\}. \tag{1.157}$$

The choice of jammer pulse parameter ρ that maximizes the bit error probability for Rayleigh fading channels is $\rho = 1$ for all $\overline{E}_b/N_J$. Thus *the constant power jammer is the worst jammer in a Rayleigh fading channel*. This is primarily due to the fact that Rayleigh fading has already created the same impact as that of a worst case pulse jammer with a resulting uncoded bit error bound that changes slowly with increasing values of $\overline{E}_b/N_J$. Changes in the signal-to-noise ratio caused by a pulse jammer no longer result in large changes in the uncoded bit error probability and the constant power jammer turns out to be the worst case. Coding and interleaving still showed dramatic improvements but this time to overcome the effects of Rayleigh fading. For other fading statistics it is not clear what is the worst pulse jammer parameter. Each case would have to be analyzed separately.

1.7 SLOW FADING MULTIPATH CHANNELS

In many radio channels signals reflect off the surface of water, buildings, trees, etc., causing multiple signal terms at the receiver. The atmosphere also causes reflections where sometimes the reflected signals are used as the primary means of sending signals from transmitters to receivers. Examples include shortwave ionospheric radio communication at HF (3 MHz to 30 MHz), tropospheric scatter radio communication at UHF (300 MHz to 3000 MHz) and SHF (3000 MHz to 30,000 MHz), and ionospheric forward scatter radio communication at VHF (30 MHz to 300 MHz). These fading multipath channels are usually modeled as having a randomly time-varying filter together with noise and interference [33].

Examination of the DS/BPSK system in a fading multipath channel begins by defining the DS/BPSK signal in (1.4) with an arbitrary phase term θ, i.e.,

$$x(t; \theta) = c(t) d(t)\sqrt{2S} \cos[\omega_0 t + \theta]. \tag{1.158}$$

The simplest multipath example is where there is an unfaded direct path signal and one reflected path signal. The received signal has the form

$$y(t) = x(t; 0) + \alpha x(t - \tau; \theta) + J(t). \tag{1.159}$$

Here α is the reflected signal amplitude term, τ is its delay relative to the direct signal, and θ is its phase relative to the direct signal. Here $x(t) = c(t)s(t)$ is the DS/BPSK signal with no fading and $J(t)$ is the jamming signal.

The usual DS/BPSK receiver would multiply $y(t)$ with the PN waveform $c(t)$ and then compute the cosine component of the carrier,

$$r_0 = \int_0^{T_b} r(t)\phi_c(t)\, dt \qquad (1.160)$$

where

$$r(t) = c(t)y(t)$$
$$= s(t) + c(t)\alpha x(t - \tau; \theta) + c(t)J(t) \qquad (1.161)$$

and

$$\phi_c(t) = \sqrt{\frac{2}{T_b}}\, \cos \omega_0 t$$

$$0 \le t \le T_b. \qquad (1.162)$$

Evaluating (1.160) using (1.161) and (1.162) gives

$$r_0 = d_0\sqrt{E_b} + n'_0 + n_0 \qquad (1.163)$$

where

$$n_0 = \int_0^{T_b} c(t)J(t)\phi_c(t)\, dt \qquad (1.164)$$

is a Gaussian random variable with variance $N_J/2$ and

$$n'_0 = \int_0^{T_b} c(t)\alpha x(t - \tau; \theta)\phi_c(t)\, dt$$

$$= \alpha d_0\sqrt{2S} \int_\tau^{T_b} c(t)c(t - \tau)\cos[\omega_0(t - \tau) + \theta]\phi_c(t)\, dt \qquad (1.165)$$

where d_0 is the value of the data signal $d(t)$ during the interval $(0, T_b)$.

Suppose in (1.159) the multipath delay τ satisfies

$$\tau \ge T_c \qquad (1.166)$$

where T_c is the PN chip time. Then for each t, $c(t)$ and $c(t - \tau)$ are independent and n' is a sum of independent random variables which can be approximated as a Gaussian random variable with variance no greater than $\alpha^2 S/(2W_{ss})$. Since typically $S \ll J$ is assumed, for all realistic values of α, this multipath noise term n'_0 is negligible compared to the noise term n_0 which is due to jamming. Thus we have the approximation

$$r_0 \cong d_0\sqrt{E_b} + n_0 \qquad (1.167)$$

for multipath delay τ satisfying (1.166).

When the multipath delay τ is greater than the chip time T_c, there is negligible degradation due to multipath. This also applies when the direct path experiences slowly varying frequency non-selective fading as discussed in the previous section. In general, however, it is possible to do better. Assuming the channel is slowly varying so that the multipath parameters α,

τ, and θ are known to the receiver, the receiver can multiply $y(t)$ with $c(t - \tau)$ and find the cosine component corresponding to the coordinate

$$\phi_c(t - \tau; \theta) = \sqrt{\frac{2}{T_b}} \cos[\omega_0(t - \tau) + \theta]$$

$$\tau \leq t \leq T_b + \tau. \tag{1.168}$$

This results in the cosine component relative to the multipath signal of the form (ignoring the direct path noise term)

$$r_1 \cong d_0 \alpha \sqrt{E_b} + n_1 \tag{1.169}$$

where

$$n_1 = \int_{\tau}^{T_b + \tau} c(t - \tau) J(t) \phi_c(t - \tau; \theta) \, dt \tag{1.170}$$

is a Gaussian random variable with variance $N_J/2$. Thus there are two outputs of the channel given by r_0 and r_1.

Except for some unrealistic waveforms for the jammer, the correlation between n_0 and n_1 is zero and thus these are independent Gaussian random variables. The optimum decision rule[5] based on r_0 and r_1 is to compare

$$r = r_0 + \alpha r_1$$

$$= d_0(1 + \alpha^2)\sqrt{E_b} + n_0 + \alpha n_1 \tag{1.171}$$

with zero as in (1.19). The bit error probability is thus

$$P_b = Q\left(\sqrt{\frac{(1 + \alpha^2)2E_b}{N_J}}\right). \tag{1.172}$$

This bit error probability is better than using a conventional DS/BPSK receiver which only uses r_0 in its decision.

Condition (1.166) for multipath delay results in a diversity system where two independent channel outputs are available. Thus DS/BPSK spread-spectrum signals not only provide protection against jamming but also can resolve multipath and take advantage of the natural diversity available.

For a multipath channel with L paths and a resulting channel output

$$y(t) = \sum_{l=1}^{L} \alpha_l x(t - \tau_l; \theta_l) + J(t) \tag{1.173}$$

it is possible to compute at the receiver the L outputs

$$r_l = \int_{\tau_l}^{T_b + \tau_l} c(t - \tau_l) y(t) \phi_c(t - \tau_l; \theta_l) \, dt \tag{1.174}$$

where $\phi_c(t - \tau; \theta)$ is given in (1.168). This assumes the receiver has complete knowledge of the multipath statistics which include amplitudes $\{\alpha_l\}$,

---

[5] This can be obtained by using the maximum-likelihood rule of comparing $p(r_0, r_1 | d = -1)$ with $p(r_0, r_1 | d = 1)$.

delays $\{\tau_l\}$, and carrier phases $\{\theta_l\}$. If now

$$|\tau_i - \tau_j| \geq T_c \quad \text{for all } i \neq j \tag{1.175}$$

then we have the approximation

$$r_l \cong d_0 \alpha_l \sqrt{E_b} + n_l$$
$$l = 1, 2, \ldots, L \tag{1.176}$$

where $\{n_l\}$ are independent Gaussian random variables with variance $N_J/2$. Here the optimum decision rule is given by (1.19) with

$$r = \sum_{l=1}^{L} \alpha_l r_l \tag{1.177}$$

and the bit error probability is

$$P_b = Q\left(\sqrt{ \frac{\left(\sum_{l=1}^{L} \alpha_l^2 \right) 2E_b}{N_J} } \right) \tag{1.178}$$

where E_b is the energy of any single multipath signal when the amplitude term is unity ($\alpha = 1$). The energy due to all multipath terms is

$$E_T = \left(\sum_{l=1}^{L} \alpha_l^2 \right) E_b \tag{1.179}$$

and (1.178) can be rewritten as

$$P_b = Q\left(\sqrt{ \frac{2E_T}{N_J} } \right). \tag{1.180}$$

Next suppose that each multipath signal has a slowly varying independent fade. The L receiver outputs are then given by

$$r_l \cong d_0 A_l \sqrt{E_b} + n_l$$
$$l = 1, 2, \ldots, L \tag{1.181}$$

where $\{A_l\}$ are independent fade random variables. Conditioned on

$$A = (A_1, A_2, \ldots, A_L) \tag{1.182}$$

the bit error probability and its Chernoff bound are given by

$$P_b(A) = Q\left(\sqrt{ \frac{\left(\sum_{l=1}^{L} A_l^2 \right) 2E_b}{N_J} } \right)$$

$$\leq \tfrac{1}{2} e^{-\left(\sum_{l=1}^{L} A_l^2 \right) E_b/N_J}$$

$$= \frac{1}{2} \prod_{l=1}^{L} e^{-A_l^2 E_b/N_J}. \tag{1.183}$$

Assuming A_l has probability density function $p_l(\cdot)$ for each $l = 1, 2, \ldots, L$ the average bit error bound is

$$P_b \leq \frac{1}{2} \prod_{l=1}^{L} \left\{ \int_0^\infty e^{-a^2 E_b/N_J} p_l(a) \, da \right\} \tag{1.184}$$

which for Rayleigh amplitudes with

$$\sigma_l^2 = \int_0^\infty a^2 p_l(a) \, da; \qquad l = 1, 2, \ldots, L \tag{1.185}$$

becomes

$$P_b \leq \frac{1}{2} \prod_{l=1}^{L} \left(\frac{1}{1 + \overline{E}_l/N_J} \right) \tag{1.186}$$

where $\overline{E}_l$ is the average signal energy in the l-th multipath signal given by

$$\overline{E}_l = \sigma_l^2 E_b; \qquad l = 1, 2, \ldots, L. \tag{1.187}$$

Note that when all the multipath energy terms are identical the exact expression for P_b is given by (1.146) with $d = L$. The general exact expression will be shown in (1.201).

In many channels, such as the tropospheric scatter channel, it is more appropriate to view the received signal as consisting of a continuum of multipath components. Such channels are usually characterized with channel output (see Proakis [33]),

$$y(t) = \int_{-\infty}^{\infty} H(\tau; t) x(t - \tau; (t \leftrightarrow \theta)(t - \tau)) \, d\tau + J(t) \tag{1.188}$$

where $H(\tau; t)$ is a randomly time-varying filter.

Associated with the randomly time-varying filter $H(\tau; t)$ are two basic parameters; T_m denotes the total multipath delay spread of the channel and B_d denotes the Doppler spread of the channel. From these define

$$\Delta f_c = \frac{1}{T_m} \tag{1.189}$$

as the "coherence bandwidth" and

$$\Delta t_c = \frac{1}{B_d} \tag{1.190}$$

as the "coherence time" of the channel. Roughly, if two CW signals of frequency separation greater than Δf_c were transmitted through the channel, then the output signals at the two carrier frequencies would have independent channel disturbances (phase and envelope). Similarly, when a single CW signal is transmitted through the channel, its output sampled at time separations greater than Δt_c would have independent channel disturbances at the sample times.

For the uncoded DS/BPSK signal one data bit is transmitted every T_b seconds. Assume

$$T_b \gg T_m \qquad (1.191)$$

so that intersymbol interference between data bits can be ignored. Also assume the channel is slowly varying so that

$$T_b \ll \Delta t_c. \qquad (1.192)$$

Thus the channel disturbance is almost constant during a data bit time T_b. Finally, since our signal is a wideband signal of bandwidth W_{ss} assume

$$W_{ss} \gg \Delta f_c. \qquad (1.193)$$

This model assumes many independent scatters are causing the continuum of multipath components. Thus the resulting channel output signal term is the sum of many independent scatters which justifies assuming it is a Gaussian random process. At any time it has a Rayleigh envelope probability distribution and an independent phase uniformly distributed over $[0, 2\pi]$. Skywave propagation where an HF signal is reflected off the ionosphere is an example where this model applies. If, however, there also exists a strong unfaded signal component such as a groundwave at HF, which appears in shorter ranges between transmitter and receivers, the signal out of the channel is assumed to have Rician fading statistics. In the following assume Rayleigh fading only.

Since the transmitted signal $x(t)$ has bandwidth W_{ss} centered at carrier frequency ω_0 radians, it can be represented in terms of samples of the inphase and quadrature components of the signal at sample times $\{n/W_{ss}: n = \ldots, -1, 0, 1, 2, \ldots\}$. Thus the channel output can be modelled as

$$y(t) = \sum_{n=-\infty}^{\infty} A_n(t) x\left(t - \frac{n}{W_{ss}}; \theta_n(t)\right) + J(t) \qquad (1.194)$$

where at any time t the envelope terms are independent Rayleigh random variables and all phase terms are independent and uniformly distributed over $[0, 2\pi]$. Also, since the total multipath spread is T_m, for all practical purposes we can truncate the number of terms to

$$L = W_{ss} T_m + 1. \qquad (1.195)$$

The assumption regarding the slowly varying nature of the channel where (1.192) holds means that $A_n(t)$ and $\theta_n(t)$ are constant during the bit time T_b. Thus

$$y(t) = \sum_{n=1}^{L} A_n x\left(t - \frac{n}{W_{ss}}; \theta_n\right) + J(t) \qquad (1.196)$$

and the same situation as before with the finite number of distinct multipath components shown in (1.173) occurs here.

For this case the receiver can compute outputs for individual multipath signals as follows,

$$r_n = \int_{n/W_{ss}}^{T_b + n/W_{ss}} c\left(t - \frac{n}{W_{ss}}\right) y(t) \phi_c\left(t - \frac{n}{W_{ss}}; \theta_n\right) dt;$$

$$n = 1, 2, \ldots, L \tag{1.197}$$

provided the phase terms $\{\theta_n\}$ are known at the receiver. Also assuming envelopes $\{A_n\}$ are known, the optimum receiver compares

$$r = \sum_{n=1}^{L} A_n r_n \tag{1.198}$$

to zero to make the binary decision $\hat{d}$.

Next examine the optimum receiver structure implied by (1.197) and (1.198). Note that (1.197) can be rewritten as

$$r_n = \int_0^{T_b} y\left(t + \frac{n}{W_{ss}}\right) c(t) \phi_c(t; \theta_n) \, dt$$

$$n = 1, 2, \ldots, L \tag{1.199}$$

and so (1.198) becomes

$$r = \int_0^{T_b} \left[\sum_{n=1}^{L} y\left(t + \frac{n}{W_{ss}}\right) A_n c(t) \phi_c(t; \theta_n) \right] dt. \tag{1.200}$$

Figure 1.17 shows a block diagram for the optimum demodulator.

The ideal tapped delay line receiver of Figure 1.17 attempts to collect coherently the signal energy from all the received signal paths that fall within the span of the delay line and carry the same information. Because its actions act like a garden rake this has been coined the "Rake receiver" [23].

The bit error bound using the ideal Rake receiver is given by (1.186). An exact expression is given by (see Proakis [3], Chapter 7)

$$P_b = \frac{1}{2} \sum_{l=1}^{L} \pi_l \left[1 - \sqrt{\frac{\overline{E}_l / N_J}{1 + \overline{E}_l / N_J}} \right] \tag{1.201}$$

where

$$\pi_l = \prod_{\substack{k=1 \\ k \neq l}}^{L} \left(\frac{\overline{E}_l}{\overline{E}_l - \overline{E}_k} \right)$$

$$l = 1, 2, \ldots, L. \tag{1.202}$$

The ideal Rake receiver assumes complete knowledge of the phase and envelope terms which appear in the correlation functions $\{A_n c(t) \phi_c(t; \theta_n)\}$. When the fading is slow this estimate is quite good.

A simpler form of the optimum receiver can be obtained using complex baseband representations for the radio signals on a carrier frequency of ω_0 radians. In general, the radio signal $f(t)$ with a carrier frequency ω_0 has representation

$$f(t) = g(t)\cos[\omega_0 t + \eta(t)]$$
$$= \text{Re}\{\mathscr{F}(t)e^{j\omega_0 t}\} \tag{1.203}$$

where $g(t)$ and $\eta(t)$ are real-valued functions and

$$\mathscr{F}(t) = g(t)e^{j\eta(t)} \tag{1.204}$$

is the complex baseband signal representing the radio signal $f(t)$. Using script letters to represent complex baseband signals

$$\begin{aligned} x(t;\theta) &= \text{Re}\{\mathscr{X}(t;\theta)e^{j\omega_0 t}\}, \\ \phi(t;\theta) &= \text{Re}\{\Phi(t;\theta)e^{j\omega_0 t}\}, \\ J(t) &= \text{Re}\{\mathscr{J}(t)e^{j\omega_0 t}\}, \\ y(t) &= \text{Re}\{\mathscr{Y}(t)e^{j\omega_0 t}\} \end{aligned} \tag{1.205}$$

where

$$\mathscr{X}(t;\theta) = c(t)\,d(t)\sqrt{2S}\,e^{j\theta},$$

$$\Phi(t;\theta) = \sqrt{\frac{2}{T_b}}\,e^{j\theta},$$

$$\mathscr{Y}(t) = \sum_{n=1}^{L} A_n c\left(t - \frac{n}{W_{ss}}\right)d\left(t - \frac{n}{W_{ss}}\right)\sqrt{2S}\,e^{j\theta} + \mathscr{J}(t). \tag{1.206}$$

Assuming the carrier frequency is much greater than the signal bandwidth, the form for r in (1.200) is given by

$$r = \text{Re}\left\{\int_0^{T_b}\left[\sum_{n=1}^{L}\mathscr{Y}\left(t + \frac{n}{W_{ss}}\right)A_n c(t)\Phi^*(t;\theta_n)\right]dt\right\}$$

$$= \text{Re}\left\{\int_0^{T_b}\left[\sum_{n-1}^{L}\sqrt{\frac{2}{T_b}}\,\mathscr{Y}\left(t + \frac{n}{W_{ss}}\right)A_n e^{-j\theta_n}c(t)\right]dt\right\}. \tag{1.207}$$

The receiver structure of Figure 1.17 can implement this complex form of the optimum receiver by replacing $y(t)$ by $\mathscr{Y}(t)$ and $A_n c(t)\Phi(t;\theta_n)$ by $A_n e^{-j\theta_n}c(t)$ for each n.

We now examine a way of estimating $A_n e^{-j\theta_n}$ which is required in a practical Rake receiver. Note that

$$\mathscr{Y}\left(t + \frac{l}{W_{ss}}\right)c(t) = c(t)\sum_{n=1}^{L}A_n c\left(t - \frac{n-l}{W_{ss}}\right)d\left(t - \frac{n-l}{W_{ss}}\right)\sqrt{2S}\,e^{j\theta_n}$$

$$+ c(t)\mathscr{J}(t)$$

$$= A_l e^{j\theta_l}\sqrt{2S}\,d(t) + c(t)\mathscr{J}(t)$$

$$+ \sum_{\substack{n-1 \\ n\neq l}}^{L}A_n c(t)c\left(t - \frac{n-l}{W_{ss}}\right)d\left(t - \frac{n-l}{W_{ss}}\right)\sqrt{2S}\,e^{j\theta_n}.$$

$$\tag{1.208}$$

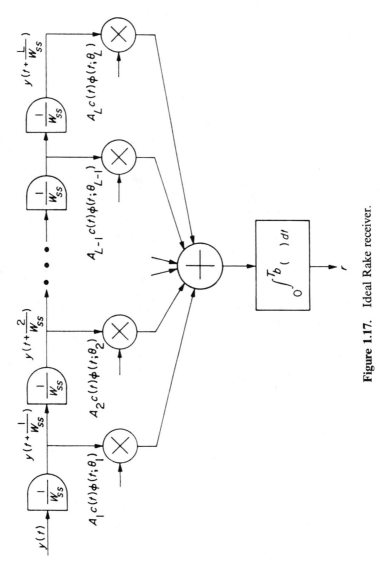

Figure 1.17. Ideal Rake receiver.

Recall that $c(t)$ is independent of $c(t - (n - l)/W_{ss})$ for each value of t when $n \neq l$ and thus,

$$\int_0^{T_b} \mathcal{Y}\left(t + \frac{l}{W_{ss}}\right) c(t)\, dt = A_l e^{j\theta_l} \sqrt{2S}\, d_0 \cdot T_b + n_l \qquad (1.209)$$

where d_0 is the data bit in $(0, T_b)$ and n_l is a Gaussian random variable. This suggests that the estimate for $A_n e^{-j\theta_n}$ be given by the conjugate of (1.209). This estimate, however, includes the data bit d_0. Assuming $A_n e^{-j\theta_n}$ remains unchanged over $2T_b$ seconds, an estimate can be based on the previous T_b second channel output signal. This estimate is shown in the complex form of the Rake receiver illustrated in Figure 1.18.

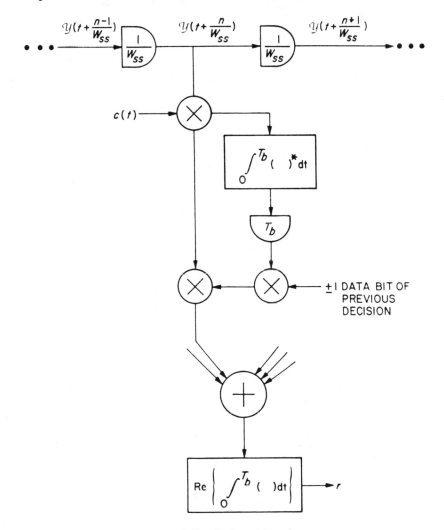

Figure 1.18. Rake with estimates.

Thus far the results in this section have applied only to uncoded DS/BPSK signals in slow fading multipath channels. With coding and ideal interleavers and deinterleavers the channel disturbance can be assumed to be independent for each transmitted coded bit. Assuming the soft decision metric of (1.120) where r is given by (1.207) and $x \in \{-1, 1\}$ the pairwise error bound for two sequences x and $\hat{x}$ is given by (1.131) where

$$D(\lambda) = E\{e^{\lambda r(\hat{x} - x)} | x\}, \quad \hat{x} \neq x. \tag{1.210}$$

From (1.181) and (1.198)

$$r = \sum_{l=1}^{L} A_l r_l$$

$$= \sum_{l=1}^{L} \left(d_0 A_l^2 \sqrt{E_s} + A_l n_l \right) \tag{1.211}$$

where $\{n_l\}$ are i.i.d. zero-mean Gaussian random variables with variance $N_J/2$. Thus

$$D(\lambda) = \prod_{l=1}^{L} E\{ e^{-2\lambda[A_l^2 \sqrt{E_s} + A_l n_l]} \}$$

$$= \prod_{l=1}^{L} E\{ E\{ e^{-2\lambda[A_l^2 \sqrt{E_s} + A_l n_l]} | A_l \} \}$$

$$= \prod_{l=1}^{L} E\{ e^{-2\lambda A_l^2 \sqrt{E_s}} E\{ e^{-2\lambda A_l n_l} | A_l \} \}$$

$$= \prod_{l=1}^{L} E\{ e^{-A_l^2 (2\lambda \sqrt{E_s} - \lambda^2 N_J)} \}$$

$$= \prod_{l=1}^{L} \left(\frac{1}{1 + 2\sigma_l^2 (2\lambda \sqrt{E_s} - \lambda^2 N_J)} \right) \tag{1.212}$$

where the amplitudes $\{A_l\}$ are again assumed Rayleigh distributed with variance as in (1.185). Then, define parameter

$$D = \min_{\lambda \geq 0} \prod_{l=1}^{L} \left(\frac{1}{1 + 2\sigma_l^2 (2\lambda \sqrt{E_s} - \lambda^2 N_J)} \right)$$

$$= \min_{s \geq 0} \prod_{l=1}^{L} \left(\frac{1}{1 + 2(\bar{E}_l / N_0)(2s - s^2)} \right) \tag{1.213}$$

where $\bar{E}_l$ is given by (1.187) with E_b replaced by E_s.

Suppose $\bar{E}_T$ is the total average energy and each multipath energy term is the same. That is,

$$\bar{E}_l = \frac{\bar{E}_T}{L} \quad l = 1, 2, \ldots, L. \tag{1.214}$$

Then $s = 1$ minimizes the bound and

$$D = \left[\frac{L}{L + \bar{E}_T/N_J} \right]^L \tag{1.215}$$

For the hard decision channel with the usual unweighted metric we have simply

$$D = \sqrt{4\varepsilon(1 - \varepsilon)} \tag{1.216}$$

where now $\varepsilon = P_b$ given by (1.201) and (1.202).

1.8 OTHER CODING METRICS FOR PULSE JAMMING

In the previous section, the ideal Rake receiver was approximated by a receiver that estimated the slowly varying multipath fading envelopes. Earlier, we had examined a similar ideal case in Section 1.4 where we considered various decoding metrics for DS/BPSK signals against pulse jamming. The ideal metric for this case is the soft decision metric with jammer state information given by

$$m(y, x; Z) = c(Z)yx \tag{1.217}$$

where Z is the jammer state random variable with probabilities

$$\Pr\{ Z = 1 \} = \rho$$
$$\Pr\{ Z = 0 \} = 1 - \rho \tag{1.218}$$

and $c(Z)$ is a weighting function such that $c(0)$ is chosen as large as possible and $c(1) = 1$. Under these conditions, it was previously shown in Chapter 4, Volume I, that the channel parameter for this case is given by

$$D = \min_{\lambda \geq 0} \rho \exp\left\{ -2\lambda E_s + \lambda^2 E_s N_J/\rho \right\}$$
$$= \rho e^{-\rho E_s/N_J} \tag{1.219}$$

which when translated to its equivalent computational cutoff rate R_0 via (1.92) is illustrated as curve (1) in Figure 1.8.

When using the same soft decision metric, i.e., (1.217) without jammer state information, we observed in Chapter 4, Volume I, that the appropriate metric weighting then becomes $c(0) = c(1) = 1$. For this case, we obtain (see Chapter 4, Volume I)

$$D = \min_{\lambda \geq 0} e^{-2\lambda E_s} \left[\rho e^{\lambda^2 E_s N_J/\rho} + 1 - \rho \right] \tag{1.220}$$

which has the equivalent computational cutoff rate $R_0 = 0$ *for all* E_s/N_0 as illustrated by curve (3) in Figure 1.8. The implication of this result is that the soft decision metric without jammer state information has no protection against a jammer who concentrates his available power in a very narrow pulse. Stated another way, a receiver that uses a soft decision metric with no jammer state information has, in the presence of an optimized pulse jammer, an unbounded probability of error. One way of getting around this intolerable degradation in the absence of jammer state information is to use hard decision decoding. This was shown, however, to result in a relatively poor performance, as can be witnessed by examining curve (4) in Figure 1.8.

In this section, we first examine other metrics that allow soft decision decoding to be used with an optimized pulse jammer and no jammer state information. Following this, we reexamine the behavior of soft decision decoding with jammer state information and the necessary modifications to the metric when, indeed, the estimate of the jammer state Z is not perfect, e.g., in the presence of background noise.

Perhaps the simplest modification of the soft decision receiver to allow operation in the presence of a pulse jammer is to *clamp* the input to the decoder at either a fixed or variable level, the latter case requiring an intelligent controller in the decoder. Clamping the decoder input level provides a means for limiting the potentially large excursions in this level caused by narrow pulse jammers and as such prevents the decoder's decision metric from being dominated by these extreme signal level swings when the jammer is present. Mathematically speaking, the operation of clamping is equivalent to first passing the decoder input y through a zero mean non-linearity with transfer function

$$y' = \begin{cases} k\sqrt{E_s}; & y > k\sqrt{E_s} \\ y; & |y| \le k\sqrt{E_s} \\ -k\sqrt{E_s}; & y < -k\sqrt{E_s} \end{cases} \qquad (1.221)$$

where y' now represents the actual decoder input and $k \ge 0$ is a parameter which is either fixed or is allowed to vary with E_s/N_J and the pulse jammer's strategy, i.e., ρ.

In the absence of jammer state information, El-Wailly [34] has shown that using the normalized maximum-likelihood metric $m(y', x) = y'x/N_J$, the channel parameter D is given by

$$D = \min_{\lambda \ge 0} e^{-\lambda 2E_s/N_J}\Bigg[(1 - \rho) + \rho\Big\{Q\Big[\sqrt{2\rho E_s/N_J}\,(k - 1)\Big]e^{2\lambda(1-k)E_s/N_J}$$
$$+ Q\Big[\sqrt{2\rho E_s/N_J}\,(k + 1)\Big]e^{2\lambda(1+k)E_s/N_J}$$
$$+ e^{\lambda^2 E_s/(\rho N_J)}[Q(A) - Q(B)]\Big\}\Bigg],$$

$$(1.222)$$

where $Q(x)$ is again the Gaussian probability integral and

$$A = (1 - k)\sqrt{2\rho E_s/N_J} - \lambda\sqrt{2E_s/(\rho N_J)}\,,$$
$$B = (1 + k)\sqrt{2\rho E_s/N_J} - \lambda\sqrt{2E_s/(\rho N_J)}\,. \qquad (1.223)$$

Intuitively, one would expect that a clamped (fixed or variable) soft decision receiver characterized by (1.221) would outperform a hard decision decoding metric and at the same time not be totally vulnerable to the optimum pulse jammer as is the unclamped soft decision metric (i.e., $k = \infty$). The degree to which this observation is true is demonstrated by the following illustrations and discussion.

Figure 1.19 is an illustration of D as computed from (1.222) versus E_s/N_J for a fixed clamping level of $k = 1$ and the worst case jammer who chooses his ρ to maximize D for each value of E_s/N_J. Also illustrated for purposes of comparison is the corresponding result for the hard decision decoding metric (with no jammer state information) which from Chapter 4, Volume I, is given by

$$D = \sqrt{4\varepsilon(1-\varepsilon)}\,; \quad \varepsilon = \rho Q\left(\sqrt{2\rho E_s/N_J}\right). \tag{1.224}$$

As a reminder, the curve for the unclamped soft decision decoder (with no jamming state information) for worst case pulse jamming would simply be a horizontal line corresponding to $D = 1$ for all E_s/N_J.

When the clamping level is allowed to vary, then assuming that the receiver contains an intelligent controller that can choose the value of k so as to minimize D for each value of ρ, Figure 1.20 illustrates the corresponding performance in terms of a three-dimensional plot of D versus E_s/N_J and ρ. Alternately, suppose it is assumed that the receiver has no intelli-

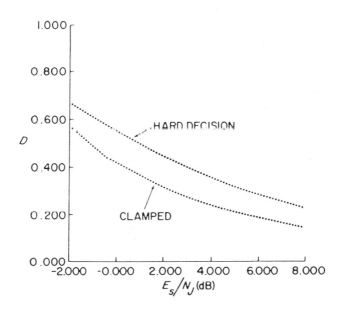

Figure 1.19. Channel parameter versus signal-to-noise ratio for worst case jammer (no jammer state information). (Reprinted from [35].)

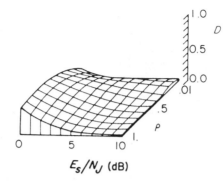

Figure 1.20. Channel parameter versus ρ and signal-to-noise ratio for best clamping level. (Reprinted from [35].)

gence, i.e., it fixes k at some value between .01 and 1.0, but the jammer, on the other hand, is intelligent and assumed to know k, in which case, he chooses ρ to maximize D. For this scenario, Figure 1.21 illustrates D versus E_s/N_J and k. Clearly, much larger values of D result for this case as compared with those in Figure 1.20.

We now return to a consideration of soft decision decoding metrics with jammer state information that derive their benefit from the fact that the contribution to the metric in those time intervals where the jammer is absent is very heavily weighted compared to that in the intervals where the jammer is known to be present. Up until now, we have assumed that knowledge of the presence or absence of the jammer during a given time interval was perfect. With, for example, additive background noise, the estimation of Z, the jammer state parameter, will not be perfect. We now examine how to suitably modify the soft decision decoding metric and the impact on its performance as a result of having imperfect jammer state information.

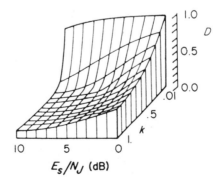

Figure 1.21. Channel parameter versus signal-to-noise ratio and clamping level for worst case jammer. (Reprinted from [35].)

DS/BPSK transmission over a channel with background additive white Gaussian noise of power spectral density N_0 and a pulse jammer with parameters $N_J = J/W$ and ρ, is equivalent to sending a BPSK signal over an additive white Gaussian noise channel with noise spectral density

$$N_{0e} = \begin{cases} N_0; & \text{if } Z = 0 \\ N_0' \triangleq N_0 + \dfrac{N_J}{\rho}; & \text{if } Z = 1 \end{cases} \qquad (1.225)$$

where Z is again the jammer state parameter with probabilities as in (1.218). If perfect jammer state information were possible, then the metric of (1.217) would be used with

$$c(0) = \frac{1}{N_0}$$

$$c(1) = \frac{1}{N_0'}. \qquad (1.226)$$

Since, as mentioned above, the background noise perfect jammer state information is not possible, then the metric of (1.217) is still appropriate except that now Z is replaced by an estimate $\hat{Z}$ of the jammer state. The manner in which the estimate $\hat{Z}$ is obtained and its statistics are the subject of the following discussion.

As in the Rake receiver illustrated in Figure 1.18, an estimate $\hat{Z}$ of the jammer state might be based on the previous BPSK transmission.[6] Another approach would be to base the estimate on the amplitude of the received signal in the current interval as follows:

$$\hat{Z} = \begin{cases} 0; & \text{if } |y| \leq \beta\sqrt{E_s} \\ 1; & \text{if } |y| > \beta\sqrt{E_s}. \end{cases} \qquad (1.227)$$

Basically, this receiver assumes that when the channel output y is close to one of the two signal terms, $\sqrt{E_s}$ or $-\sqrt{E_s}$, then the channel contains only background noise. Substituting (1.226) in (1.217) then gives the metric examined by El-Wailly and Costello [35]:

$$m(y, x; \hat{Z}) = \begin{cases} \dfrac{yx}{N_0}; & \text{if } \hat{Z} = 0 \\ \dfrac{yx}{N_0'}; & \text{if } \hat{Z} = 1. \end{cases} \qquad (1.228)$$

It should be noted that the metric of (1.228) can also be regarded as one that assumes no jammer state information since indeed the jammer state information comes directly from measurements on the observable (i.e., the channel output) rather than from an external source. Thus, combining

---

[6] This estimate can be done before deinterleaving at the receiver to make the channel memoryless.

(1.227) and (1.228), we get the alternate form

$$m(y, x; \hat{Z}) = \begin{cases} \dfrac{yx}{N_0}; & \text{if } |y| \le \beta\sqrt{E_s} \\[2mm] \dfrac{yx}{N_0'}; & \text{if } |y| > \beta\sqrt{E_s} \end{cases} \tag{1.229}$$

For the metric of (1.229), El-Wailly and Costello [35] have shown that the channel parameter D is given by

$$\begin{aligned} D = \min_{\lambda \ge 0} (1 - \rho)\Big\{ & I\Big[-\infty, (1 - \beta)\sqrt{E_s}, N_0', N_0\Big] \\ & + I\Big[(1 - \beta)\sqrt{E_s}, (1 + \beta)\sqrt{E_s}, N_0, N_0\Big] \\ & + I\Big[(1 + \beta)\sqrt{E_s}, +\infty, N_0', N_0\Big]\Big\} \\ & + \rho\Big\{ I\Big[-\infty, (1 - \beta)\sqrt{E_s}, N_0', N_0'\Big] \\ & + I\Big[(1 - \beta)\sqrt{E_s}, (1 + \beta)\sqrt{E_s}, N_0, N_0'\Big] \\ & + I\Big[(1 + \beta)\sqrt{E_s}, +\infty, N_0', N_0'\Big]\Big\}, \end{aligned} \tag{1.230}$$

where

$$I[L, H, \hat{N}, N] = e^{-(E_s/\hat{N})[2\lambda - (N/\hat{N})\lambda^2]}[F(L) - F(H)], \tag{1.231}$$

and

$$F(t) = Q\Big[t\sqrt{2/N} - \lambda\sqrt{(2E_s/\hat{N})(N/\hat{N})}\Big]. \tag{1.232}$$

Figure 1.22 illustrates the performance of a receiver using the metric of (1.229) in terms of D versus ρ with β as a parameter for $E_s/N_J = 5$ dB and a background-to-jammer-noise-spectral-density ratio $N_0/N_J = 1.0$. Also shown for comparison are the corresponding results for the ideal soft decision metric with unknown and perfectly known jamming state information. We observe that the performance corresponding to (1.230) is much improved over that of the unknown jammer state case but not as good as when the jammer state is perfectly known. Also, a value of $\beta = 2$ seems to give best performance.

Suppose that we again use the metric of (1.228) but now the jammer state estimate is provided by means external to the decoder. In particular, assume that the probability of error in Z is described by the false alarm and missed detection probabilities

$$P_{FA} = \Pr\{\hat{Z} \ne Z\} \text{ when } Z = 0$$

$$P_{MD} = \Pr\{\hat{Z} \ne Z\} \text{ when } Z = 1 \tag{1.233}$$

i.e., the probability of estimating the state of the jammer depends on the state of the jammer. Typically, we would want to choose $P_{MD} \ll P_{FA}$ since

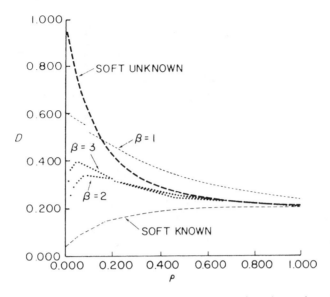

Figure 1.22. Channel parameter vs. ρ for system with estimate based on signal level and with $E_s/N_J = 5$ dB and $N_0/N_J = 1.0$. (Reprinted from [35].)

the effect on the metric of missing a jammed channel symbol is much more severe than that produced by assuming the jammer is present when indeed he is not. For this scenario, the channel parameter D has been shown to be given by [35]

$$D = \min_{\lambda \geq 0} (1 - \rho)(1 - P_{FA})e^{-[2\lambda - \lambda^2]E_s/N_0}$$

$$+ (1 - \rho)P_{FA}e^{-[2\lambda - (N_0/N_0')\lambda^2]E_s/N_0'}$$

$$+ \rho(1 - P_{MD})e^{-[2\lambda - \lambda^2]E_s/N_0'}$$

$$+ \rho P_{MD}e^{-[2\lambda - (N_0'/N_0)\lambda^2]E_s/N_0}. \qquad (1.234)$$

which is illustrated in Figure 1.23 versus ρ and P_{FA} for $P_{MD} = 10^{-8}$, $E_s/N_J = 5$ dB, and $N_0/N_J = 1.0$.

Finally, it should be obvious that one could employ a metric that combines the advantages of clamping the decoder input level with a jammer state estimate provided by external means. Such metrics have been considered in [35] and their performance analyzed by the general analysis techniques of Chapter 4, Volume I.

In conclusion, we leave the reader with the thought that while many other metrics for the pulse jamming channel with additive background noise are theoretically possible, from a practical standpoint, one wants to select a metric that is easy to implement and robust in the sense that the worst case

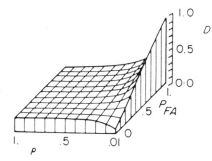

Figure 1.23. Channel parameter versus ρ and P_{FA} for $P_{MD} = 10^{-8}$, $E_s/N_J = 5$ dB, and $N_0/N_J = 1.0$. (Reprinted from [35].)

jammer does not do much more harm than the baseline constant power jammer.

1.9 DISCUSSION

With DS/BPSK signals we have shown that any jammer signal, after despreading at the receiver, can be approximated as additive Gaussian noise. Thus there is an equivalent Gaussian noise channel with BPSK data modulation. For uncoded BPSK data modulation with BPSK PN spreading, however, it was possible for the jammer power in the BPSK signal coordinate (cosine) to be as much as 3 dB more than expected with natural noise of the same power. To combat this potential degradation BPSK data modulation with QPSK PN spreading can be used. In this case the uncoded bit error probability is (see (1.59) and (1.72))

$$P_b = Q\left(\sqrt{\frac{E_b}{\sigma^2}}\right) \tag{1.235}$$

where

$$\sigma^2 = N_J/2 \tag{1.236}$$

is the equivalent two-sided jammer noise spectral density.

Although the jammer statistics can be approximated as Gaussian regardless of the waveform used by the jammer, its power may be varied in time. This results in a Gaussian noise channel where the noise variance varies in time. Denote this variance sequence as $\{\sigma_n^2\}$ where σ_n^2 is the noise variance during the n-th BPSK modulated data symbol.

By concentrating its power on a few uncoded data bits, the jammer can dramatically increase the average bit error probability for the same time-averaged power level. To combat this jammer strategy some form of coding is required. For a sequence of L-coded symbols $x_1, x_2, \ldots, x_L$ where $x_n \in \{1, -1\}$ the ideal decoder metric is (see [33])

$$m(y, x) = \sum_{n=1}^{L} \frac{y_n x_n}{\sigma_n^2} \qquad (1.237)$$

where $y_1, y_2, \ldots, y_L$ is the sampled BPSK correlator output. With this type of decoder metric the degradation due to time-varying jammer power (pulse jamming) can be effectively neutralized.

The ideal soft decision metric (1.237) assumes complete knowledge of the jammer power level during each coded BPSK modulation symbol time. That is, it assumes knowledge of the noise variances $\{\sigma_n^2\}$. Section 1.7 examined various alternative metrics where this jammer state information is not available or only partially available or estimated. Various quantized forms of this metric have also been considered in this chapter.

DS/BPSK waveforms can also be used effectively in fading channels where a more complex receiver structure is required. Here, too, decision and decoding metrics are used that require complete or estimated information regarding the channel.

The discussion of this chapter focussed on BPSK data modulation where coding consists of transmitting a constrained sequence of binary (± 1) symbols each modulated on a BPSK waveform. In some applications it is useful to use orthogonal binary sequences. In particular, $M = 2^L$ orthogonal binary sequences of L bit length can be generated by defining a sequence of $m \times m$ binary matrices $\{H_m\}$ as,

$$H_{m+1} = \begin{bmatrix} H_m & H_m \\ H_m & -H_m \end{bmatrix} \qquad (1.238)$$

where

$$H_1 = \begin{bmatrix} 1 & 1 \\ 1 & -1 \end{bmatrix}. \qquad (1.239)$$

The $M = 2^L$ rows of H_L are L bit length orthogonal binary sequences.

Orthogonal binary sequences can be viewed as an orthogonal code. Alternatively the entire L bit sequence of BPSK waveforms can be considered to be a single M-ary waveform. The set of waveforms form M orthogonal signals which can be demodulated coherently or non-coherently. These M-ary waveforms can be regarded as basic signals for an M-ary input modulation. In some applications DS/BPSK spread-spectrum signals are converted to these M-ary modulations where demodulation is done non-coherently.

In the next chapter we examine M-ary orthogonal signals that are non-coherently demodulated. There the orthogonal signals consist of CW carriers of different frequencies referred to as frequency-shift-keying (FSK) modulation. Many of the results there apply to the M-ary waveforms described above which consist of orthogonal binary sequences modulated onto BPSK waveforms.

1.10 REFERENCES

[1] *Proceedings of the 1973 Symposium on Spread Spectrum Communications*, Naval Electronics Laboratory Center, San Diego, CA, March 13–16, 1973.

[2] *Spread Spectrum Communications*, Lecture Series No. 58, Advisory Group for Aerospace Research and Development, North Atlantic Treaty Organization, July 1973 (AD 766914).

[3] R. C. Dixon, ed., *Spread Spectrum Techniques*, New York: IEEE Press, 1976.

[4] R. C. Dixon, *Spread Spectrum Systems*, New York: John Wiley, 1976.

[5] L. A. Gerhardt and R. C. Dixon, Special Issue on Spread Spectrum Communications, *IEEE Trans. Commun.*, COM-25, August 1977.

[6] D. J. Torrieri, *Principles of Military Communication Systems*, Dedham, MA: Artech House, 1981.

[7] C. E. Cook, F. W. Ellersick, L. B. Milstein, and D. L. Schilling, eds., Special Issue on Spread Spectrum Communications, *IEEE Trans. Commun.*, COM-30, May 1982.

[8] J. K. Holmes, *Coherent Spread Spectrum Systems*, New York: John Wiley, 1982.

[9] G. C. Clark, Jr., and J. B. Cain, *Error-Correction Coding for Digital Communications*, New York: Plenum Press, 1981.

[10] *MILCOM Conference Record*, 1982 IEEE Military Communications Conference, Boston, MA, October 17–20, 1982.

[11] Proceedings of the *1983* Spread Spectrum Symposium, *Long Island, NY. Sponsored by the Long Island Chapter of the IEEE Commun. Soc., 807 Grundy Ave., Holbrook, NY.*

[12] R. A. Scholtz, "The spread spectrum concept," *IEEE Trans. Commun.*, COM-25, pp. 748-755, August 1977.

[13] M. P. Ristenbatt and J. L. Daws, Jr., "Performance criteria for spread spectrum communications," *IEEE Trans. Commun.*, COM-25, pp. 756–763, August 1977.

[14] C. L. Cuccia, "Spread spectrum techniques are revolutionizing communications," *MSN*, pp. 37–49, September 1977.

[15] R. E. Kahn, S. A. Gronemeyer, J. Burchfield, and R. C. Kunzelman, "Advances in packet radio technology," *Proc. IEEE*, vol. 66, pp. 1468–1496, November 1978.

[16] A. J. Viterbi, "Spread spectrum communications—Myths and realities," *IEEE Commun. Mag.*, vol. 17, pp. 11–18, May 1979.

[17] R. L. Pickholtz, D. L. Schilling, and L. B. Milstein, "Theory of spread spectrum communications—A tutorial," *IEEE Trans. Commun.*, COM-30, pp. 855–884, May 1982.

[18] C. E. Cook and H. S. Marsh, "An introduction to spread spectrum," *IEEE Commun. Mag.*, vol. 21, pp. 8–16, March 1983.

[19] M. Spellman, "A comparison between frequency hopping and direct sequence PN as antijam techniques," *IEEE Commun. Mag.*, vol. 21, pp. 37–51, March 1983.

[20] P. A. Kullstam, "A Theoretical Investigation of Spread Spectrum Modulation Concepts and Performance," Ph.D. Thesis, Catholic University of America, May 1977.

[21] P. A. Kullstam, "Spread spectrum performance analysis in arbitrary interference," *IEEE Trans. Commun.*, COM-25, pp. 848–853, August 1977.

[22] W. Feller, *An Introduction to Probability Theory and Its Applications, Vol. II*, New York: John Wiley, 1966.

[23] R. Price, "The detection of signals perturbed by scatter and noise," *IRE Trans. Inform. Theory*, PGIT-4, pp. 163–170, September 1954.

[24] R. Price and P. E. Green, Jr., "A communication technique for multipath channels," *Proc. IRE*, vol. 46, pp. 555–570, March 1958.

[25] R. Price and P. E. Green, Jr., "Signal processing in radar astronomy—Communication via fluctuating multipath media," MIT Lincoln Laboratory, Lexington, Mass., Tech. Report No. 234, October 1960.

[26] T. Kailath, "Correlation detection of signals perturbed by a random channel," *IRE Trans. Inform. Theory*, IT-6, pp. 361–366, June 1960.

[27] T. Kailath, "Channel characterization: Time-variant dispersive channels," Chap. 6, in *Lectures on Communication System Theory*, E. Baghdady, ed., New York: McGraw-Hill, 1961.

[28] G. L. Turin, "On optimal diversity reception," *IRE Trans. Inform. Theory*, IT-7, pp. 154–166, July 1961.

[29] G. L. Turin, "On optimal diversity reception II," *IRE Trans. Commun. Systems*, vol. CS-12, pp. 22–31, March 1962.

[30] R. Price, "Error probabilities for adaptive multichannel reception of binary signals," *IRE Trans. Inform. Theory*, IT-8, pp. 305–316, September 1962.

[31] D. Chase, "Digital signal design concepts for a time-varying Rician channel," *IEEE Trans. Commun.*, COM-24, pp. 164–172, February 1976.

[32] J. F. Pieper, J. G. Proakis, R. R. Reed, and J. K. Wolf., "Design of efficient coding and modulation for a Rayleigh fading channel," *IEEE Trans. Inform. Theory*, IT-24, pp. 457–468, July 1978.

[33] J. G. Proakis, *Digital Communications*, New York: McGraw-Hill, 1983.

[34] F. El-Wailly, "Convolutional Code Performance Analysis of Jammed Spread Spectrum Channels Without Side Information," Ph.D. dissertation, Illinois Institute of Technology, 1982.

[35] F. El Wailly, and D. J. Costello, "Analysis of coded spread spectrum soft decision receivers: Part I – Direct sequence modulation with pulse jamming, Part II–Frequency hopped modualtion with partial band jamming," submitted to *IEEE Trans. Inform. Theory*.

Chapter 2

NON-COHERENT
FREQUENCY-HOPPED SYSTEMS

In Chapter 1 we considered direct-sequence (DS) spread-spectrum (SS) communication systems with coherent phase-shift-keyed (PSK) modulation. These systems use a pseudonoise (PN) sequence to directly spread the data-modulated carrier, producing a continuous $(\sin x/x)^2$ power spectrum (assuming the usual rectangular modulation pulse shape) with bandwidth W_{ss}. Because of practical considerations (it is difficult to synchronize the PN generator in the receiver to sub-nanosecond accuracy), current DS systems typically operate with PN rates of 100 Mchips/sec or less, implying that W_{ss} is limited to several hundred MHz.

We will now analyze the other principal category of SS systems, namely, frequency hopping (FH) with non-coherent M-ary frequency-shift-keyed (MFSK) modulation. This is essentially a conventional MFSK scheme in which the carrier frequency is pseudorandomly hopped over W_{ss} under the control of a PN sequence. Specifically, as illustrated in Figure 2.1, successive (not necessarily disjoint) k-chip segments of the PN sequence drive a frequency synthesizer which hops the carrier over 2^k frequencies. On a given hop, the signal bandwidth is identical to conventional MFSK, which is typically much smaller than W_{ss}; however, averaged over many hops, the FH/MFSK spectrum occupies the entire SS bandwidth. Current technology permits FH bandwidths of the order of several GHz, which is an order of magnitude larger than implementable DS bandwidths. Because they must operate over such wide bandwidths, FH synthesizers generally do not maintain phase coherence over successive hops: consequently, coherent data demodulation techniques are possible only within each hop. As mentioned above and shown in Figure 2.1, this chapter will consider only non-coherent energy detection.

In other publications, the term "fast frequency hopping" (FFH) has occasionally been used to denote systems with relatively high hop rates, independent of the data rate. However, synthesizer technology is progress-

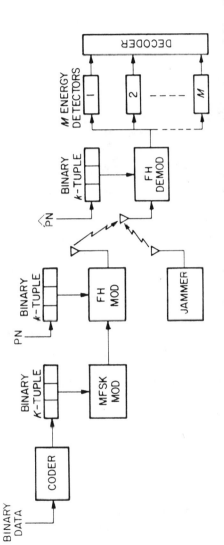

Figure 2.1. Functional block diagram of FH/MFSK system perturbed by jamming. Transmit one of $M = 2^K$ tones; carrier is hopped to one of 2^k frequencies determined by k-chip segments of PN code; dehopping requires derived PN reference ($\widehat{PN}$); non-coherent detection.

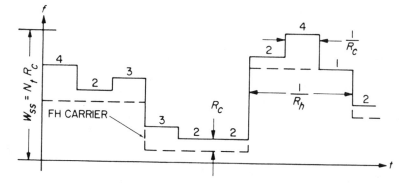

Figure 2.2. Example of SFH/MFSK signal, with $K = 2$ or $M = 4$ symbols $(1, 2, 3, 4)$, $R_c = R_s = R_b/2 = 3R_h$, and N_t equally spaced FH tone frequencies.

ing so rapidly that today's FFH system will invariably become tomorrow's slow frequency-hopped (SFH) system under this terminology. For example, in the recent past it was common for FH systems to operate at several hundred hops/sec, and 10–20 Khops/sec was considered to be state-of-the-art. Yet prototype synthesizer implementations have now been developed that deliver several hundred Khops/sec, and rates greater than 1 Mhop/sec should be realizable in the near term. To avoid technology-dependent terminology, we will define an FFH system to be one in which the hop rate R_h is an integer multiple of the MFSK symbol rate R_s, while the term SFH denotes the reverse condition.

Another terminological ambiguity is the widespread use of the word "chip" to refer to an individual FH/MFSK tone of shortest duration, which should not be confused with the PN chips that drive the frequency synthesizer. In an FFH system where there are multiple hops per M-ary symbol, each hop is a chip, whereas, in an SFH system as shown in Figure 2.2, a chip denotes an M-ary symbol. The chip rate $R_c = \max(R_h, R_s)$ is the highest FH system clock rate; the energy detectors in Figure 2.1 generate outputs at this rate. In the FFH mode, the M-ary symbol metrics are usually formed by linearly combining the R_h/R_s detected chip energies for each symbol, resulting in a non-coherent combining loss that will be discussed in detail later.

With non-coherent detection, the MFSK tones on a given hop must be separated in frequency by an integer multiple of R_c to provide orthogonality. This implies that a transmitted symbol will not produce any crosstalk in the other $M - 1$ energy detectors, and, if the M-ary band contains additive white Gaussian noise (AWGN), the components of that noise in each detector output will be uncorrelated. Figure 2.3(a) depicts a common implementation in which the entire SS band is partitioned into $N_t = W_{ss}/R_c$ equally spaced FH tones; these are then grouped into $N_b = N_t/M$ adjacent,

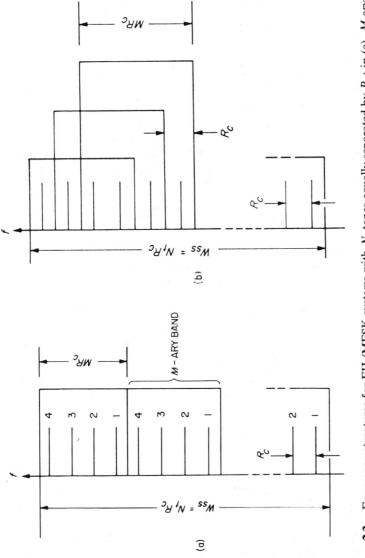

Figure 2.3. Frequency structures for FH/MFSK systems with N_t tones equally separated by R_c: in (a), M-ary bands are contiguous and non-overlapping, whereas in (b), these bands are shifted by R_c, which scrambles the FH tone/M-ary symbol mapping ($M = 4$ here).

non-overlapping M-ary bands, each with bandwidth MR_c. Under this arrangement, the PN binary k-tuples direct the frequency synthesizer to any of $N_b = 2^k$ carrier frequencies, and each FH tone is assigned to a specific, hop-invariant M-ary symbol. It is conceivable that a sophisticated jammer could exploit this fixed assignment scheme. One method of scrambling the FH tone/M-ary symbol mapping from hop to hop is to allow the synthesizer to hop the carrier over all but $M - 1$ of the N_t available frequencies so that adjacent M-ary bands are only shifted by R_c as shown in Figure 2.3(b). A more jam-resistant (and more expensive) approach is to use M distinct frequency synthesizers to individually hop the M-ary symbols, destroying the contiguous nature of an M-ary band.

There are situations in which it is desirable to avoid certain regions of the radio frequency (RF) band (e.g., fading or narrowband jamming), and here FH enjoys a distinct advantage over DS systems. The synthesizer algorithm which maps the k-chip PN segments into specific carrier frequencies can be modified to eliminate these undesirable bands, resulting in a discontinuous spectrum.

In addition to its anti-jam (AJ) capability, an SS signal is generally difficult to detect and even harder to decipher by an unauthorized receiver: this characteristic is usually referred to as a low probability of intercept (LPI). Most interceptors operate as energy detectors, and they have to monitor the received signal long enough to achieve a sufficiently high signal-to-noise ratio (SNR) for reliable detection in the presence of background noise. The LPI advantage of an SS signal is that its power is spread over a bandwidth considerably larger than conventional transmissions, significantly increasing the noise in a receiver that is not privy to the despreading sequence. In the past, when implementable FH systems operated at low hop rates, their transmissions could conceivably be detected by narrowband monitors capable of following the pseudorandom frequency variations (these so-called "frequency followers" could be used to drive repeat-back jammers which could effectively defeat the AJ capability of FH systems). Now, with hop rates of 100 KHz or more, the current generation of FH systems no longer has this vulnerability: an FH interceptor must detect multiple hops of the transmitted signal with a front end wide enough to accommodate the entire SS bandwidth. And, since FH signals can occupy much larger SS bandwidths than DS signals, they have a corresponding LPI advantage.

As with all SS systems, an FH receiver must derive a synchronized replica of the received spreading sequence to perform properly. And the despreading operation, in which the received signal is despread by this SS reference, simultaneously spreads any accompanying interference over W_{ss} or more, so that most of the interfering power can be eliminated by a narrowband filter matched to the data modulation bandwidth. As noted in Volume I, this AJ capability of an SS system is often measured by its processing gain (PG), a term which has been saddled with several conflicting definitions in the literature. For example, in his pioneering text on SS systems, Dixon at one point defines PG as the ratio of the SNR's at the output and input of the despreader [1, p. 8]; elsewhere (e.g., [1, p. 7]), he uses the definition we prefer, PG = W_{ss}/R_b ((3.4) in Chapter 3, Volume I), which has the advantage that it does not depend on the particular choice of modulation or coding. Adding to this confusion, for FH systems, Dixon defines PG to be "the number of available frequency choices" [1, p. 29], which could refer to either N_t or N_b in our notation. This last definition is actually consistent with (3.4) in Volume I under certain constraints: the FH/MFSK scheme described earlier has $N_t = W_{ss}/R_c$ equally spaced tones so that PG = $N_t R_c/R_b$. For the special case of SFH with uncoded binary FSK (BFSK, or MFSK with $M = 2$), $R_c = R_b$ so that PG reduces to N_t; this holds even if the SS bandwidth is not contiguous.

Although our definition of PG is universally valid for all SS systems, the interpretation of W_{ss} is not always obvious. An instructive case in point is a generalization of the FH/MFSK implementation of Figure 2.3. Consider the structure of Figure 2.4, where each M-ary band again has tones with uniform spacing R_c, but the bands themselves are now uniformly separated by an arbitrary amount ΔW, which is algebraically negative if the bands overlap. For clarification, $\Delta W = 0$ in Figure 2.3(a), while $\Delta W = -(M - 1)R_c$ in Figure 2.3(b). In the overlapping band case, with $-MR_c < \Delta W < 0$, the spectrum averaged over many hops is approximately rectangular with

$$W_{ss} = N_b M R_c - (N_b - 1)|\Delta W|. \tag{2.1}$$

However, when $\Delta W > 0$, there are unused gaps in the multihop spectrum so that the occupied SS bandwidth that contributes to the PG is given by

$$W_{ss} = N_b M R_c \tag{2.2}$$

under the simplified approximation of a piecewise rectangular spectrum.

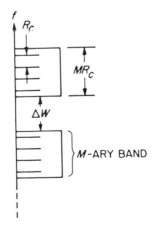

Figure 2.4. A generalization of frequency structure of FH/MFSK systems (shown for $M = 4$) in Figure 2.3: each M-ary band contains tones with uniform separation R_c, but adjacent bands are spaced ΔW apart.

All of this underscores our contention that while PG is a gross measure of the SS advantage relative to conventional modulation techniques, it is more accurate and meaningful to quantitatively characterize the performance of a specific system over a particular jamming channel by the bit error probability (P_b) as a function of the received SNR (see (3.3), Volume I). The remainder of this chapter is devoted to this type of characterization for non-coherent FH/MFSK systems. Where possible, exact closed form performance expressions will be derived; however, in most cases involving coded communication links, exponentially tight upperbounds on P_b will be computed in the interest of mathematical expediency. Exact analyses of coded systems generally prove to be intractable, necessitating the use of computer simulations or numerical integrations. While these approaches can generate more accurate results than bounding techniques, they are inconvenient, requiring a new computer calculation whenever even a single system parameter is changed. Furthermore, the functional dependence of a closed form P_b bound on the various system parameters can provide useful insights into system sensitivity to variations in these parameters. Of course, where simulation results are available, comparisons will be made to indicate the accuracy of our bounds.

Our examination of the performance of non-coherent FH/MFSK communications begins with uncoded signals received over the familiar additive white Gaussian noise (AWGN) channel. This could represent the situation in which a relatively unsophisticated jammer injects spectrally flat noise over the entire SS bandwidth into the FH receiver, and will be used as a benchmark against which more intelligent jammers and FH countermeasures can be compared.

As in the DS case, we will conservatively adopt the worst case perspective that the jammer has *a priori* knowledge of all relevant signalling parameters, with the critical exception of real time PN spreading sequence synchronization. Specifically, a smart jammer is assumed to have the ability to optimize its strategy to exploit information about W_{ss}, M, N_t, N_b, R_b, R_h, the location of the FH tones and the M-ary bands, the channel code and decoding algorithm, the detection metric, the signal power, and the nominal P_b. Also, reflecting realistic jamming scenarios, we will neglect the received thermal or non-hostile background noise in deference to the typically dominant jamming power.

We will consider the two principal types of intelligent but non-adaptive FH jamming threats, namely, partial-band noise and multitone interference. It will be shown that when such jammers optimize their system parameters based on their assumed knowledge of the FH/MFSK target, the resulting performance degradation can be severe, particularly at low error probabilities. To counteract this threat, the FH system must incorporate some coding redundancy which allows the received data decisions to be based on multiple hops. One particularly simple coding scheme that we will consider is time diversity or repetition coding, which can also be used in a concatenation structure to augment the effectiveness of the more powerful block and convolutional codes.

In analyzing these coded systems, we will focus on a detection metric that assumes that the receiver can determine with certainty whether a given hop is jammed. This is the so-called perfect jamming state knowledge or side-information case discussed in Chapters 3 and 4, Volume I, and it is a common assumption in many published articles on pulsed jamming of DS systems and the dual case of partial-band jamming of FH systems (e.g., [2, p. 288]). However, we should caution that in practice a receiver's derived jamming state side information is subject to error, and such a metric is non-robust in the sense that it depends on the received signal, thermal noise, and jamming power levels. While a few other, more robust metrics are considered at the end of this chapter, our main intent here is not to exhaustively consider all possible detection metrics, but rather to use the principal selected metric to illustrate techniques for analyzing the performance of a variety of coded FH/MFSK systems in the presence of different jamming strategies.

We will also not consider the class of adaptive jammers known as repeat-back or frequency-follower jammers. As technology moves to higher hop rates, this threat becomes less viable since it requires that the jammer intercept the FH signal, detect the frequency of the M-ary band, and synthesize an appropriate narrowband signal, all within each hop dwell time.

2.1 BROADBAND NOISE JAMMING

Consider a non-coherent FH/MFSK communication link with received power S transmitting data at a bit rate R_b: the received energy per bit is

$$E_b = S/R_b. \tag{2.3}$$

Suppose this signal is jammed by Gaussian noise with received power J and an approximately rectangular spectrum which coincides with the FH bandwidth W_{ss}. This is equivalent to an AWGN channel, with an effective noise power spectral density

$$N_J = J/W_{ss}. \tag{2.4}$$

The SS system can therefore be characterized by the SNR

$$E_b/N_J = (S/J)(W_{ss}/R_b). \tag{2.5}$$

Identifying PG $= W_{ss}/R_b$, S/J as the signal-to-jamming power ratio prior to despreading, and E_b/N_J as the despread SNR in the information bandwidth, we see that this particular example agrees with Dixon's definition of PG as the ratio of the despread-to-spread SNR's.

We will initially restrict our attention to uncoded SFH systems with an alphabet of $M = 2^K$ orthogonal signals, each containing K bits of information. The chip, M-ary symbol, bit, and hop rates satisfy

$$R_c = R_s = R_b/K \geq R_h. \tag{2.6}$$

The received symbol energy is then

$$E_s = S/R_s = KE_b. \tag{2.7}$$

Without loss of generality, assume that symbol 1 is sent on a given data transmission. Referring to the block diagram of Figure 2.1, the output e_1 of the first energy detector, normalized for convenience by $1/N_J$, is a non-central chi-square random variable with 2 degrees of freedom, whose probability density function is given by

$$p(e_1) = \begin{cases} \exp(-e_1 - E_s/N_J)I_0\left(2\sqrt{e_1 E_s/N_J}\right); & e_1 \geq 0 \\ 0; & e_1 < 0 \end{cases} \tag{2.8}$$

where $I_0(\cdot)$ is the zeroeth-order modified Bessel function of the first kind. Since the symbols are separated in frequency by the symbol rate R_s, the other $M - 1$ energy detectors contain no signal components. Their outputs are identically distributed central chi-square random variables with 2 degrees of freedom:

$$p(e_i) = \begin{cases} \exp(-e_i); & e_i \geq 0 \\ 0; & e_i < 0 \end{cases} \quad (i = 2, \ldots, M). \tag{2.9}$$

Because the M signals are orthogonal, the e_i's are all statistically independent.

The system performance in this example is the same as for non-coherent detection of conventional (unhopped) M-ary orthogonal signals in AWGN. The derivation of this performance has been widely documented (e.g., [3, Section 8.10]): however, it is reviewed here for the sake of completeness. The M-ary symbol error probability (P_s) is given by

$$P_s = \Pr\left\{ \bigcup_{i=2}^{M} (e_i \geq e_1) \right\} \tag{2.10a}$$

$$= 1 - \Pr\left\{ \bigcap_{i=2}^{M} (e_i < e_1) \right\}$$

$$= 1 - \int_0^\infty de_1\, p(e_1)\left[\int_0^{e_1} de_2\, p(e_2) \right]^{M-1}$$

$$= 1 - e^{-E_s/N_J} \int_0^\infty du\, u\, e^{-u^2/2} I_0\left(u\sqrt{\frac{2E_s}{N_J}} \right)\left[\int_0^u dv\, v\, e^{-v^2/2} \right]^{M-1}$$

$$\tag{2.10b}$$

where (2.8) and (2.9) were used with the change of variables $u = \sqrt{2e_1}$ and $v = \sqrt{2e_2}$. But

$$\left[\int_0^u dv\, v\, e^{-v^2/2} \right]^{M-1} = \left(1 - e^{-u^2/2} \right)^{M-1}$$

$$= \sum_{j=0}^{M-1} (-1)^j \binom{M-1}{j} e^{-ju^2/2} \qquad (2.11)$$

so that (2.10b) reduces to

$$P_s = 1 - e^{-E_s/N_J} \sum_{j=0}^{M-1} (-1)^j \binom{M-1}{j}$$

$$\times \underbrace{\int_0^\infty du\, u\, e^{-(j+1)u^2/2} I_0\!\left(u\sqrt{\frac{2E_s}{N_J}} \right)}_{\left(\dfrac{1}{j+1}\right) e^{E_s/(j+1)N_J}}. \qquad (2.12)$$

The elimination of the integral in (2.12) is based on the identity

$$\int_0^\infty de_1\, p(e_1) = 1 \qquad (2.13)$$

with appropriate parameter changes. Letting $i = j + 1$, (2.12) becomes

$$P_s = 1 - \frac{1}{M} \sum_{i=1}^{M} (-1)^{i-1} \binom{M}{i} e^{-(E_s/N_J)(1-1/i)}$$

$$= \frac{1}{M} \sum_{i=2}^{M} (-1)^i \binom{M}{i} e^{-(E_s/N_J)(1-1/i)}. \qquad (2.14)$$

For larger values of K (i.e., K's of the order of 5 or more), the summation of (2.14) involves an exponentially large number of terms with alternating signs, each of which is composed of the product of a factor $\binom{M}{i}$ which can be very large in the neighborhood $i \sim M/2$, and a very small factor, $\exp[-(E_s/N_J)(1 - 1/i)]$. Consequently, evaluation of this expression on a digital computer often yields unsatisfactory results. A more practical approach is to apply numerical integration techniques to the integral of (2.10b) and (2.11):

$$P_s = 1 - e^{E_s/N_J} \int_0^\infty du\, u\, e^{-u^2/2} \left(1 - e^{-u^2/2} \right)^{M-1} I_0\!\left(u\sqrt{\frac{2E_s}{N_J}} \right). \qquad (2.15)$$

However, for large values of E_s/N_J (i.e., small P_s), the value of the integral is close to 1, and the expression is again inherently inaccurate. To reduce the computational burden, make the variable change $x = u^2/2$:

$$P_s = 1 - e^{-E_s/N_J} \int_0^\infty dx\, e^{-x} \left(1 - e^{-x} \right)^{M-1} I_0\!\left(2\sqrt{\frac{xE_s}{N_J}} \right). \qquad (2.16)$$

Now, integrating (2.16) by parts, and introducing the Marcum Q-function [4] defined by

$$Q_M(\sqrt{2\gamma},\sqrt{2x}) = \int_x^\infty du\, e^{-(u+\gamma)}I_0(2\sqrt{u\gamma}) \qquad (2.17)$$

with

$$Q_M(\sqrt{2\gamma},0) = 1$$

we can write

$$P_s = (M-1)\int_0^\infty dx\left[1 - Q_M\left(\sqrt{\frac{2E_s}{N_J}},\sqrt{2x}\right)\right]e^{-x}(1-e^{-x})^{M-2}.$$

$$(2.18)$$

In practice, (2.18) can be evaluated accurately by using about 300 equally spaced increments for x between 0 and 30; iterative formulas exist for computing $Q_M(\cdot,\cdot)$.

As an alternative to computing P_s exactly using the techniques discussed above, it is often more convenient and insightful to evaluate the simple, closed form expression for its union upperbound: from (2.10a),

$$P_s \le (M-1)\text{Pr}\{e_2 \ge e_1\}$$

$$= \left(\frac{M-1}{2}\right)\exp(-E_s/2N_J) \qquad (2.19)$$

where we have applied (2.14) with $M = 2$. Note that the upperbound in (2.19) is the leading (i.e., $i = 2$) term in the summation of (2.14), and is satisfied with equality for $M = 2$.

When a symbol error occurs, the decision is equally likely to favor any of the $M-1$ incorrect orthogonal symbols. Since $M/2$ of these incorrect binary K-tuples will produce an error in a given bit, the probability of a bit error is given by [3, (8.14)]

$$P_b = \frac{M}{2(M-1)}P_s \qquad (2.20a)$$

$$= \frac{1}{2(M-1)}\sum_{i=2}^M (-1)^i\binom{M}{i}e^{-(KE_b/N_J)(1-1/i)} \qquad (2.20b)$$

$$\le 2^{K-2}e^{-KE_b/2N_J} \qquad (2.20c)$$

where (2.20c) is satisfied with equality when $K = 1$.

The performance results above are illustrated in Figure 2.5. It is evident that as E_b/N_J becomes large enough for a given K or M, the union bound approaches arbitrarily close to the exact bit error probability, reflecting the dominance of the leading term in the summation of (2.20b). Also, the performance improves as K increases, which is to be expected since MFSK modulation is equivalent to using an (M, K) binary block orthogonal code (i.e., a code which maps blocks of K bits into $M = 2^K$ symbol codewords).

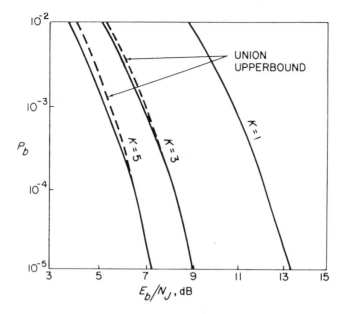

Figure 2.5. Performance of conventional MFSK ($M = 2^K$) system over AWGN channel, with bit energy E_b and noise power density N_J.

The broadband noise jamming results of (2.20) will be used as a reference point for comparing the performance of more sophisticated jammers and FH schemes in the sections that follow.

2.2 WORST CASE JAMMING

In this section we will examine the degradation in performance of uncoded SFH/MFSK systems that occurs when a smart jammer has *a priori* knowledge of all FH system parameters except for the PN code (the jammer can even know which PN code is being used provided that it cannot achieve real time PN synchronization), and devises an optimum strategy to exploit this. We will consider the two most effective jamming strategies against FH systems: partial-band noise and multitone jamming. In the multitone category, we will analyze several distinct approaches, and identify the best of these.

Much of the credit for the initial work in this area belongs to Houston [5], and we will refer frequently to his results below.

2.2.1 Partial-Band Noise Jamming

Suppose a Gaussian noise jammer chooses to restrict its total power J (referenced to the FH receiver input) to a fraction ρ ($0 < \rho \leq 1$) of the full SS bandwidth W_{ss}. As shown in Figure 2.6, the jamming noise power is

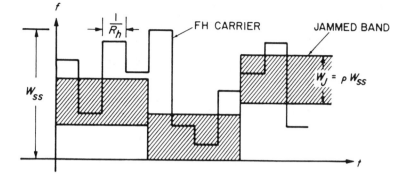

Figure 2.6. Partial-band noise jamming of FH system: jammer concentrates power in fraction $\rho \in (0, 1]$ of SS bandwidth, and hops noise band to prevent FH band avoidance countermeasure.

spread uniformly over $W_J = \rho W_{ss}$, resulting in an increased power density

$$N_J' = \frac{J}{W_J} = \frac{J}{\rho W_{ss}} = \frac{N_J}{\rho},\qquad (2.21)$$

and a correspondingly degraded SNR level

$$\frac{E_b}{N_J'} = \frac{\rho E_b}{N_J}\qquad (2.22)$$

in the jammed band, where N_J is still defined by (1.4).

Recall that an FH system can in principle avoid certain frequency bands that it determines are particularly noisy. Consequently, we assume as in Figure 2.6 that the jammer hops the jammed band over W_{ss}, slowly relative to the FH dwell time $1/R_h$, but often enough to deny the FH system the opportunity to detect that it is being jammed in a specific portion of W_{ss} and take remedial action. Also, to simplify the analysis, we will assume that shifts in the jammed band coincide with carrier hop transitions, so that the channel is stationary over each hop. Furthermore, we will assume that on a given hop, each M-ary band lies entirely inside or outside W_J. This last restriction is common to most analytical treatments of partial-band FH jamming and the dual case of pulsed DS jamming. (Since the FH carrier hops in and out of W_J, partial-band jamming can be regarded as pulsed jamming with non-uniformly spaced pulses of duration $1/R_h$, under the simplifying assumptions made above.) The performance computed based on these assumptions is actually somewhat pessimistic, as Viterbi has noted [6, p. 14]: on a given M-ary symbol transmission, if only part of the M-ary band is jammed, and/or if it is only jammed over part of the symbol band, less noise is intercepted by the energy detectors, thereby reducing the probability of error. We should add that it does not matter whether the jammed band W_J is a single contiguous region as suggested by Figure 2.6: the analysis below is transparent to partitions of W_J so long as each of them satisfies the assumptions above.

Because of the pseudorandom hopping, it is reasonable to model the FH/MFSK system in partial-band noise as a two-state channel, independent from hop to hop. With probability ρ, an M-ary transmission is jammed and the conditional P_b is determined by the SNR ratio of (2.22); but, since we are neglecting thermal noise, with probability $(1 - \rho)$, the transmission is noiseless and an error-free decision is made. Then the average error rate is simply

$$P_b = \rho P_b\left(\frac{\rho E_b}{N_J}\right),\qquad(2.23)$$

where the term on the right denotes the expression of (2.20b) with E_b/N_J replaced by $\rho E_b/N_J$.

What (2.23) tells us is that if ρ is reduced, the probability that an M-ary transmission is jammed is decreased, but jammed signals suffer a higher conditional error rate: the net effect may degrade the average FH/MFSK performance, depending on the values of M and E_b/N_J. The utility of jamming only part of the RF band is illustrated in Figure 2.7 for $M = 2$. Suppose S, J, W_{ss}, and R_b combine to make $E_b/N_J = 10.9$ dB. In broadband noise ($\rho = 1$), the resulting $P_b = 10^{-3}$.

If the jammer concentrates the same noise power over half the SS band ($\rho = 1/2$), only half the transmissions are jammed, but these have a

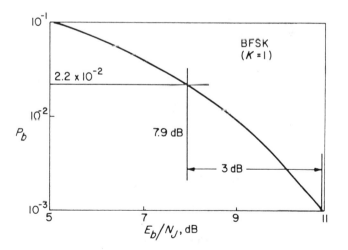

Figure 2.7. Illustration of partial-band jamming advantage against FH/MFSK systems. Referring to (2.20b) and (2.23), if $M = 2$ and $E_b/N_J = 10.9$ dB, when $\rho = 1$ then $P_b = 10^{-3}$; however, if $\rho = 1/2$, then conditional SNR ratio in jammed band is 7.9 dB, so that conditional $P_b = 2.2 \times 10^{-2}$, resulting in average $P_b = 1.1 \times 10^{-2}$.

conditional $P_b = 2.2 \times 10^{-2}$, which results in an average $P_b = 1.1 \times 10^{-2}$ according to (2.23). So in this example, reducing ρ to $1/2$ degrades the performance more than an order of magnitude because of the steepness of the P_b curve in the selected region. (The results below indicate that the worst performance for these parameters occurs at $\rho = .16$, for which $P_b = 3.0 \times 10^{-2}$.)

Figure 2.8 illustrates the performance of an FH/BFSK system in partial-band noise for several partial-band jamming factors ρ. For small enough E_b/N_J, it is evident that broadband noise jamming ($\rho = 1$) is the most effective. In general, for any value of E_b/N_J, there is an optimum value of $\rho \in (0, 1]$ from the jammer's viewpoint which maximizes P_b, and this is denoted by ρ_{wc} (for worst case jamming). The performance in worst case partial-band noise is the upper envelope (or supremum) of the family of P_b curves for fixed values of ρ: as shown in Figure 2.8, when E_b/N_J exceeds a threshold level, $\rho_{wc} < 1$ indicating a partial-band jamming advantage, and the performance curve is a straight line. Of course, in practice, it may be difficult for the jammer to match ρ to the actual E_b/N_J.

The worst case partial-band noise jammer chooses ρ to maximize the P_b for a given M and E_b/N_J. From (2.20b) and (2.23), the resulting average

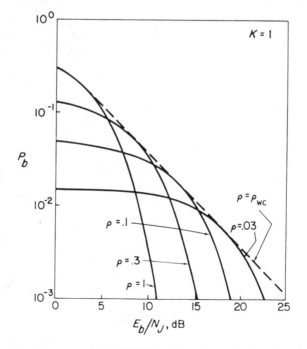

Figure 2.8. Performance of FH/BFSK system in partial-band noise for several fixed values of ρ. The performance in worst case partial-band noise is realized when the jammer chooses $\rho = \rho_{wc}$ to maximize P_b for a given E_b/N_J. Note that ρ_{wc} decreases as E_b/N_J gets larger.

performance can be expressed as

$$P_b = \max_{0 < \rho \leq 1} \left[\frac{\rho}{2(M-1)} \sum_{i=2}^{M} (-1)^i \binom{M}{i} e^{-(\rho K E_b/N_J)(1-1/i)} \right]. \quad (2.24)$$

For $M = 2$, this maximization is a simple mathematical calculation; for larger values of M, it must be evaluated numerically. The results have the form [5, (15) and (16)]

$$P_b = \begin{cases} \dfrac{1}{2(M-1)} \displaystyle\sum_{i=2}^{M} (-1)^i \binom{M}{i} e^{-(KE_b/N_J)(1-1/i)}, \\[2mm] \quad \text{and } \rho_{wc} = 1; \ \dfrac{E_b}{N_J} \leq \gamma \\[4mm] \dfrac{\beta}{E_b/N_J}, \text{ and } \rho_{wc} = \dfrac{\gamma}{E_b/N_J}; \ \dfrac{E_b}{N_J} \geq \gamma \end{cases} \quad (2.25)$$

where the parameters β and γ are tabulated for $1 \leq K \leq S$ below and ρ_{wc} denotes the jammer's optimum ρ.

(2.25) demonstrates that so long as E_b/N_J is not unusually small, worst case partial-band jamming converts the exponential relationship between P_b and E_b/N_J in (2.20) into an inverse linear dependence. As shown in Figure 2.9, the resulting degradation can be severe for small P_b's: for example, the loss is 14.7 dB at $P_b = 10^{-3}$ for $K = 1$, and increases with K. At P_b's of 10^{-5} and lower, this gap exceeds 30 dB for any K, illustrating the effectiveness of worst case partial-band noise jamming against uncoded FH/MFSK signals at typical operating points.

(2.25) indicates that ρ_{wc} becomes very small for large E_b/N_J's; that is, a worst case noise jammer concentrates its power in a small portion of W_{ss} at low P_b's. The signals do not get jammed most of the time, but those that do are likely to result in errors. This is an indication that some form of coding redundancy that causes data decisions to depend on multiple symbol transmissions can reduce the effectiveness of partial-band jamming; the degree to which this statement is true will become evident in Section 1.3.

Table 2.1
Parameters associated with performance of uncoded FH/MFSK signals in worst case partial-band noise, as defined in (2.25).

K	β	γ, dB
1	$e^{-1} = .3679$	3.01
2	.2329	.76
3	.1954	− .33
4	.1812	− .59
5	.1759	−1.41

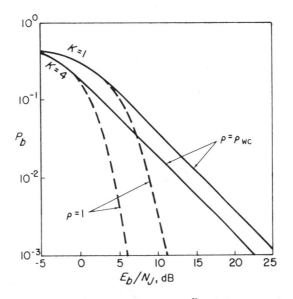

Figure 2.9. Degradation in FH/MFSK ($M = 2^K$) performance due to worst case ($\rho = \rho_{wc} \leq 1$) partial-band noise relative to broadband ($\rho = 1$) noise jamming.

2.2.2 Multitone Jamming

A second, sometimes more effective, class of intelligent FH jamming than partial-band noise is multitone or multiple CW tone interference. In this category, the jammer divides its total received power J into Q distinct, equal power, random phase CW tones. These are distributed over the spread-spectrum bandwidth W_{ss} according to one of several strategies illustrated in Figure 2.10 and discussed in detail below. One of the reasons that multitone jamming can be more effective against FH/MFSK signals than partial-band noise is that CW tones are the most efficient way for a jammer to inject energy into the non-coherent detectors.

The analysis below involves several simplifying assumptions to allow us to focus on the issues of interest. We continue to neglect receiver thermal noise under the assumption that it is dominated by the jamming interference. As shown in Figure 2.10, we assume that each jamming tone coincides exactly in frequency with one of the N_t available FH slots, with at most one tone per slot. Furthermore, while the multitone jammer may periodically rearrange the location of its tones to thwart any FH avoidance measures, it is assumed that such changes coincide with hop transitions. While these artificial constraints could never be achieved in practice, like the earlier partial-band jamming assumptions, they simplify the analysis and yield somewhat pessimistic performance results. For example, if a jamming tone is offset in frequency from an FH slot, less energy gets into the adjacent MFSK detector reducing the jamming effectiveness, and the performance must be averaged over the frequency offset distribution.

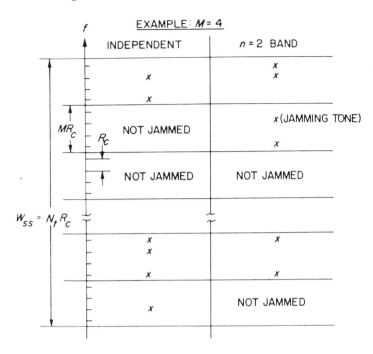

Figure 2.10. Multitone jamming strategies: independent multitone jamming distributes the tones pseudorandomly over all N_t FH frequencies; band multitone jamming places $n \in [1, M]$ tones in each jammed M-ary band.

We denote the fraction of the FH slots jammed by

$$\rho \equiv \frac{Q}{N_t} \tag{2.26}$$

analogous to the fraction of the SS bandwidth jammed in the partial-band noise case. However, there was a uniformity in the partial-band jamming scheme that does not generally carry over to the multitone case. In the former, an entire M-ary band was assumed to be evenly degraded by Gaussian noise if it was jammed at all. Under the various multitone scenarios, the number of symbols in an M-ary band that can be hit by jamming tones can range anywhere between zero and all M. A single jamming tone hitting any of the $M - 1$ untransmitted symbols can produce an error, provided that the jamming tone power J/Q is slightly greater than the signal power S, in the absence of additive noise. Since the jammer is presumed to not know the hopping sequence, it is to its advantage to place its tones in as many M-ary bands as possible, even if most of these jammed bands contain a single jamming tone. Consequently, with regard to multitone jamming, a more significant parameter than ρ is

$$\mu = \Pr\{\text{any symbol in an } M\text{-ary band is jammed}\} \tag{2.27}$$

Note that $\mu = \rho$ in the partial-band noise case.

If in fact it can be shown that the best multitone strategy is to maximize μ, the jammer should leave at least $M - 1$ unjammed FH slots between each of its Q available tones (assuming $Q \leq N_t/M$) so that no M-ary band can contain more than one jamming tone. Of course, this assumes that the FH system adopts the distinct M-ary band structure of Figure 2.3(a) or (b), with all MFSK symbols occupying adjacent FH slots. Without loss of generality, the multitone analysis below assumes the non-overlapping band structure of Figure 2.3(a) for conceptual convenience, although the results would be unchanged for the arrangement of Figure 2.3(b).

The "band multitone" strategy of Figure 2.10 is a generalization of these arguments in which a jammed band contains exactly n jamming tones, under the implied assumption that Q/n is an integer. If the jammed bands are selected pseudorandomly, we have

$$\mu = \frac{QM}{nN_t} = \frac{M}{n}\rho.$$ (2.28)

In previous publications, multitone jamming analyses of non-coherent FH/MFSK systems have restricted the jamming tones to be contiguous in the frequency domain, with uniform spacing R_c or MR_c [5, p. 53] and [7]. Indeed, Trumpis refers to this structure as "partial-band multitone jamming" by analogy to the noise jamming case. In our notation, these two schemes fall into the band multitone jamming category with $n = 1$ or M. If a large μ is desirable, $n = M$ band multitone jamming should prove to be relatively ineffective since it does not judiciously allocate its available power to the largest number of bands.

A simpler strategy that bypasses the assumptions that permit the band multitone implementation is to pseudorandomly distribute the Q jamming tones uniformly over the N_t available FH slots. Under this so-called "independent multitone" scheme shown in Figure 2.10, a given FH slot is jammed with probability ρ, and we assume that other slots are independently jammed with the same probability. In fact, the independence assumption is not always justified. Conditioned on one FH slot being jammed, another given slot is jammed with probability $(Q - 1)/(N_t - 1)$, which is essentially equal to ρ in (2.26) implying independent, equally likely jamming of both slots provided $N_t \geq Q \gg 1$. Since our performance analysis focusses on the dehopped M-ary band containing the transmitted data symbol, we can assume that each of the M slots in that band is independently tone jammed with probability ρ by extending the provision above to

$$N_t \geq Q \gg M - 1.$$ (2.29)

Under this assumption,

$$\mu = 1 - (1 - \rho)^M$$ (2.30)

which reduces to the value of μ for $n = 1$ band multitone jamming when ρ is small; this suggests that these two strategies may be equally effective for $\rho \ll 1$.

2.2.2.1 Random Jamming Tone Phase

For all of the multitone strategies under consideration, there is the possibility that a given MFSK received signal will itself be jammed. When this occurs, the relative phase of the data and jamming tones impacts the energy detector output for that frequency, a factor first analyzed by Trumpis [7, pp. 15–16]. We have already assumed that there is no frequency offset between the data and jamming tones. We now additionally assume that the phase difference ϕ is uniformly distributed over $[0, 2\pi)$.

The received signal power is S: denote the received power in each jamming tone by

$$\frac{J}{Q} = \frac{S}{\alpha}, \tag{2.31}$$

where α is a parameter to be optimized by the jammer. Then a received data signal which has been tone jammed has the composite form

$$\sqrt{2S} \sin \omega_0 t + \sqrt{\frac{2S}{\alpha}} \sin(\omega_0 t + \phi) = \sqrt{2S'} \sin(\omega_0 t + \phi'), \tag{2.32}$$

where the phasor representation of (2.32) in Figure 2.11 indicates that the total resultant power is

$$S' = S\left[\left(1 + \frac{1}{\sqrt{\alpha}}\cos\phi\right)^2 + \left(\frac{1}{\sqrt{\alpha}}\sin\phi\right)^2\right]$$

$$= S\left(1 + \frac{2}{\sqrt{\alpha}}\cos\phi + \frac{1}{\alpha}\right). \tag{2.33}$$

The energy detector output for that M-ary symbol (or chip, in the uncoded case) is

$$\frac{S'}{R_c} = E_c\left(1 + \frac{2}{\sqrt{\alpha}}\cos\phi + \frac{1}{\alpha}\right), \tag{2.34}$$

where the symbol (chip) energy $E_c = S/R_c = S/R_s$. Table 2.2 extends (2.34) to the other three cases required to determine the system performance with multitone jamming in the absence of thermal noise.

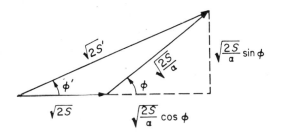

Figure 2.11. Phasor representation of tone jammed MFSK data signal.

Table 2.2
Normalized energy detector outputs.

	If tone jammed	If not jammed
Transmitted M-ary symbol	$1 + \dfrac{2}{\sqrt{\alpha}}\cos\phi + \dfrac{1}{\alpha}$	1
Any of the $(M - 1)$ other symbols	$\dfrac{1}{\alpha}$	0

The table reveals the range of α for which symbol errors can occur. If none of the M dehopped symbols is jammed, there can be no error. If the data symbol is not jammed and *any* of the other symbols is hit with a jamming tone, an error will *always* be made if $\alpha < 1$, and *never* for $\alpha > 1$ (ties that occur when $\alpha = 1$ can be resolved with an M-sided coin flip, or simply assigned to the error side of the ledger). Note that in the uncoded case, with hard decisions made on each symbol, there is no advantage to hitting more than one of the untransmitted symbols. If only the data symbol is hit, its energy detector output is greater than zero except for a singular value of ϕ depending on α; since that singular point occurs with probability zero, an error cannot occur. Finally, if the transmitted symbol and any of the other symbols are simultaneously jammed, an error can occur with probability

$$\Pr\left\{\cos\phi < -\frac{\sqrt{\alpha}}{2}\right\} = \frac{1}{\pi}\cos^{-1}\left(\frac{\sqrt{\alpha}}{2}\right), \qquad (2.35)$$

which is non-zero for $0 \le \alpha < 4$. This is a surprising result. Most multitone jamming analyses simply assume axiomatically that the jammer should set $\alpha = 1_-$, so that each jamming tone has power slightly in excess of S. Houston pointed out that the jammer could do better by optimizing α, but he neglected the impact of the random jammer phase and restricted α to the range $(0, 1)$ [5, pp. 53–54]. Only Trumpis [7, p. 16] recognized that an error could still occur even when the power in each jamming tone was up to 6 dB *below* the received signal power. It remains to be seen whether in fact a worst case jammer would ever select $\alpha > 1$ for any of the multitone strategies, and this will be resolved in the next two sections.

2.2.2.2 Band Multitone Jamming

The simplest scheme to analyze is $n = 1$ band multitone jamming. The random jamming phase has no effect on the system performance, and it demonstrates the basic approach to determining the worst case α for the other multitone strategies.

From Table 2.2 and the discussion above, a symbol error can only occur if the M-ary band containing the transmitted data is jammed, and then only if the data symbol itself is not hit and $\alpha < 1$:

$$P_s = \mu\left(\frac{M-1}{M}\right). \tag{2.36}$$

Since we know that

$$N_t = \frac{W_{ss}}{R_s} = \frac{KW_{ss}}{R_b} \tag{2.37}$$

(2.3), (2.4), (2.28), and (2.31) allow us to rewrite μ in the form

$$\mu = \frac{\alpha M}{nKE_b/N_J} \tag{2.38}$$

for multitone band jamming with arbitrary n.

Recalling the relation between P_s and P_b (see (2.20a)), and setting $n = 1$, the performance for a given α is specified by

$$P_b = \frac{\alpha M}{2KE_b/N_J} \tag{2.39}$$

The worst case performance is achieved by maximizing α. But, in addition to the constraint $\alpha < 1$, there is the requirement that the number of jamming tones not exceed the number of M-ary bands, i.e., $\mu \leq 1$. Therefore, the worst case $n = 1$ band multitone jammer sets α equal to

$$\alpha_{wc} = \begin{cases} \dfrac{K}{M}\left(\dfrac{E_b}{N_J}\right); & \dfrac{E_b}{N_J} < \dfrac{M}{K} \\[3mm] 1_-; & \dfrac{E_b}{N_J} \geq \dfrac{M}{K} \end{cases} \tag{2.40}$$

for which (2.39) reduces to

$$P_b = \begin{cases} \dfrac{1}{2}; & \dfrac{E_b}{N_J} \leq \dfrac{2^K}{K} \\[3mm] \dfrac{2^{K-1}}{KE_b/N_J}; & \dfrac{E_b}{N_J} \geq \dfrac{2^K}{K} \end{cases} \tag{2.41}$$

since $M = 2^K$.

Under the worst case partial-band noise scenario, the relationship between P_b and E_b/N_J became inverse linear for P_b's below a threshold that depended on K (see (2.25)); here that same type of relationship arises *for all* $P_b < 1/2$, independent of K. Both worst case jammers are significantly more effective than broadband noise against uncoded FH/MFSK signals with non-coherent detection, especially for large E_b/N_J, as shown in Figure 2.12. However, of the two, it is clear that the multitone structure is a better

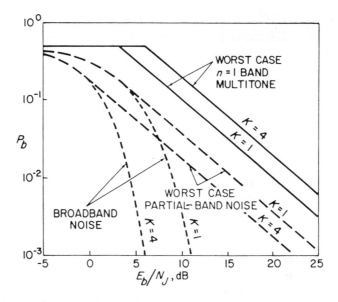

Figure 2.12. Advantage of $n = 1$ band multitone strategy over partial-band noise and broadband noise jamming of uncoded FH/MFSK signals.

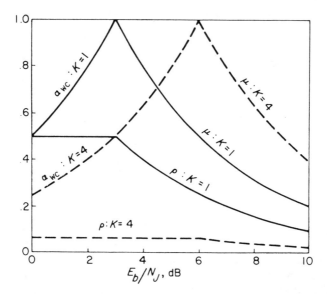

Figure 2.13. Parameters characterizing worst case $n = 1$ band multitone jamming of uncoded FH/MFSK signals.

jamming strategy, particularly for larger values of K. The performance in additive noise improves with K due to the block orthogonal coding gain implicit in the MFSK modulation; *for multitone jamming the performance degrades with K* since the critical parameter μ is proportional to $2^K/K$. In the inverse linear performance regions, the advantage of worst case $n = 1$ band multitone jamming over partial-band noise is ((2.25) and (2.41))

$$\Delta(E_b/N_J) = 10\log_{10}\left(\frac{2^{K-1}}{K\beta}\right)\,\text{dB} \qquad (2.42)$$

where β has been computed by Houston [5, (16); or see our corrected Table 2.1 above]. In particular, $\Delta(E_b/N_0) = 4.3$ dB at $K = 1$, and 10.5 dB at $K = 4$. Since β decreases monotonically with K, the effectiveness of the worst case partial-band noise and $n = 1$ band multitone jamming schemes continues to diverge as K increases.

Figure 2.13 illustrates that for a given value of K, when E_b/N_J falls below the threshold specified in (2.40), the entire SS bandwidth is saturated with one jamming tone per M-ary band. In this saturation region, $\mu = 1$ and $\rho = 1/M$ (refer to (2.28) and (2.38)), and the jamming tone power rises above S. For large values of E_b/N_J, $\alpha_{\text{wc}} = 1$ and μ and ρ asymptotically approach zero.

We now turn to the general class of band multitone jamming with $n > 1$ tones per jammed band. Because a jammed band will contain at least two interfering tones, there is now the possibility of simultaneously hitting the data symbol and at least one other M-ary symbol in the dehopped band. Consequently, the random jamming phase arguments summarized in Table 2.2 are applicable here.

The band containing a transmitted data symbol will be jammed with probability μ defined in (2.38). Conditioned on this event, the probability that one of the n jamming tones hits the data symbol is

$$\frac{\binom{M-1}{n-1}}{\binom{M}{n}} = \frac{n}{M}. \qquad (2.43)$$

As discussed in Section 2.2.2.1, a symbol error occurs only if (i) the data symbol is not jammed, any of the other symbols is hit, and $\alpha < 1$, or (ii) the data symbol is jammed along with any of the other symbols, the random jamming phase lies in the range defined by (2.35), and $\alpha < 4$:

$$P_s = \mu\left[\left(1 - \frac{n}{M}\right)u_{-1}(1-\alpha) + \frac{n}{M\pi}\cos^{-1}\left(\frac{\sqrt{\alpha}}{2}\right)\right] \qquad (2.44)$$

where $u_{-1}(\cdot)$ is the unit step function. Invoking (2.38) and constraining

$\mu \leq 1$, we find that the worst case performance is given by

$$P_b = \frac{M^2}{2nK(M-1)E_b/N_J}$$

$$\times \max_{0 < \alpha \leq \min\left(4, \frac{nKE_b/N_J}{M}\right)} \left\{ \alpha \left[\left(1 - \frac{n}{M}\right) u_{-1}(1-\alpha) + \frac{n}{M\pi} \cos^{-1}\left(\frac{\sqrt{\alpha}}{2}\right) \right] \right\}.$$

$$(2.45)$$

Before performing the maximization over α in (2.45), it is instructive to determine the behavior of the term $\alpha \cos^{-1}(\sqrt{\alpha}/2)$ over the range $0 < \alpha \leq 4$. As shown in Figure 2.14, it has a single maximum at $\alpha = 2.52$, and decreases monotonically on either side of that peak. The other term that must be maximized in (2.45), $\alpha u_{-1}(1-\alpha)$, increases monotonically with α over its range $(0, 1)$.

Depending on the range of E_b/N_J and n/M, the worst case performance is specified by

(i) $\dfrac{E_b}{N_J} \leq \dfrac{M}{nK}$

$$\alpha_{wc} = \left(\frac{nK}{M}\right)\frac{E_b}{N_J} \Rightarrow \mu = 1, \rho = \frac{n}{M}$$

$$\leq 1$$

$$P_b = \frac{M}{2(M-1)}\left\{1 - \frac{n}{M}\left[1 - \frac{1}{\pi}\cos^{-1}\left(\frac{\sqrt{\alpha_{wc}}}{2}\right)\right]\right\} \qquad (2.46a)$$

(ii) $\dfrac{E_b}{N_J} \geq \dfrac{M}{nK}$

Because the function being maximized in (2.45) is discontinuous at $\alpha = 1$, it is easier to maximize it separately over the ranges $0 < \alpha < 1$ and $1 \leq \alpha \leq (nK/M)(E_b/N_J)$, and then to determine the larger of these two maxima for the given ratio n/M. Since $(\alpha/\pi)\cos^{-1}(\sqrt{\alpha}/2)$ peaks at .525 when $\alpha = 2.52$, (2.45) reduces to

$$P_b = \frac{M^2}{2nK(M-1)E_b/N_J}\max\left(1 - \frac{2n}{3M}, \frac{\beta n}{M}\right) \qquad (2.46b)$$

where

$$\beta \equiv \begin{cases} \dfrac{nKE_b/N_J}{M\pi}\cos^{-1}\left(\dfrac{1}{2}\sqrt{\dfrac{nKE_b/N_J}{M}}\right); & \dfrac{E_b}{N_J} \leq 2.52\dfrac{M}{nK} \\[4mm] .525; & \dfrac{E_b}{N_J} \geq 2.52\dfrac{M}{nK}. \end{cases}$$

Defining

$$\gamma \equiv \left(\beta + \tfrac{2}{3}\right)^{-1} \in [.839, 1] \qquad (2.46c)$$

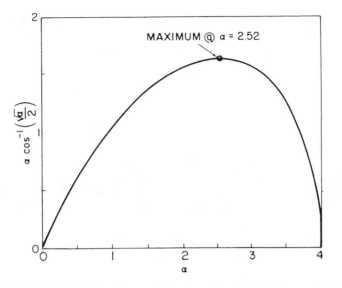

Figure 2.14. The behavior of one of the terms involved in the maximization of P_b in (2.45).

we must consider two ranges of n/M for the given range of E_b/N_J:

(a) $n \leq \gamma M$

$$\alpha_{wc} = 1_- \Rightarrow \mu = \frac{M}{nKE_b/N_J}, \; \rho = \frac{1}{KE_b/N_J}$$

$$P_b = \frac{M(M - \frac{2}{3}n)}{2nK(M-1)E_b/N_J}. \tag{2.46d}$$

(b) $\gamma M \leq n \leq M$

$$\alpha_{wc} = \min\left(2.52, \frac{nKE_b/N_J}{M}\right) \geq 1$$

$$\mu = \min\left(1, \frac{2.52M}{nKE_b/N_J}\right)$$

$$\rho = \min\left(\frac{n}{M}, \frac{2.52}{KE_b/N_J}\right)$$

$$P_b = \frac{M\beta}{2K(M-1)E_b/N_J}. \tag{2.46e}$$

One of the most interesting conclusions presented by (2.46a–e) is that there exist conditions for which it is in fact advantageous to allocate less power to each jamming tone than the received signal power. This occurs only for sufficiently large values of E_b/N_J and $n \sim M$.

In particular, consider the special case of $n = M$. Here we know with certainty that all M symbols of a jammed band will be hit, so that the only error mechanism is the random jamming phase effect. This is reinforced by (2.44), which reduces to

$$P_s = \frac{\alpha \cos^{-1}(\sqrt{\alpha}/2)}{\pi K E_b/N_J}.$$ (2.47)

We have already seen in Figure 2.14 that this function of α peaks at $\alpha = 2.52$. So we are not surprised that α_{wc} can exceed 1, nor that it saturates at 2.52 for $E_b/N_J \geq 2.52/K$ as shown in Figure 2.15. The impact of allowing α_{wc} to exceed 1 is to increase the likelihood μ that a band is jammed (see Figure 2.16), even though the larger value of α decreases the range of jamming phase ϕ for which an error can occur (refer to (2.35)). Figure 2.17 shows that the resulting uncoded FH/MFSK performance is 2 dB worse in the inverse linear region than the same jammer with $\alpha_{wc} = 1_-$.

The advantage of using $\alpha > 1$ does not only apply to the limiting case of $n = M$ band multitone jamming where all errors are attributed to the random phase effect. (2.46e) shows that α_{wc} peaks at 2.52 for large enough E_b/N_J whenever $n \geq .84M$: interestingly, in this region, the asymptotic P_b is independent of n. By contrast, had we restricted α to be less than 1, the worst case asymptotic P_b would have been that of (2.46d). The jamming advantage realized by permitting $\alpha_{wc} \geq 1$ is the ratio of these two P_b expressions

$$\frac{.525 \frac{n}{M}}{1 - \frac{2n}{3M}}; \qquad \frac{n}{M} \geq .84.$$ (2.48)

For example, at $M = 16$ that advantage is 0.4 dB at $n = 14$ and 1.2 dB at $n = 15$, as well as the previously mentioned 2.0 dB at $n = M = 16$.

A final observation from Figure 2.17 is that unlike $n = 1$ band multitone jamming (refer to Figure 2.12), when $n = M$ the performance improves with increasing K or M, just as it does for broadband or worst case partial-band noise. Actually, (2.46e) shows that this effect occurs over the range $.84M \leq n \leq M$ for large enough E_b/N_J, since $M/K(M - 1)$ increases with $K \geq 1$.

For $K = 1$ and 4, Figure 2.18 illustrates the relative effectiveness of band multitone jamming for $n = 2$ and M. Figure 2.19 shows the variation in performance with n for $K = 4$. The implication of these two graphs is that the performance improves with n for fixed K, and degrades with K for fixed n. In fact, these conclusions are essentially correct. Consider the large E_b/N_J asymptotic expressions for P_b in (2.41), (2.46d), and (2.46e): these all have the form

$$P_b = \left(\frac{M}{2K}\right) \frac{\zeta}{E_b/N_J}$$ (2.49)

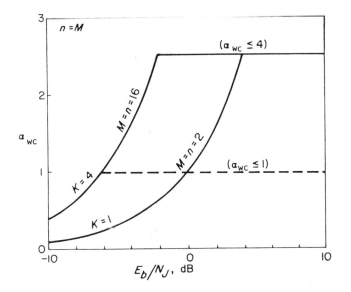

Figure 2.15. Optimum power allocation for $n = M$ band multitone jammer.

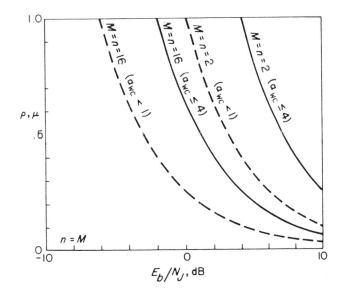

Figure 2.16. The probability ρ that a given FH slot is jammed, and the probability μ that a given M-ary band is jammed, for worst case $n = M$ band multitone jamming.

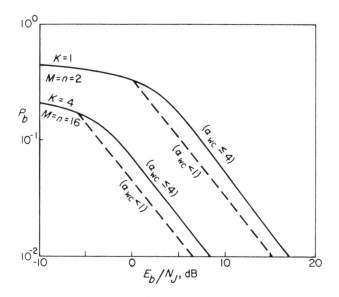

Figure 2.17. Performance of uncoded FH/MFSK system with $n = M$ band multitone interference when jamming tone powers are optimized (solid lines), or are constrained to always exceed the received signal power (dotted lines).

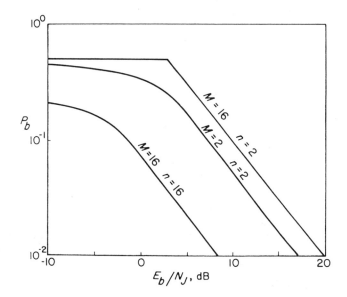

Figure 2.18. Relative effectiveness of band multitone jamming for several values of M and n.

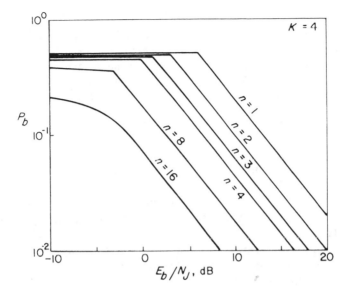

Figure 2.19. Performance improves with n for band multitone jamming, for a given K.

where ζ is given by

$$\zeta = \begin{cases} 1; & n = 1 \\ \dfrac{\left(\dfrac{M}{n} - \dfrac{2}{3}\right)}{M - 1}; & 2 \leq n \leq .84M \text{ (provided } K \geq 2) \\ \dfrac{.525}{M - 1}; & .84M \leq n \leq M. \end{cases} \quad (2.50)$$

(2.49) and (2.50) reveal a singular exception to our premature conjectures: for $n - 2$ the asymptotic P_b is minimized at $K = 2$. With this one anomaly, the relative effectiveness of band multitone jamming for large E_b/N_J can be summarized as follows:

(i) for a given K, the performance improves with n;
(ii) for a given n, the performance degrades with K;
(iii) for a given $n/m \geq .84$, the performance improves with K.

From the jammer's point of view, the $n = 1$ band multitone strategy is preferred, while the $n = M$ case is relatively ineffective.

2.2.2.3 Independent Multitone Jamming

As we noted in our introduction to multitone jamming, the particular appeal of the so-called independent category is its simplicity. Its effectiveness is not contingent on the jammer knowing the location of the tone frequencies in each M-ary band, which could be independently hopped over the entire spread-spectrum bandwidth, and there is no need to regulate the number of jamming tones on each band. Utilizing a seemingly inelegant strategy, the Q jamming tones are distributed uniformly over the N_t available frequencies. If $Q \gg M - 1$ (see (2.29)), each of the FH slots that comprise a given M-ary band are independently jammed with probability ρ, which reduces to

$$\rho = \frac{\alpha}{KE_b/N_J} \tag{2.51}$$

based on (2.5), (2.26), (2.31), and (2.37). We will now analyze the effectiveness of independent multitone jamming, and compare it with the other jamming schemes.

Referring to Table 2.1, a symbol error can occur only if at least one of the $M - 1$ untransmitted symbols in the M-ary band containing the data is jammed. Then, conditioned on this event, a symbol error is made if the data tone is not jammed and $\alpha < 1$, or the data tone is jammed but the relative jamming tone phase ϕ satisfies (2.35) with $\alpha < 4$. By the assumptions implicit in the independent multitone strategy, the likelihood of the data tone being jammed is independent of the prerequisite that one or more of the other untransmitted tones is jammed:

$$P_s = \left[1 - (1 - \rho)^{M-1}\right]\left[(1 - \rho)u_{-1}(1 - \alpha) + \frac{\rho}{\pi}\cos^{-1}\left(\frac{\sqrt{\alpha}}{2}\right)u_{-1}(4 - \alpha)\right] \tag{2.52}$$

Applying (2.20a) and (2.51), with the restriction that $\rho \leq 1$, the bit error rate in worst case jamming is specified by

$$P_b = \frac{M}{2(M - 1)} \max_{0 < \alpha < \min(4, KE_b/N_J)} \left[1 - (1 - \rho)^{M-1}\right]$$
$$\times \left[(1 - \rho)u_{-1}(1 - \alpha) + \frac{\rho}{\pi}\cos^{-1}\left(\frac{\sqrt{\alpha}}{2}\right)\right] \tag{2.53}$$

where ρ is a function of α defined by (2.51).

For $E_b/N_J < \gamma$ (tabulated below), Figure 2.20 shows that $\alpha_{wc} < 1$, but the maximization in (2.53) must be computed separately for each combination of K and E_b/N_J. However, Table 2.3 shows that this computationally difficult region corresponds to $P_b > .37$ for any K, so it is not of practical interest. When $E_b/N_J \geq \gamma$, $\alpha_{wc} = 1_-$, $\rho \leq 1/K\gamma \equiv \rho_0 < 1\ \forall K$, and (2.53) reduces to

$$P_b = \frac{M}{2(M - 1)}\left(1 - \frac{2}{3}\rho\right)\left[1 - (1 - \rho)^{M-1}\right]. \tag{2.54}$$

Note that when $\alpha = \alpha_{wc} = 1_-$ and $KE_b/N_J \gg 1$, (2.51) says that $\rho \ll 1$ (see Figure 2.20 as well): then (2.54) reduces to

$$P_b \cong \frac{M\rho}{2} = \frac{2^{K-1}}{KE_b/N_J} \tag{2.55}$$

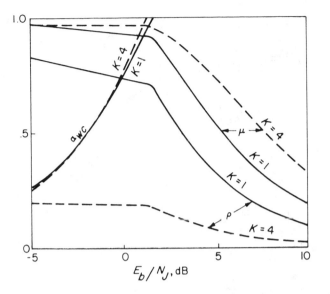

Figure 2.20. Worst case independent multitone jamming parameters: α_{wc} is ratio of signal-to-jamming tone powers, ρ is probability that a given FH slot is jammed, and μ is probability that at least one slot in an M-ary band is jammed.

which is identical to (2.41) for $n = 1$ band multitone jamming. So, as illustrated in Figure 2.21, for the low error rates that typify most practical communication systems, these two jamming strategies are equally effective against uncoded non-coherent FH/MFSK signals. Recall that we suggested this hypothesis at the beginning of this section based on identical expressions for μ when ρ is small (refer to (2.30) and the discussion that follows).

A summary of the relative effectiveness of all of the worst case partial-band noise and multitone jammers is shown in Figure 2.22 for $K = 4$. The $n = 1$ band and independent multitone strategies are the most effective in the absence of coding; partial-band noise is as effective as band multitone jamming for $n \sim M/2$; and $n = M$ band multitone jamming is the least

Table 2.3
Parameters associated with performance of uncoded FH/MFSK signals in worst case independent multitone jamming.

K	γ, dB	$\rho_0 \equiv \dfrac{1}{K\gamma}$	$P_b\|_{\rho = \rho_0}$
1	1.54	.70	.37
2	.45	.45	.39
3	.54	.29	.42
4	1.26	.19	.45

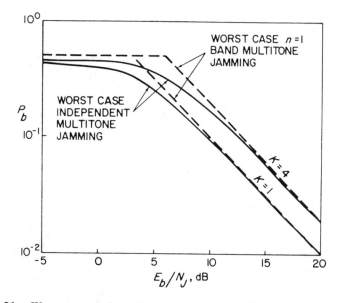

Figure 2.21. Worst case independent and $n = 1$ band multitone jammers are equally effective against uncoded FH/MFSK signals for low P_b's.

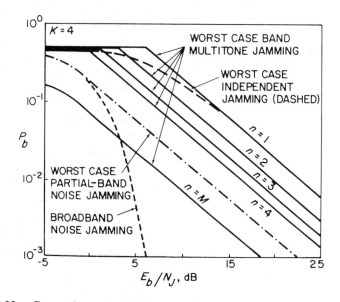

Figure 2.22. Comparison of relative effectiveness of various worst case jamming strategies against uncoded, non-coherently detected FH/16-ary FSK signals.

effective (even worse than broadband noise for low E_b/N_J). Furthermore, all of the worst case jammers result in the inverse linear performance characteristic for large enough E_b/N_J.

2.3 CODING COUNTERMEASURES

Up to this point, each M-ary symbol was sent entirely on a single hop. And we have seen that sophisticated non-adaptive (i.e., non-repeat back) jammers can severely degrade the performance of such systems. These jammers all concentrate their available power over portions of the total spread-spectrum bandwidth W_{ss}; although only a small fraction of the data may be hit, those data suffer a relatively high conditional error rate.

An effective countermeasure against this form of concentrated jamming is to introduce coding redundancy, so that if a given hop is jammed other hopefully more reliable received data can be used to resolve the contaminated data. By basing data decisions on multiple hops, the jammer will have to deconcentrate its power so as to hit a larger fraction of the transmitted data. The ultimate goal is to force the jammer to retreat back towards the original broadband (unconcentrated) noise jamming strategy. In this manner, we would expect to recover the desired exponential performance characteristic.

In computing the performance of a coded system for which each data decision depends on multiple uses of a noisy channel, it is generally impractical to derive closed form exact error rate expressions. Such computations typically involve the integrals of special functions which are at best cumbersome or must be evaluated numerically. Furthermore, these calculations must often be repeated when a single system parameter is changed. Consequently, it is more expedient to use Chernoff upperbounding techniques which involve the statistics of a single channel use. This approach often yields closed form expressions which can produce useful insights. Furthermore, these bounds are exponentially accurate in the region of greatest interest, namely, small P_b's. Our analysis will be based extensively on such bounds; where exact results have been documented, they will be used to determine the accuracy of our bounds.

2.3.1 Time Diversity

One of the simplest yet effective coding techniques is to subdivide each information symbol into equal energy subsymbols which are then transmitted over independent channel states. This is referred to as diversity transmission or a repetition code.

In the context of the FH/MFSK structure, each M-ary symbol is partitioned into m subsymbols or "chips" (not to be confused with the PN spreading code chips) with energy $E_c = KE_b/m$. As illustrated in Figure

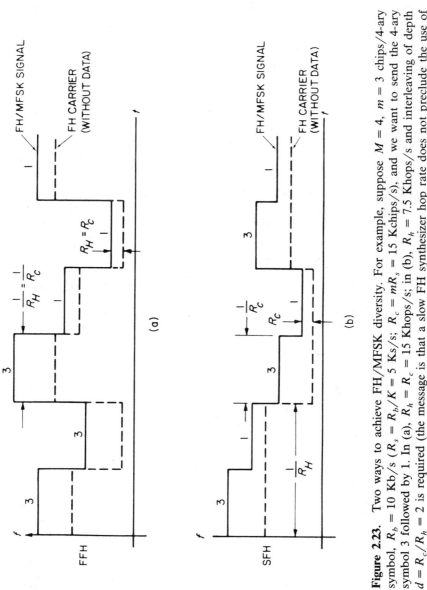

Figure 2.23. Two ways to achieve FH/MFSK diversity. For example, suppose $M = 4$, $m = 3$ chips/4-ary symbol, $R_b = 10$ Kb/s ($R_s = R_b/K = 5$ Ks/s; $R_c = mR_s = 15$ Kchips/s), and we want to send the 4-ary symbol 3 followed by 1. In (a), $R_h = R_c = 15$ Kchips/s; in (b), $R_h = 7.5$ Khops/s and interleaving of depth $d = R_c/R_h = 2$ is required (the message is that a slow FH synthesizer hop rate does not preclude the use of diversity).

2.23, each chip is transmitted on a different hop using fast frequency hopping (FFH) or slow frequency hopping (SFH) with pseudorandom interleaving. The intention is that each chip comprising an M-ary symbol will have an independent chance of being jammed. Because these chips are distributed in time, this technique is often termed "time diversity," although "jamming-state diversity" would be an equally descriptive label. The resulting chip rate $R_c = mR_b/K$; recall that we use the term FFH to denote the condition $R_c = R_h$ (the hop rate), whereas SFH implies $R_c > R_h$.

For each received M-ary chip, after the FH carrier has been demodulated, the received signal is input to M non-coherent detectors which measure the energy at each of the possible tone frequencies over the chip duration $1/R_c$ (refer to Figure 2.1). To ensure that none of the signal energy is detected by the $M - 1$ energy detectors at the untransmitted frequencies, the spacing between adjacent FH slots (within an M-ary band) must be an integer multiple of R_c; if the minimum separation is used, then the number of available frequencies

$$N_t = \frac{W_{ss}}{R_c} = \frac{KW_{ss}}{mR_b} \qquad (2.56)$$

decreases with the amount of diversity m per M-ary symbol[1] (see Figure 2.3). In addition to the M detected energies per chip, the receiver may have some channel state side information available to aid the data decision process [8]. We will assume that *the receiver knows with certainty whether each hop is jammed or not*. One way to derive this information in practice is to implement automatic gain control (AGC) in the receiver, which may be monitored to determine whether jamming power is corrupting a given hop [2, p. 288]. However, we will instead use an approach suggested by Trumpis [7]: since only one of the energy detector outputs will be high[2] on a given chip transmission in the absence of jamming, we will *declare a chip to be jammed when two or more energy detector outputs are high*.

Since the channel statistics can change at the pleasure of the jammer, any decision metric will be suboptimum for some form of jamming. For simplicity, we will use the following procedure based on the soft decision energy detector outputs for the M tones and m diversity transmissions relevant to a given M-ary symbol, along with the corresponding jamming state side information. If *any* of the m chips is not jammed, an error-free M-ary decision can be made; otherwise, select the largest of the metrics

$$\left\{ \Lambda_i \equiv \sum_{j=1}^{m} e_{ij}; \ 1 \le i \le M \right\} \qquad (2.57)$$

---

[1]Viterbi and Jacobs [2] and others prefer the parameter $L = m/K$, which is the diversity per bit.

[2]Actually, since we are neglecting receiver thermal noise, the other $M - 1$ detected energies will be identically zero.

where e_{ij} is the energy detector output for the i-th M-ary symbol on the j-th diversity transmission. The linear sum metric of (2.57) is maximum likelihood for a Rayleigh fading channel [9, pp. 533–540], and it is asymptotically optimum for the additive white Gaussian noise channel in the limit of large E_b/N_J. However, in general it is suboptimum, and we will later see that there is a signal-to-noise ratio reduction as a result of the non-coherent energy combining implicit in the Λ_i's.

2.3.1.1 Partial-Band Noise Jamming

We will now analyze the performance of the FH/MFSK system with diversity utilizing the soft decision metric with side information in a worst case partial-band noise environment (refer back to Section 2.2.1). Suppose symbol 1 is sent: conditioned on the j-th chip being jammed, e_{1j} is a non-central chi-square random variable while the remaining $M-1$ e_{ij}'s are central chi-square random variables, each with two degrees of freedom. The probability density functions of the e_{ij}'s have the form of (2.8) and (2.9), with E_s/N_J replaced by $E_c/N_J' = \rho K E_b/m N_J$ (see (2.21)).

Let H_j denote the event that the j-th chip is hit by partial-band noise, with $H \equiv (H_1, H_2, \ldots, H_m)$ representing the event that all m diversity transmissions are jammed; since the H_j's are independent,

$$\Pr\{H\} = \rho^m. \tag{2.58}$$

With perfect side information, an M-ary symbol error requires that H occur:

$$P_s = \rho^m \Pr\left\{ \bigcup_{i=2}^{M} (\Lambda_i \geq \Lambda_1) \middle| H \right\}. \tag{2.59}$$

Adopting the bounding approach of [2], the M-ary problem of (2.59) is first reduced to a binary problem by applying the union bounding technique and noting that the Λ_i's are identically distributed for $i \neq 1$:

$$P_s \leq (M-1)\rho^m \Pr\{\Lambda_2 \geq \Lambda_1 | H\}. \tag{2.60}$$

This expression is then reduced to the statistics of a single diversity transmission by using (2.57) and the Chernoff bound [2, (9)], with $L = m$ and E_b replaced by $E_s = KE_b$]:

$$\Pr\{\Lambda_2 \geq \Lambda_1 | H\} = \Pr\left\{ \sum_{j=1}^{m} (e_{2j} - e_{ij}) \geq 0 \middle| H \right\}$$

$$\leq \frac{1}{2}\left[\overline{e^{\lambda(e_{2j} - e_{1j})}}^{|H_j} \right]^m; \qquad \lambda \geq 0$$

$$= \frac{1}{2}\left[\frac{e^{-(\lambda/1+\lambda)(\rho K E_b/m N_J)}}{1 - \lambda^2} \right]^m; \quad 0 \leq \lambda < 1 \quad (2.61)$$

which should be minimized over the Chernoff parameter λ. The factor of $1/2$ in the Chernoff bound is justified under certain conditions [10] which are satisfied for partial-band noise jamming (see Appendix 2A). Using (2.20a) and optimizing over λ and ρ for worst case jamming,

$$P_b \le \frac{M}{4} \max_{0<\rho\le1} \min_{0\le\lambda<1} \left[\left(\frac{\rho}{1-\lambda^2} \right) e^{-(\lambda/1+\lambda)(\rho KE_b/mN_J)} \right]^m \quad (2.62)$$

for arbitrary K, E_b/N_J, and m. Computing the joint extrema over ρ and λ, there are two distinct solutions:

(i) $m \ge KE_b/3N_J$

$$P_b \le \frac{M}{4} \left[\left(\frac{1}{1-\lambda^2} \right) e^{-2\beta(\lambda/1+\lambda)} \right]^m \quad (2.63)$$

where

$$\rho_{wc} = 1, \ \beta \equiv \frac{KE_b}{2mN_J},$$

and

$$2\lambda = \sqrt{1 + 6\beta + \beta^2} - (1 + \beta)$$

(ii) $1 \le m \le KE_b/3N_J$

$$P_b \le \frac{M}{4} \left(\frac{4mN_J}{eKE_b} \right)^m \quad \text{and} \quad \rho_{wc} = \frac{3m}{KE_b/N_J}. \quad (2.64)$$

Note that as m increases for a given KE_b/N_J, the jammer's need to hit all m diversity transmissions forces it to broaden the jammed frequency band;

Table 2.4

Parameters derived by Trumpis [7] for exact performance of FH/MFSK signals with diversity m in worst case partial-band noise (see (2.65)).

	$K = 1$		$K = 2$		$K = 3$		$K = 4$		$K = 5$	
m	β	γ, dB	β	γ, dB	β	γ, dB	β	γ, dB	β	γ, dB
2	.4168	6.7	.5959	4.2	.8575	2.8	1.0796	2.0	1.3659	1.4
3	.5210	8.8	.8265	6.1	1.2565	4.6	1.8198	3.7	2.5937	3.0
4	.6797	10.2	1.1401	7.4	1.8320	5.9	2.8251	4.8	4.1493	4.1
6	1.2392	12.1	2.2245	9.2	3.8464	7.6	6.4370	6.5	10.3502	5.7
8	2.3584	13.5	4.4101	10.5	8.0165	8.9	14.0997	7.7	24.0142	6.8
10	4.6110	14.5	8.8385	11.5	16.5774	9.8	30.2829	8.7	61.6169	7.8

when $m > KE_b/3N_J$, the worst case partial-band noise jammer is simply the broadband noise jammer with $\rho = 1$.

The accuracy of these upperbounds can be investigated by comparing them with exact expressions derived by Trumpis [7], involving the optimization of a sum of generalized Laguerre polynomials over ρ, which must be performed by a computer for each combination of K and m (the computation is lengthy for larger values of K and m since the number of terms in the summation is $(M - 1)[\frac{1}{2}M(m - 1) + 1] \sim m2^{2K-1}$). For small E_b/N_J, $\rho_{wc} = 1$ and P_b must be computed numerically for each value of E_b/N_J; however, in the region of interest where P_b is low, Trumpis showed that

$$P_b = \beta\left(\frac{mN_J}{KE_b}\right)^m \quad \text{and} \quad \rho_{wc} = \frac{\gamma}{E_b/N_J} \qquad (2.65)$$

$$\text{provided} \quad E_b/N_J \geq \gamma$$

where β and γ are given in Table 2.4.

The accuracy of the upperbounds for several values of K and m is illustrated in Figures 2.24–2.26. For large enough E_b/N_J, it is seen that the separation between the bounds and the exact performance approaches an asymptotic limit. We can compute this asymptotic difference analytically using (2.64) and (2.65) for large E_b/N_J. For a given K and m, (2.64)

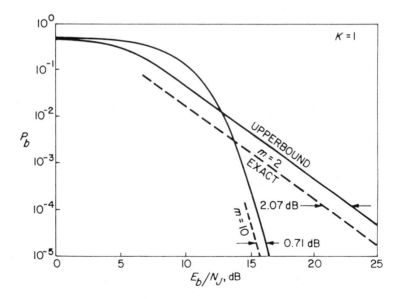

Figure 2.24. Comparison of upperbounds on performance of FH/BFSK $(K = 1)$ signals with diversity m chips/bit in worst case partial-band noise against exact P_b (dashed curves).

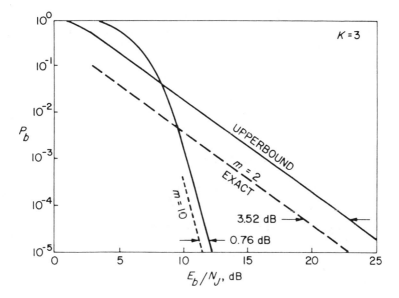

Figure 2.25. Same as Figure 2.24, except FH/MFSK signals with $K = 3$ or $M = 8$, and m chips/8-ary symbol.

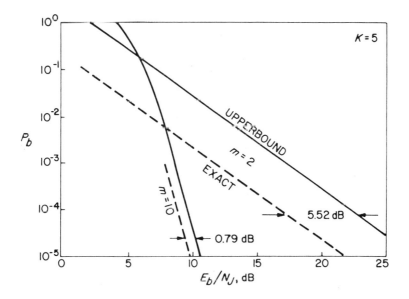

Figure 2.26. Same as Figures 2.24 and 2.25, except $K = 5$ or $M = 32$ with m diversity chips/32-ary symbol.

Table 2.5

Asymptotic accuracy of performance upperbound in (2.64) for low P_b

m	$\Delta E_b/N_J$, dB		
	$K = 1$	$K = 3$	$K = 5$
2	2.07	3.52	5.52
10	.71	.76	.79

overbounds the signal-to-noise level required to achieve a particular P_b:

$$\left(\frac{E_b}{N_J}\right)_1 \equiv \frac{4m}{eK}\left(\frac{M}{4P_b}\right)^{1/m}. \tag{2.66}$$

The actual required signal-to-noise ratio is given by (2.65):

$$\left(\frac{E_b}{N_J}\right)_2 = \frac{m}{K}\left(\frac{\beta}{P_b}\right)^{1/m}. \tag{2.67}$$

The bound accuracy may be specified by the ratio

$$\Delta E_b/N_J \equiv \left(\frac{E_b}{N_J}\right)_1 \Bigg/ \left(\frac{E_b}{N_J}\right)_2 = \frac{4}{e}\left(\frac{M}{4\beta}\right)^{1/m} \tag{2.68}$$

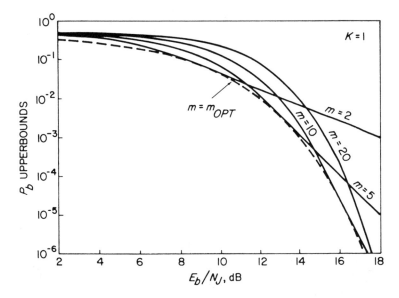

Figure 2.27. Variation in performance upperbounds for FH/BFSK signals in worst case partial-band noise with diversity m chips/bit; optimum diversity (m_{opt}) performance is lower envelope, shown as dotted curve.

which is given in Table 2.5 for selected values of K and m. We see that the bound is quite accurate for larger values of m; for example, it is only $3/4$ dB above the exact result for $m = 10$, with minor variation over K. However, for small m the bound is pessimistic by several dB, and the accuracy degrades with increasing K.

Returning to the upperbounds now, (2.63) and (2.64) are used to illustrate the variation in performance with diversity m for a given K in Figures 2.27–2.29. Notice that the curves cross each other, so that for fixed K and E_b/N_J, the performance actually degrades if m is too large due to non-coherent combining of the detected chip energies in the metric of (2.57). For example, consider the performance curves for $K = 3$ in Figure 2.28: if $E_b/N_J = 12$ dB, P_b is lower at $m = 10$ than at $m = 5$ or 20. The existence of an optimum amount of diversity for a given K and E_b/N_J is further illustrated in Figure 2.30. (Although the diversity m is treated like a continuous variable, it should be remembered that it only has meaning for integer values. Since the minima in Figure 2.30 are quite broad, there is essentially no loss in performance when the continuous value of m_{opt}, defined below in (2.70), is truncated to the nearest integer.) The loss in performance for $m > m_{opt}$ due to non-coherent combining is clearly evident in Figure 2.30.

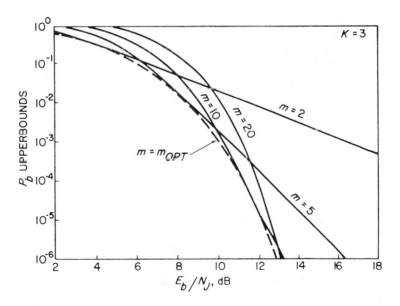

Figure 2.28. Same as Figure 2.27, except FH/MFSK signals with $K = 3$ or $M = 8$, and m diversity chips/8-ary symbol.

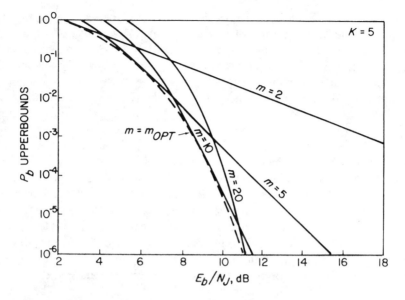

Figure 2.29. Same as Figures 2.27 and 2.28, except that $K = 5$ or $M = 32$ and m diversity chips/32-ary symbol.

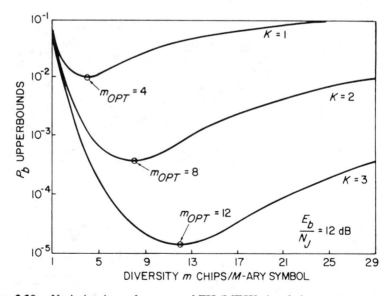

Figure 2.30. Variation in performance of FH/MFSK signals in worst case partial-band noise with amount of diversity for $E_b/N_J = 12$ dB. The P_b upperbound is minimized at $m = m_{opt}$, and the performance degrades for larger values of m due to non-coherent combining of the diversity chip energies.

Referring again to Figures 2.27–2.29, the performance in worst case jamming with optimized diversity is the lower envelope of the set of curves with parameter m, shown as a dotted curve in each graph. Analytically, we are looking for the solution

$$P_b \leq \frac{M}{4} \min_{m \geq 1} \max_{0 < \rho \leq 1} \min_{0 \leq \lambda < 1} \left[\left(\frac{\rho}{1 - \lambda^2} \right) e^{-(\lambda/1 + \lambda)(\rho K E_b / m N_J)} \right]^m \quad (2.69)$$

which implies that we want the smaller of the minima of (2.63) and (2.64) over $m \geq 1$. We find that the desired solution derives from (2.64), provided E_b/N_J is large enough:

$$\left. \begin{array}{c} P_b \leq 2^{K-2} e^{-K E_b / 4 N_J} \\[2mm] m_{\text{opt}} = \dfrac{K E_b}{4 N_J} \\[2mm] \rho_{\text{wc}} = \frac{3}{4} \end{array} \right\} \quad \text{provided} \quad \frac{E_b}{N_J} \geq \frac{4}{K}. \quad (2.70)$$

The constraint in (2.70) simply guarantees that $m_{\text{opt}} \geq 1$; for smaller values of E_b/N_J, $m_{\text{opt}} = 1$ and (2.25) yields the exact P_b (which is obviously discontinuous from the upperbound of (2.70) at its lower limit on E_b/N_J).

The result of (2.70) was first documented in [2], where the extrema m_{opt} and ρ_{wc} were termed "quasi-optimum" since they are based on an upperbound. It is plotted in Figure 2.31 for $1 \leq K \leq 5$, which shows that the performance with optimum diversity in worst case partial-band noise im-

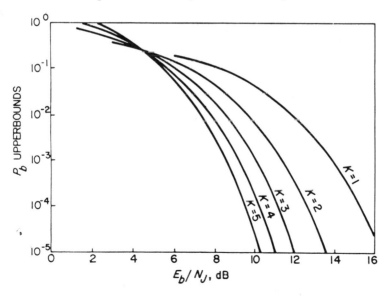

Figure 2.31. Performance of FH/MFSK signals with optimum diversity in worst case partial-band noise ($\rho_{\text{wc}} = 3/4$, $\forall K$ and $E_b/N_J \geq 4/K$).

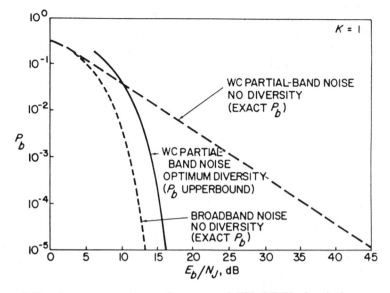

Figure 2.32. Improvement in performance of FH/BFSK signals in worst case (WC) partial-band noise due to optimum diversity; recovery from no diversity case (dashed curve) is significant at low P_b's (e.g., better than 29 dB at $P_b = 10^{-5}$), and performance is within 3 dB of broadband noise (dotted curve).

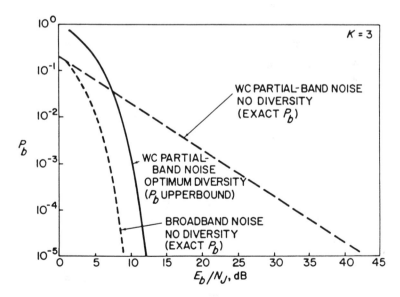

Figure 2.33. Same as Figure 2.32, except FH/MFSK signals with $K = 3$ ($M = 8$); optimum diversity recovers about 31 dB relative to no diversity case at $P_b = 10^{-5}$ (benchmark).

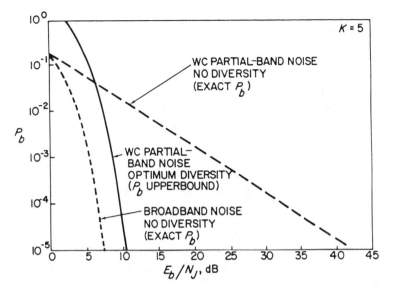

Figure 2.34. Same as Figures 2.32 and 2.33, except $K = 5$ ($M = 32$); optimum diversity recovers over 32 dB relative to no diversity case at $P_b = 10^{-5}$.

proves with K, as it did with no diversity in worst case partial-band and broadband noise (see Figure 2.9). Most important, we see in (2.70) that optimum diversity restores the desired exponential performance behavior characteristic of brute force broadband noise jamming. As shown in Figures 2.32–2.34, optimum diversity recovers most of the advantage that a worst case partial-band noise jammer enjoys against uncoded FH/MFSK signals. For example, Table 2.6 contrasts the performance with no diversity and optimum diversity at the benchmark $P_b = 10^{-5}$, based on the exact expressions of 2.25 and the upperbound of (2.70):

Table 2.6

Effectiveness of diversity against worst case partial-band noise: $\Delta E_b/N_J$ is improvement with optimum diversity at $P_b = 10^{-5}$.

| | $P_b = 10^{-5}$ | | | | |
| | $m = 1$ | | $m = m_{opt}$ | | |
K	E_b/N_J, dB (Exact)	ρ_{wc}	E_b/N_J, dB (Upperbound)	ρ_{wc}	$\Delta E_b/N_J$, dB (Lowerbound)
1	45.66	5.43×10^{-5}	16.36	.75	29.30
2	43.67	5.12×10^{-5}	13.62	.75	30.05
3	42.91	4.74×10^{-5}	12.12	.75	30.79
4	42.56	4.84×10^{-5}	11.11	.75	31.45
5	42.42	4.57×10^{-5}	10.36	.75	32.06

In particular, Table 2.6 demonstrates that optimum diversity recovers more than 29 dB at $K = 1$ (even more for larger K) at $P_b = 10^{-5}$. Furthermore, comparing the upperbounds of (2.20c) and (2.70), we see that the performance with optimum diversity in worst case partial-band noise is only about 3 dB worse than with no diversity in broadband noise.

(2.70) shows that the desired amount of diversity m_{opt} can become large for sufficiently low P_b's, especially for larger alphabet sizes M, under worst case partial-band noise jamming conditions (e.g., $m_{\text{opt}} = 20$ at $P_b = 10^{-5}$ when $K = 5$). This is an indication that reliable communication over such channels demands a substantial amount of coding redundancy. Since time diversity corresponds to a primitive repetition code, it is conceivable that more powerful codes may achieve the required redundancy with improved efficiency; this will be examined in Section 2.3.2.

A surprising result is that with optimum diversity, $\rho_{\text{wc}} = \frac{3}{4}$ independent of K and E_b/N_J, provided that the constraint in (2.70) is satisfied. This invariance can probably be attributed to the form of the P_b upperbound in (2.69); an exact joint determination of m_{opt} and ρ_{wc} would in all likelihood exhibit some complex variation with K and E_b/N_J. Table 2.6 reminds us that in the absence of diversity, at low P_b's, an effective partial-band noise jammer concentrates its available power over a small fraction of W_{ss} (e.g., $\rho_{\text{wc}} \sim 5 \times 10^{-5}$ at $P_b = 10^{-5}$ for $1 \leq K \leq 5$). Optimum diversity forces the jammer to retreat back towards the broadband ($\rho = 1$) strategy, with the resultant loss of most of its prior advantage (see Figures 2.32–2.34).

It should be noted that the solution to the joint optimization of the P_b bound in (2.69) over m and ρ is *not a saddlepoint*; that is, if we reverse the order of that optimization and consider instead the expression

$$P_b \leq \frac{M}{4} \max_{0 < \rho \leq 1} \min_{m \geq 1} \min_{0 \leq \lambda < 1} \left[\left(\frac{\rho}{1 - \lambda^2} \right) e^{-(\lambda/1 + \lambda)(\rho K E_b / m N_J)} \right]^m \quad (2.71)$$

the result would differ from (2.70) [11]. Let us consider the implications of (2.69). The underlying assumption is that the three extrema are computed for arbitrary but fixed values of K (or M) and E_b/N_J. The first minimization over λ simply guarantees the tightest Chernoff upperbound, and the optimum λ is in general a function of m and ρ:

$$\lambda = \tfrac{1}{2} \left[\sqrt{1 + 6\beta + \beta^2} - (1 + \beta) \right] \quad (2.72)$$

where

$$\beta \equiv \frac{\rho K E_b}{2 m N_J}$$

which is a variation of (2.63). Next, the P_b bound with λ given by (2.72) is maximized over ρ for any m, and the optimum ρ depends on m, as does λ above. In particular, if $E_b/N_J \geq 3m/K$ so that $\rho_{\text{wc}} \leq 1$, the solution is (2.64) with the optimum λ in (2.72) reducing to $1/2$; for smaller E_b/N_J,

$\rho_{wc} = 1$ and (2.72) reduces to (2.63). The implication is that no matter which value of m is selected by the communicator, the jammer is privy to that parameter, either through surveillance or espionage, and can *subsequently* choose ρ_{wc} accordingly. Since the communicator must reveal his strategy first while the jammer gets the last move in this game of electronic countermeasure (ECM) and counter-countermeasure (ECCM), the jammer realizes an advantage in principle. Under these playing rules, the best strategy for the communicator is to choose m according to (2.70), provided $E_b/N_J \geq 4/K$ so that $m_{opt} \geq 1$; then ρ_{wc} in (2.64) reduces to 3/4.

Suppose instead that the jammer must declare his choice of ρ first, and the communicator can subsequently select m knowing ρ. Under this reversal of the game rules to the detriment of the jammer, the optimized performance is given by (2.71), with λ still defined by (2.72). The solution is then $m_{opt} = \rho_{wc} = 1$, corresponding to the performance of an uncoded FH/MFSK system in broadband Gaussian noise [11]. The performance improvement is illustrated in Figures 2.32–2.34.

This introduces some game theoretic considerations regarding the selection of m and ρ. The arguments are easier to follow in the context of a specific example, so suppose we want to achieve a P_b of 10^{-5} with $K = 4$. Based on (2.70), the communicator would design for $E_b/N_J = 11.11$ dB with $m = 13$. For these values of K, E_b/N_J, and m, (2.64) (or (2.70)) shows that the jammer will maximize the P_b upperbound at 10^{-5} if he uses $\rho = 3/4$; any other choice of ρ would result in a lower P_b upperbound. So the advantage to the communicator of designing his system parameters based on (2.70) is that it guarantees a minimum performance level for any partial-band noise jammer.

Suppose the communicator anticipates that the jammer will use $\rho = 3/4$, or discovers this by monitoring the interference or utilizing covert methods. He may elect to simplify his system implementation by eliminating the diversity ($m = 1$): then, the exact performance is given by ((2.24) without the maximization over ρ)

$$P_b = \frac{\rho}{2(M-1)} \sum_{i=2}^{M} (-1)^i \binom{M}{i} e^{-(\rho K E_b/N_J)(1-1/i)} \qquad (2.73)$$

with $K = 4$, $M = 16$, and $\rho = 3/4$. The desired P_b of 10^{-5} can now be realized with E_b/N_J reduced to 9.23 dB, representing a savings of about 1.9 dB (remember that we are comparing an exact result with an upperbound) over the $m = 13$ case. The risk in such a strategy is that the jammer may somehow discover that the communicator is not using diversity and retaliate by reducing ρ to about 5×10^{-5}, which would require $E_b/N_J = 42.56$ dB to achieve the desired P_b. So, in electing to use $m = 1$, the communicator could save less than 1.9 dB at the cost of a possible degradation of more than 31.5 dB. The prudent approach is to accept the guaranteed performance of (2.70).

2.3.1.2 Band Multitone Jamming

We now turn to the performance of FH/MFSK signals with m-diversity in a band multitone jamming environment. Recall from Section 2.22, and particularly Figure 2.10, that this a highly structured jamming strategy in which an M-ary band is jammed with probability μ (refer to (2.28)), and a jammed band contains exactly n jamming tones, each with power S/α (see (2.31)), where α is a parameter to be optimized by the jammer. With m-diversity, the expression for μ in (2.38) has the modified form

$$\mu = \frac{\alpha m M}{n K E_b / N_J} \tag{2.74}$$

since $R_s = mR_b/K$ in (2.37).

For the special case of $n = 1$ and $K = 1$, we can readily compute the exact performance. An error will occur whenever the m diversity transmissions are all jammed, the untransmitted symbol is hit each time, and $\alpha < 1$ ((2.36) raised to the m-th power):

$$P_b = \left(\frac{\mu}{2}\right)^m = \left(\frac{\alpha m}{E_b/N_J}\right)^m. \tag{2.75}$$

This is maximized at $\alpha = 1_-$ provided E_b/N_J is large enough so that $\mu \le 1$; therefore the performance of FH/BFSK signals with diversity m in worst case $n = 1$ band multitone jamming is exactly specified by

$$P_b = \begin{cases} \left(\dfrac{m}{E_b/N_J}\right)^m \text{ and } \alpha_{\text{wc}} = 1_-; & \dfrac{E_b}{N_J} \ge 2m \\[3mm] 2^{-m} \text{ and } \alpha_{\text{wc}} = \dfrac{E_b/N_J}{2m}; & \dfrac{E_b}{N_J} < 2m. \end{cases} \tag{2.76}$$

This is plotted in Figure 2.35 for various values of m, including the uncoded $m = 1$ case. Note that (2.76) is valid $\forall m \ge 1$, and is identical to (2.40) and (2.41) for $K = m = 1$.

The slope discontinuity in these curves is characteristic of band multitone jamming because of the constraint that each jammed M-ary band contain *exactly* n jamming tones (recall Figures 2.12, 2.18, and 2.19). The horizontal portion of each curve in Figure 2.35 is the "saturation region" where $\mu = 1$; that is, each binary ($M = 2$) band contains its quota of one jamming tone. In this region, the power in each jamming tone exceeds S and continues to rise (α_{wc} falls in (2.76)) as E_b/N_J decreases.

There is somewhat of a paradox here. The $n = 1$ band multitone scenario is highly structured. It concedes the jammer knowledge of the location of the M-ary bands, information that is used to undermine the receiver's awareness of which hops are jammed by distributing the jamming power so as to hit the maximum number of hops. Yet, unlike the less structured partial-band noise behavior illustrated in Figure 2.30, the constraint im-

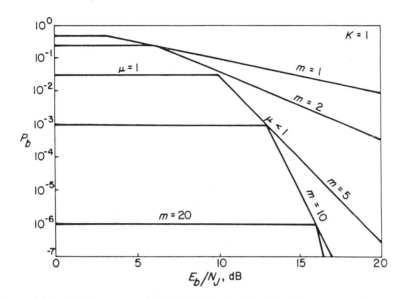

Figure 2.35. Performance of FH/BFSK signals with diversity m chips/bit in worst case $n = 1$ band multitone jamming. In the horizontal regions, the probability μ that a binary band is jammed is 1.

posed by $n = 1$ band multitone jamming ultimately leads to error-free performance as m becomes large. For small values of m, each jamming tone contains the same power as the signal tones, and the $Q = J/S$ jamming tones are less than the number of M-ary bands. However, as m increases so does the chip rate R_c, which causes the number of M-ary bands to decrease (refer to Figure 2.10). Eventually, the number of M-ary bands equals J/S, and the saturation condition $\mu = 1$ is reached. In this region, under the assumption of perfect receiver jamming state side information, P_b is simply the probability 2^{-m} that the incorrect binary symbol is hit on all m diversity transmissions.

This surprising behavior is better illustrated in Figure 2.36, which demonstrates the variation in performance with diversity for several SNR's. There is an initial minimum P_b in the region $\mu < 1$ or $m < E_b/2N_J$; from (2.76), this occurs at

$$m = m_{\text{opt}} = \frac{E_b/N_J}{e} < \frac{E_b/N_J}{2}$$

where

$$P_b = e^{-e^{-1}E_b/N_J} \tag{2.77}$$

and

$$\alpha_{\text{wc}} = 1_-$$

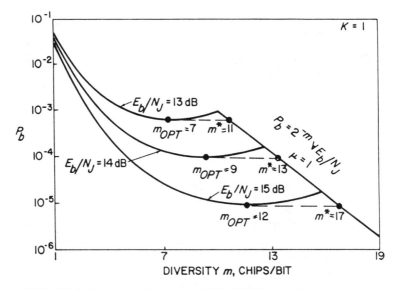

Figure 2.36. Variation in performance of FH/BFSK signals in worst case $n = 1$ band multitone jamming with diversity. Note that P_b decreases as 2^{-m} for large m, independent of E_b/N_J. For fixed E_b/N_J, the initial minimum occurs at $m = m_{opt}$; however, lower P_b's are later achieved for $m > m^*$ (refer to (2.79)).

provided

$$E_b/N_J \geq e \Rightarrow m_{opt} \geq 1.$$

For smaller values of E_b/N_J, $m_{opt} = 1$ and the performance is given by (2.40) and (2.41):

$$P_b = \begin{cases} \dfrac{1}{E_b/N_J} \text{ and } \alpha_{wc} = 1_-; & 2 \leq \dfrac{E_b}{N_J} \leq e \\[3mm] \dfrac{1}{2} \text{ and } \alpha_{wc} = \dfrac{E_b}{2N_J}; & \dfrac{E_b}{N_J} \leq 2. \end{cases} \qquad (2.78)$$

However, in the saturation region, P_b falls below that defined in (2.77) for $m > m^*$, where

$$m^* \equiv \frac{E_b/N_J}{e \ln 2} = \frac{m_{opt}}{\ln 2} = 1.44 m_{opt} \qquad (2.79)$$

and P_b becomes arbitrarily small for larger values of m. If it is important to use minimal diversity, m_{opt} is a good choice. However, since m^* is only moderately larger, better performance can be achieved without resorting to impractically large amounts of diversity.

Figure 2.37 illustrates the extreme effectiveness of diversity for FH/BFSK signals in worst case $n = 1$ band multitone jamming. Even at the local

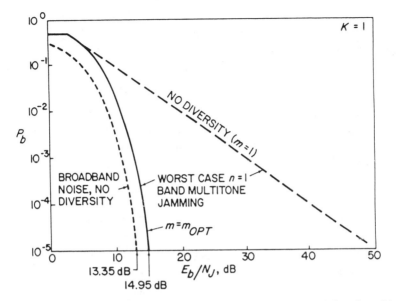

Figure 2.37. Effectiveness of diversity against worst case $n = 1$ band multitone jamming for FH/BFSK signalling. Improvement is 35 dB relative to $m = 1$ system, and performance is only 1.6 dB worse than in broadband noise, with $m = m_{opt}$ at $P_b = 10^{-5}$.

minimum, $m = m_{opt}$, the improvement relative to $m = 1$ is 35 dB at $P_b = 10^{-5}$, and the performance at that P_b is only 1.6 dB worse than in broadband noise. And for $m > m^*$, the performance is even better; in fact, for large enough values of m, P_b is lower than in broadband noise (it is not plotted since P_b becomes arbitrarily small for large m). If the $m = m_{opt}$ performance here is compared with the worst case partial-band noise/optimum diversity curve in Figure 2.32, which was 3 dB worse than the broadband noise case for small P_b's, recall that the latter curve is an upperbound while the former is exact.

Turning now to larger size alphabets ($K > 1$) but still considering only $n = 1$ band multitone interference, the performance with m-diversity and the suboptimum linear sum metric with perfect side information will now be analyzed using the union/Chernoff bounding techniques of Section 2.3.1.1. Because of the side information, it is convenient to introduce the event H_j that the j-th chip is jammed, as in the partial-band noise case. Adopting the convention of Section 2.3.1 which declares that a particular chip is jammed when two or more energy detector outputs are high on a given transmission, for $n = 1$ band multitone jamming,

$$\Pr\{H_j\} = \left(\frac{M-1}{M}\right)\mu. \tag{2.80}$$

Under the assumption that the receiver has error-free information about which of the m diversity transmissions are jammed, an M-ary symbol error requires the occurrence of $H \equiv (H_1, H_2, \ldots, H_m)$, where

$$\Pr\{H\} = \left[\left(\frac{M-1}{M}\right)\mu\right]^m = \left[\frac{(M-1)\alpha m}{KE_b/N_J}\right]^m \tag{2.81}$$

using (2.74).

Let e_{ij} denote the energy detected for the i-th M-ary symbol on the j-th diversity transmission, normalized by the chip signal energy E_c, and suppose symbol 1 is sent. Then, as in (2.60) and (2.61), the M-ary detection problem is reduced to a binary one by employing the union bound:

$$P_s \le (M-1)\Pr\{H\}\Pr\left\{\sum_{j=1}^m (e_{2j} - e_{1j}) \ge 0 \middle| H\right\}. \tag{2.82}$$

Next, the m-ary statistics are simplified to those for a single chip transmission by applying the Chernoff bound:

$$\Pr\left\{\sum_{j=1}^m (e_{2j} - e_{1j}) \ge 0 \middle| H\right\} \le \left\{\overline{e^{\lambda(e_{2j} - e_{1j})}}^{|H_j}\right\}^m; \qquad \lambda \ge 0. \tag{2.83}$$

The factor of $1/2$ that appeared in the partial-band noise bound of (2.61) is absent here because the discrete-valued e_{ij}'s do not satisfy the sufficient condition specified in [10] and Appendix 2A. Conditioned on H_j, e_{1j} and e_{2j} have the following probability distributions:

$$\Pr\{e_{1j} = 1|H_j\} = 1$$

$$\Pr\{e_{2j} = \beta|H_j\} = \begin{cases} \dfrac{1}{M-1}; & \beta = \dfrac{1}{\alpha} \\ \dfrac{M-2}{M-1}; & \beta = 0 \end{cases} \tag{2.84}$$

so that

$$\Pr\{(e_{2j} - e_{1j}) = \beta|H_j\} = \begin{cases} \dfrac{1}{M-1}; & \beta = \dfrac{1}{\alpha} - 1 \\ \dfrac{M-2}{M-1}; & \beta = -1 \end{cases}. \tag{2.85}$$

Since (2.83) (or common sense) says that $(e_{2j} - e_{1j})$ must have a non-zero probability of being greater than zero for a symbol error to occur, (2.85) shows that this requires that $\alpha < 1$. Using the familiar relationship of (2.20a) between bit and symbol error rates, and combining (2.81)–(2.83) and (2.85), we have, for arbitrary m and α,

$$P_b \le \frac{M}{2} F^m \tag{2.86}$$

where

$$F \equiv \min_{\lambda \geq 0} \left[\frac{\alpha m e^{-\lambda}}{KE_b/N_J} (M - 2 + e^{\lambda/\alpha}) \right].$$

Performing the minimization by setting $\partial F/\partial \lambda$ to zero yields

$$F = \frac{m}{KE_b/N_J} \left[\frac{\alpha(M-2)}{1-\alpha} \right]^{1-\alpha} \tag{2.87}$$

provided $M > 2$, and $\alpha \geq 1/(M-1)$ so that $\lambda \geq 0$.

For worst case jamming, we want to choose $\alpha \in (0,1)$ to maximize F. Although it is not immediately obvious from the form of (2.87), Figure 2.38 demonstrates that F has a unique maximum for each value of $K \geq 2$ over this range of α. Denote the maximizing value of α by α_0, and define

$$\beta \equiv \left(\frac{E_b}{mN_J} \right) F \bigg|_{\alpha=\alpha_0} = \frac{1}{K} \left[\frac{\alpha_0(M-2)}{1-\alpha_0} \right]^{1-\alpha_0} \tag{2.88}$$

where β and α_0 are given in Table 2.7 below. The constraint $\mu \leq 1$ must also be satisfied: using (2.74) with $n = 1$, this translates into

$$\alpha \leq \frac{K}{mM} \left(\frac{E_b}{N_J} \right). \tag{2.89}$$

Consequently, the performance in worst case $n = 1$ band multitone jam-

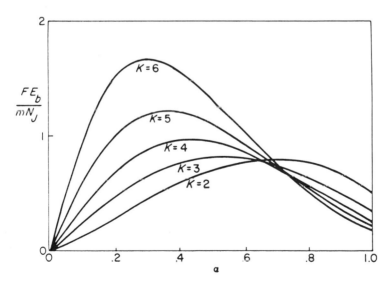

Figure 2.38. Demonstration that P_b bound for FH/MFSK signals with arbitrary diversity m and SNR E_b/N_J in $n = 1$ band multitone jamming has unique maximum over jammer parameter α (refer to (2.86)–(2.88)).

ming with diversity m is upperbounded by

$$P_b \leq \begin{cases} \dfrac{M}{2}\left(\dfrac{\beta m}{E_b/N_J}\right)^m \text{ and } \alpha_{wc} = \alpha_0; & \dfrac{E_B}{N_J} \geq \zeta m \\[4mm] \dfrac{M}{2}\left(\dfrac{1}{\alpha_{wc}M}\left[\dfrac{\alpha_{wc}(M-2)}{1-\alpha_{wc}}\right]^{1-\alpha_{wc}}\right)^m & \\[4mm] \text{and } \alpha_{wc} = \dfrac{K}{mM}\left(\dfrac{E_b}{N_J}\right); & \dfrac{mM}{K(M-1)} \leq \dfrac{E_b}{N_J} \leq \zeta m \end{cases} \qquad (2.90)$$

where $\zeta \equiv \alpha_0 M/K$ is also computed in Table 2.7.

Although (2.90) is an upperbound, it is relatively simple and easy to evaluate, and many behavioral performance characteristics can be deduced from it. It is plotted in two different formats for several values of $K \geq 2$ in Figures 2.39–2.41. Unlike the $K = 1$ curves in Figure 2.35, the P_b upperbounds for fixed m in Figures 2.39 and 2.40 have no slope discontinuities; this does not rule out such behavior in the exact bit error rate. The arguments in Section 2.3.1.2 discussing the saturation effect imposed by the constraints of band multitone jamming are valid for all values of K, so we should not be surprised that the $K = 2$ performance in Figure 2.41 improves for larger values of m. (This effect would be evident for other values of K in Figure 2.41 if we had gone beyond $m = 30$.) Mathematically, for large values of m in the saturation region of (2.90),

$$\alpha_{wc} \ll 1 \Rightarrow P_b \leq \frac{M}{2}\left(\frac{M-2}{M}\right)^m \underset{m \to \infty}{\to} 0. \qquad (2.91)$$

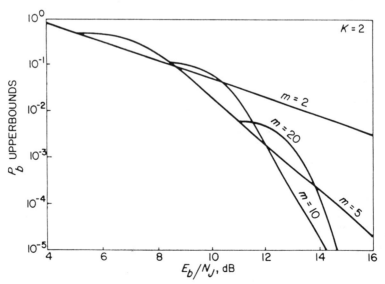

Figure 2.39. Performance upperbounds for FH/4-ary FSK signals with diversity m chips per 4-ary symbol in worst case $n = 1$ band multitone jamming.

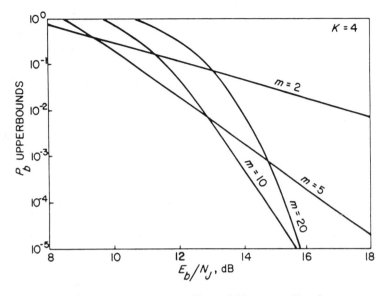

Figure 2.40. Same as Figure 2.39, except $K = 4$.

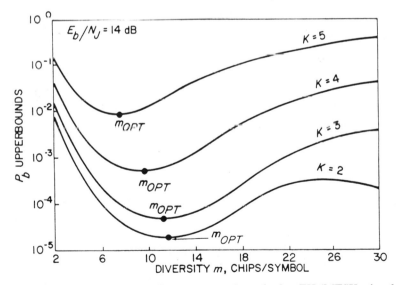

Figure 2.41. Behavior of performance upperbounds for FH/MFSK signals in worst case $n = 1$ band multitone jamming as a function of diversity m. Note that performance improves for large m in the $K = 2$ curve, and the performance degrades as K increases.

Another way of seeing this is to recall that under the perfect side information assumption, an M-ary symbol error requires that all m diversity chips be jammed, denoted by the event H in (2.81). In the saturation region,

$$\Pr\{H\} = \left(\frac{M-1}{M}\right)^m \xrightarrow[m\to\infty]{} 0 \qquad (2.92)$$

and the error rate must fall accordingly for large amounts of diversity. However, unlike the $K = 1$ case where it was possible to achieve P_b's less than the first local minimum at $m = m_{opt}$ (see Figure 2.36) with moderately larger values of m, such is not the case for $K \geq 2$. Figure 2.41 clearly implies that for practical implementations with non-binary alphabets, the best design tradeoff is to use $m = m_{opt}$. Finally, Figure 2.41 demonstrates that for $K \geq 2$, the performance with diversity degrades as K increases in worst case $n = 1$ band multitone jamming, in contrast to the behavior in worst case partial-band noise; this echoes our earlier observations for $m = 1$ signalling (e.g., refer to Figure 2.12).

With diversity transmission, where received data decisions involve multiple channel uses, it is generally difficult and cumbersome to compute the exact performance. We circumvent this problem by evaluating exponentially tight upperbounds. Still, the accuracy of these bounds remains a source of concern and raises issues of credibility. In the partial-band noise case we were able to use some exact calculations derived by Trumpis as a benchmark (recall Figures 2.25 and 2.26). Here, for $n = 1$ band multitone jamming and a soft decision non-coherent combining metric with perfect jamming state information, we can use combinatorial techniques to again compute the exact performance. However, the results are much more complex, particularly for large amounts of diversity, and they do not yield insights as readily as the closed form expression of (2.90).

Suppose symbol 1 is sent and event H occurs, which is a necessary condition for a symbol error in the perfect side information case. Then, for a given jammer parameter α, the normalized conditional soft decision symbol metrics of (2.57) are

$$\Lambda_1 = m$$
$$\Lambda_i = l_i/\alpha; \qquad 2 \leq i \leq M \qquad (2.93)$$

where l_i denotes the number of times the i-th symbol is hit by a jamming tone over the m diversity transmissions. Note that conditioned on H, each of the m hops contains only one jamming tone, which is equally likely to hit any of the $M - 1$ untransmitted symbols on that hop. This is the situation pictured in the grid at the beginning of Appendix 2B ($I \to M - 1$, $J \to m$). The conditional symbol error rate is given by

$$P_s|^H = \Pr\left\{ \bigcup_{i=2}^{M} (\Lambda_i \geq \Lambda_1) \middle| H \right\}$$
$$= \Pr\{l_{max} \geq m\alpha | H\} \qquad (2.94)$$

where

$$l_{max} \equiv \max_{2 \le i \le M} \{l_i\}. \tag{2.95}$$

The probability of any particular pattern of m jamming tones conditioned on H is $(M-1)^{-m}$. Appendix 2B computes the number of these patterns satisfying the constraint on l_{max} in (2.94); in the notation of Appendix 2B,

$$P_s|^H = (M-1)^{-m} S_{M-1,m}(k); \qquad k \equiv \lceil m\alpha \tag{2.96}$$

where $\lceil x$ is the smallest integer greater than or equal to x. Using (2.81) and (2B.8), we have

$$P_b = \frac{M}{2(M-1)} \Pr\{H\} P_s|^H = \frac{M}{2} \left(\frac{\alpha m}{KE_b/N_J} \right)^m$$

$$\times \left\{ \sum_{i=k}^{m} \binom{m}{i} (M-2)^{m-i} - \left(\frac{M-2}{2} \right) \sum_{i=k}^{\lfloor m/2 \rfloor} \binom{m}{i} \binom{m-i}{i} (M-3)^{m-2i} \right.$$

$$\left. - (M-2) \sum_{i=k}^{\lfloor m/2 \rfloor} \binom{m}{i} \sum_{j=i+1}^{m-i} \binom{m-i}{j} (M-3)^{m-i-j} \right\};$$

where $k - 1 < m\alpha \le k$, and

provided $\frac{m}{3} < k \le m$ $\tag{2.97}$

where $\lfloor x$ denotes the integer part of x, and we are adopting the convention that $\sum_{i=a}^{b} c_i = 0$ if $a > b$. For the special case of $k = 1$ or 2, (2B.6) and (2B.7) yield simpler expressions for P_b:

$$P_b = \frac{M}{2(M-1)} \left[\frac{\alpha m(M-1)}{KE_b/N_J} \right]^m; \qquad k = 1 \quad \text{and} \quad 0 < m\alpha \le 1 \tag{2.98}$$

$$P_b = \frac{M}{2} \left(\frac{\alpha m}{KE_b/N_J} \right)^m \times \begin{cases} (M-1)^{m-1} - \dfrac{(M-2)!}{(M-m-1)!}; & m < M-1 \\ (M-1)^{m-1}; & m > M-1 \end{cases}$$

provided $k = 2$ and $1 < m\alpha \le 2$. $\tag{2.99}$

So for $m \le 8$, we can compute the exact performance for any $k \in [1, m]$; for larger m, there is a computational gap for $3 \le k \le \lfloor m/3$. We could extend (2.97) to smaller values of k, but the result would be even more complex. As it is, compare (2.97) with the simplicity of the P_b bound of (2.90). And we still have to maximize these exact expressions over $\alpha < 1$ to establish the worst case jamming condition. If we treat k rather than α as the maximizing variable, it is clear from (2.97)–(2.99) that P_b is maximized over α for a given k by selecting

$$m\alpha = k. \tag{2.100}$$

Then why did we not replace $m\alpha$ by k in (2.97)–(2.99) and be done with it? The reason is that we must also satisfy the constraint $\mu \leq 1$, which implies that

$$m\alpha \leq \frac{KE_b}{MN_J}. \qquad (2.101)$$

In the singular cases for which KE_b/MN_J is precisely integer, $k \leq KE_b/MN_J$ and (2.100) is valid for all k in this range. In general, when KE_b/MN_J is non-integer, we have

$$m\alpha = \begin{cases} k; & k \leq \left\lfloor \dfrac{KE_b}{MN_J} \right\rfloor \\[3ex] \dfrac{KE_b}{MN_J}; & k = \left\lceil \dfrac{KE_b}{MN_J} \right\rceil = \left\lfloor \dfrac{KE_b}{MN_J} \right\rfloor + 1. \end{cases} \qquad (2.102)$$

The worst case jamming condition is now established by maximizing the expressions for P_b in (2.97)–(2.99) over the range $1 \leq k \leq \min(m, \lceil KE_b/MN_J \rceil)$; this operation must be performed using a computer for each combination of K, E_b/N_J, and m, further emphasizing the practical utility of the P_b upperbound. Note that since (2.97) can be evaluated only if $k = \lceil m\alpha \rceil > m/3$, (2.101) limits us to the range

$$m \gtrsim \frac{3KE_b}{MN_J}. \qquad (2.103)$$

These exact results are first used to assess the accuracy of the upper-bounds plotted in Figures 2.39–2.41. As we observed in the partial-band noise case of Figures 2.24–2.26, for worst case $n = 1$ band multitone jamming, Figures 2.42 and 2.43 demonstrate that the bounds are pessimistic by several dB for small $m \sim 2$, but are accurate to within about $1/2$ dB of the exact performance for moderate $m \gtrsim 10$. Actually, the asymptotic accuracy is about 30% better for the band multitone case relative to the partial-band noise curves. Note that for the values of m selected, we could not plot exact P_b's for SNR's that were too small to satisfy the constraint of (2.103). Figure 2.44 gives a different perspective on the accuracy of the P_b bounds. In the $K = 2$ case, we see that the exact P_b does indeed decrease for sufficiently large m, so this characteristic is not simply a bound anomaly. The piecewise linear nature of the exact curves is due to the real world constraint that m and the maximizing parameter k take on only integer values, so the worst case performance variation with diversity is not as smooth as the bounds suggest. As a matter of interest, the exact performance for $K = 1$ is superimposed on Figure 2.44. Before the saturation condition $\mu = 1$ is reached, for the selected SNR, we see that the $K = 1$ performance is only marginally better than for $K = 4$ and worse than for $K = 2$ (or $K = 3$, which was not plotted). However, because the $K = 1$

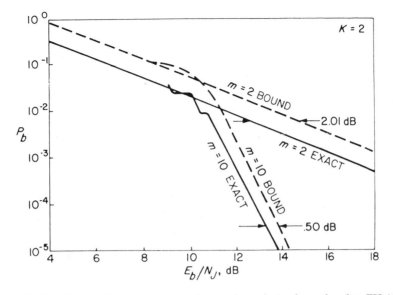

Figure 2.42. Comparison of exact and upperbound P_b formulas for FH/4-ary FSK signalling with diversity m chips per symbol in worst case $n = 1$ band multitone jamming. Accuracy of bound improves with m, and is moderately tighter than for worst case partial-band noise (Figures 2.24 and 2.25).

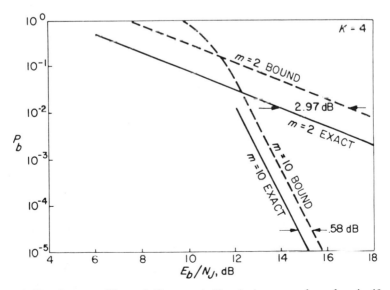

Figure 2.43. Same as Figure 2.42, except $K = 4$. Accuracy degrades significantly with K for small $m \sim 2$, but increases only slightly with K for large $m \lesssim 10$.

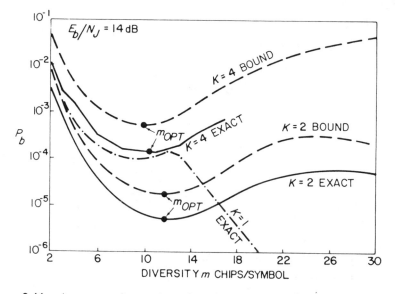

Figure 2.44. Accuracy of upperbound performance variation with diversity for FH/MFSK signals in worst case $n = 1$ band multitone jamming (refer to Figure 2.41). Performance for $K = 1$ is slightly better than $K = 4$ for small m, but improves drastically as 2^{-m} in the saturation region $m \geq E_b/2N_J$ where $\mu = 1$ (see (2.76)).

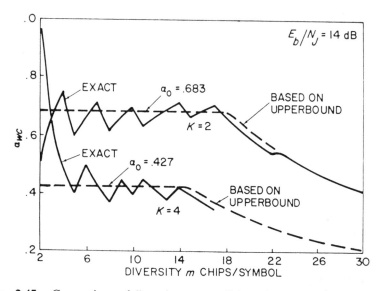

Figure 2.45. Comparison of "quasi-worst case" jamming parameter α_{wc} derived from P_b upperbound of (2.86) and (2.87) with actual worst case parameter based on (2.97)–(2.99) and (2.102) describing exact performance of FH/MFSK signals with diversity in $n = 1$ band multitone jamming.

performance improves so rapidly for $m \geq E_b/2N_J$ where $\mu = 1$ (refer to (2.76)), it is difficult to make a definite "optimum diversity" comparison with larger values of K. Figure 2.45 addresses the accuracy of the "quasi-worst case" jamming parameter α_{wc} derived from the P_b bound of (2.86) and (2.87). Prior to saturation, for smaller values of m, the exact α_{wc} fluctuates up and down due to the discrete nature of (2.102) (for each integer value of m, $\alpha_{wc} = k_{wc}/m$, where k_{wc} is the integer k that maximizes P_b in (2.97)–(2.99) subject to $\alpha m = k$), but the constant quasi-worst case $\alpha_{wc} = \alpha_0$ (see (2.90) and Table 2.7) is a good average fit to these fluctuations. In the saturation region, for sufficiently large m, both the quasi-worst case and exact parameters satisfy $\alpha_{wc} = KE_b/mMN_J$ (see (2.90) and (2.102)), so there is total agreement here.

We now consider the question of optimum diversity for $K \geq 2$. Recall that for $K = 1$, it was not clear whether to select the local minimum of the P_b versus m curves, or to use a moderately larger value of $m \geq m^*$ for which P_b became arbitrarily small (Figure 2.36 and (2.76) and (2.79)). Figure 2.44 shows that although P_b for $K \geq 2$ again decreases for sufficiently large m in the saturation region, the values of m for which this effect becomes significant are much larger than the initial local minima at m_{opt}, and are not of interest for realistic implementations. So we will look for the value of m that minimizes P_b for $\alpha = \alpha_{wc}$ in the region $\mu < 1$. For the bound of (2.90) (the relative simplicity of which we should all have a new appreciation for), this implies the restriction $E_b/N_J \geq \zeta m$; the minimum occurs at $\beta m/E_b/N_J = e^{-1}$, so that the performance with "quasi-optimum" diversity is defined by

$$P_b \leq \frac{M}{2}e^{-\delta E_b/N_J}; \quad \text{where } \delta \equiv \frac{1}{\beta e}$$

$$m_{opt} = \delta E_b/N_J$$

$$\alpha_{wc} = \alpha_0 \tag{2.104}$$

provided

$$\frac{E_b}{N_J} \geq \frac{1}{\delta} \equiv \gamma_0.$$

The constraint on E_b/N_J ensures that $m_{opt} \geq 1$. Of course, for the solution to be valid, we must verify that m_{opt} is in the pre-saturation region:

$$m_{opt} \leq \frac{E_b}{\zeta N_J}$$

$$\Downarrow$$

$$\zeta \leq \beta e. \tag{2.105}$$

Table 2.7 shows that this constraint is satisfied, and also lists the numerical values of δ and γ_0.

Table 2.7
Parameters associated with P_b upperbounds for FH/MFSK signals
with diversity in worst case $n = 1$ band multitone jamming.

K	α_0	β	ζ	βe	δ	γ_0, dB
2	.683	.7945	1.366	2.160	.4631	3.34
3	.527	.8188	1.405	2.226	.4493	3.48
4	.427	.9583	1.708	2.605	.3839	4.16
5	.356	1.2204	2.278	3.317	.3014	5.21

For the exact P_b expressions of (2.97)–(2.99) and (2.102), there is no elegant way to compute the optimum diversity. The results involve several levels of computer number crunching. For each value of K and E_b/N_J, a value of m is selected within the range specified by (2.103). For this parameter set, the value of k that maximizes P_b is determined, corresponding to α_{wc}; the corresponding worst case P_b for that value of m is stored. The procedure is repeated for other values of m until the worst case P_b is minimized, defining m_{opt}. (This exercise shows the relative convenience of the performance bound.)

The resulting performance for optimum diversity and worst case $n = 1$ band multitone jamming is compared with the much simpler upperbound result in Figure 2.46. The bound is only pessimistic by $1/2$ dB, which is a reasonable price to pay for its computational convenience. Interestingly, Figures 2.47 and 2.48 demonstrate that the quasi-optimum jamming parameters α and diversity m based on the upperbound agree quite well on the average with the exact parameters, whose variations are due to the discrete nature of the optimizations. So, for both the communicator and the jammer, the bound provides an accurate determination of the key system design parameter.

The effectiveness of optimum diversity against worst case $n = 1$ band multitone jamming is shown in Figures 2.49 and 2.50 for $K = 2$ and 4. The improvement in performance is substantial relative to the no diversity case: 36 dB at $K = 2$ and 38 dB at $K = 4$. (From the perspective of the E_b/N_J scale used in these graphs, the accuracy of the bounds seems to be quite adequate.) Recall that in the worst case partial-band noise case, optimum diversity provided performance less than 3 dB worse than in broadband noise for small P_b's. Worst case $n = 1$ band multitone jamming is more effective against FH/MFSK signals with optimum diversity for $K \geq 2$: the degradation relative to broadband noise is over 3 dB at $K = 2$, and almost 7 dB at $K = 4$.

The performance of FH/MFSK signals with optimum diversity in worst case $n = 1$ band multitone jamming is summarized in Figure 2.51 for $1 \leq K \leq 5$. As in the $m = 1$ case, P_b increases with K, except for the anomalous $K = 1$ case. For $K = 1$, the performance is plotted for the local extremum m_{opt}, but we can make P_b much smaller (easily below the $K = 2$ curve) for moderately larger amounts of diversity.

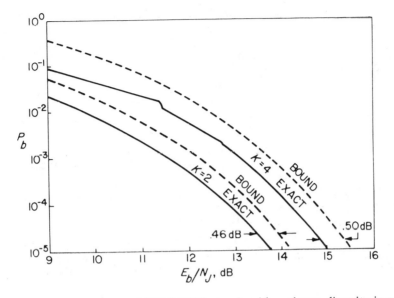

Figure 2.46. Performance of FH/MFSK signals with optimum diversity in worst case $n = 1$ band multitone jamming: comparison of upperbound with exact P_b.

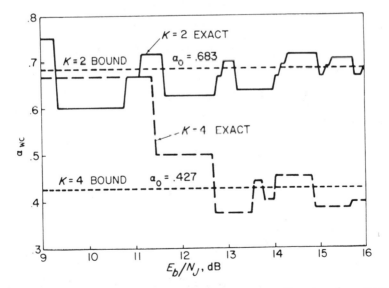

Figure 2.47. Comparison of worst case jamming power ratio α based on exact and upperbound performance of Figure 2.46. Agreement is good despite fluctuations in exact α_{wc} due to discrete nature of optimization, with exception of $K = 4$ for small E_b/N_J.

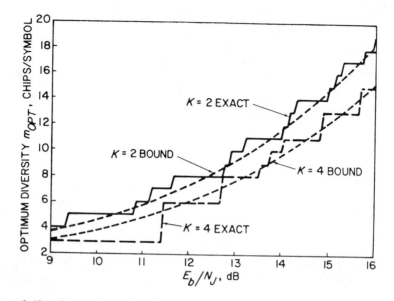

Figure 2.48. Comparison of optimum diversity for exact and upperbound performance of Figure 2.46.

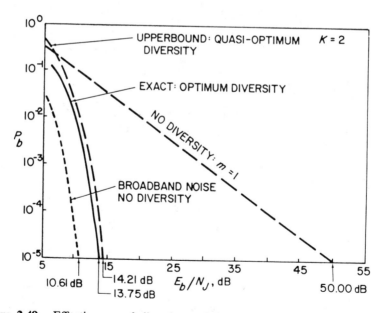

Figure 2.49. Effectiveness of diversity against worst case $n = 1$ band multitone jamming for FH/MFSK signals, with $K = 2$ above. Improvement relative to $m = 1$ case is over 36 dB, and performance is only 3.1 dB worse than in broadband noise, with $m = m_{opt}$ at $P_b = 10^{-5}$.

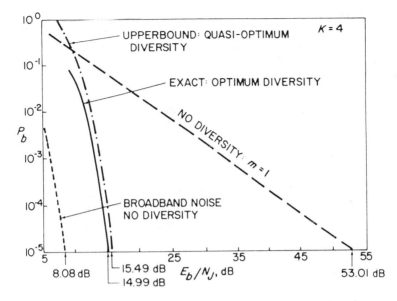

Figure 2.50. Same as Figure 2.49, except $K = 4$. Optimum diversity provides 38 dB improvement relative to $m = 1$ system, and is only 6.9 dB worse than in broadband noise at $P_b = 10^{-5}$.

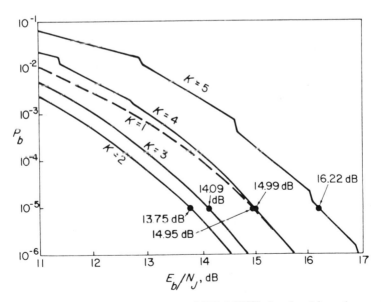

Figure 2.51. Overview of performance of FH/MFSK signals with optimum diversity in worst case $n = 1$ band multitone jamming. Note that the $K = 1$ curve is for the local minimum at m_{opt}, and arbitrarily better performance can be achieved for sufficiently large yet practically implementable values of m.

We now consider band multitone jamming for $n \in [2, M]$. To avoid the complexity of an exact performance analysis, we will again employ the union/Chernoff bound approach. Since $n \geq 2$, if an M-ary bound is jammed, we are guaranteed to have two or more high energy detector outputs. Therefore, from (2.74),

$$\Pr\{H\} = \mu^m = \left[\frac{\alpha m M}{nKE_b/N_J} \right]^m . \tag{2.106}$$

Again assuming that symbol 1 is sent, we need the statistics of the differenced energy detector output $e_{2j} - e_{1j}$ conditioned on H_j. Referring to Table 2.2, which considers the relative phase ϕ of the received signal and jammer tones, and incorporating the joint likelihood that either or both symbols 1 and 2 are hit when the M-ary band is jammed, we can write

$\Pr\{e_{2j} - e_{1j} = \beta | H_j, \phi\}$

$$= \begin{cases} \dfrac{\dbinom{M-2}{n-2}}{\dbinom{M}{n}} = \dfrac{n(n-1)}{M(M-1)} ; & \beta = -1 - \dfrac{2}{\sqrt{\alpha}}\cos\phi \\[4ex] \dfrac{\dbinom{M-2}{n-1}}{\dbinom{M}{n}} = \dfrac{n(M-n)}{M(M-1)} ; & \beta = -1 - \dfrac{1}{\alpha} - \dfrac{2}{\sqrt{\alpha}}\cos\phi \\[4ex] \dfrac{n(M-n)}{M(M-1)} ; & \beta = \dfrac{1}{\alpha} - 1 \\[4ex] \dfrac{\dbinom{M-2}{n}}{\dbinom{M}{n}} = \dfrac{(M-n)(M-n-1)}{M(M-1)} ; & \beta = -1. \end{cases}$$

$$\tag{2.107}$$

Since ϕ is assumed to be uniformly distributed over $[0, 2\pi)$,

$$\overline{e^{-(2\lambda/\sqrt{\alpha})\cos\phi}} = I_0\left(\frac{2\lambda}{\sqrt{\alpha}} \right) \tag{2.108}$$

where $I_0(\cdot)$ is the zeroeth-order modified Bessel function of the first kind. Then, as in (2.82), (2.83), and (2.86):

$$P_b \leq \frac{M}{2} F^m \tag{2.109}$$

where

$$F \equiv \min_{\lambda \geq 0} \left[\overline{\mu e^{\lambda(e_{2j} - e_{1j})}}^{|H_j|} \right]$$

$$= \min_{\lambda \geq 0} \left\{ \frac{\alpha m e^{-\lambda}}{nK(M-1)E_b/N_J} \left[n(n-1)I_0\left(\frac{2\lambda}{\sqrt{\alpha}} \right) \right. \right.$$

$$+ n(M-n)e^{-\lambda/\alpha}I_0\left(\frac{2\lambda}{\sqrt{\alpha}} \right) + n(M-n)e^{\lambda/\alpha}$$

$$\left. \left. + (M-n)(M-n-1) \right] \right\}. \tag{2.110}$$

Because of the Bessel function, we cannot express the minimization over λ in closed form. Likewise, we must resort to numerical techniques to determine the worst case jamming performance, for which F must be maximized over $\alpha \in (0, 4)$. However, it can be demonstrated that this joint extremum is unique, and occurs at $\alpha = \alpha_0$, tabulated below for selected combinations of K and n. Defining the parameter β as in (2.90), for the region $\mu \geq 1$, we have

$$P_b \leq \frac{M}{2} \left(\frac{\beta m}{E_b/N_J} \right)^m \quad \text{and} \quad \alpha_{\text{wc}} = \alpha_0; \quad \frac{E_b}{N_J} \geq \zeta m \tag{2.111}$$

where β and $\zeta \equiv \alpha_0 M/nK$ are also given in Table 2.8 for various values of K and n. Unfortunately, in the saturation region $E_b/N_J \leq \zeta m$ (or $\mu = 1$), $\alpha_{\text{wc}} = (nK/mM)E_b/N_J$ and F must be minimized numerically over $\lambda \geq 0$ for each value of E_b/N_J (this is not typically a region of practical interest unless there is some additional channel coding since the M-ary error rate is relatively high).

As usual, the easiest way to observe the variation in performance with the parameters implicit in (2.111) is to examine different graphical representations. However, since a given two-dimensional parametric plot can only illustrate the relationship between three of the independent variables P_b, E_b/N_J, K, m, and n, there are many graphs to choose from; the few selected below provide some valuable insights. Figure 2.52 is representative of the behavior of P_b as a function of m and n for fixed K and E_b/N_J. We see that there is an optimum diversity m_{opt} which increases with n. And the performance appears to improve with n for a given m, which Table 2.8 indicates is due to the fact that β decreases with n for fixed K. This suggests that it is to the jammer's advantage to keep n small so as to place a jammer tone in as many M-ary bands as possible: this is a direct consequence of the assumption that the receiver has perfect jamming state side

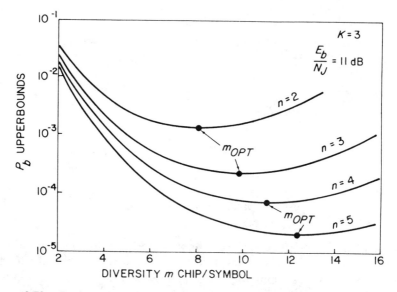

Figure 2.52. Performance of FH/8-ary FSK signals with m diversity chips/symbol in worst case band multitone jamming with n jamming tones per 8-ary band. These curves demonstrate the existence of an optimum diversity m_{opt}, and the decreased jamming effectiveness for larger n.

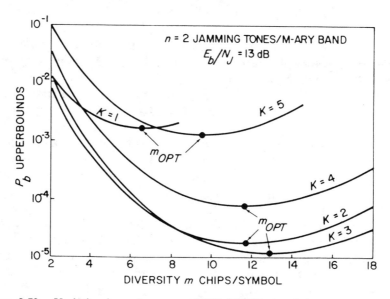

Figure 2.53. Variation in performance of FH/MFSK signals in worst case $n = 2$ band multitone jamming with alphabet size $M = 2^K$ and diversity m. Note the supremacy of the $K = 3$ case.

Table 2.8
Parameters associated with performance of FH/MFSK signals with diversity
in worst case band multitone jamming.

K	n	α_0	β	ζ	δ	γ_0, dB	$\delta\zeta$
1	2	2.395	1.1381	2.395	.3232	4.91	.774
2	2	1.072	.6305	1.072	.5835	2.34	.626
	3	1.745	.5784	1.163	.6361	1.96	.740
	4	2.395	.5691	1.197	.6465	1.89	.774
3	2	.701	.5767	.935	.6379	1.95	.596
	3	.898	.4723	.798	.7790	1.08	.622
	4	1.169	.4237	.779	.8682	.61	.676
	5	1.488	.4009	.794	.9177	.37	.729
	6	1.804	.3894	.802	.9446	.25	.758
	7	2.106	.3832	.802	.9601	.18	.770
	8	2.394	.3794	.798	.9697	.13	.774
4	2	.535	.6354	1.070	.5790	2.37	.620
	3	.625	.5023	.833	.7324	1.35	.610
	4	.716	.4297	.716	.8560	.68	.613
	5	.816	.3844	.653	.9571	.19	.625
	6	.931	.3541	.621	1.0388	− .17	.645
	7	1.064	.3335	.608	1.1031	− .43	.671
	8	1.213	.3193	.607	1.1523	− .62	.699
	12	1.827	.2933	.609	1.2545	− .98	.764
	16	2.394	.2845	.598	1.2929	− 1.12	.774
5	2	.430	.7771	1.376	.4734	3.25	.651
	32	2.395	.2276	.479	1.6162	− 2.08	.774

information. Figure 2.53 fixes n and E_b/N_J and illustrates the variation in P_b with m and K. Here we discover that m_{opt} does not vary monotonically with K. Also, the best performance with optimum diversity is realized when $K = 3$, for the selected case $n = 2$ and $E_b/N_J = 13$ dB; in fact, it will be shown below that this observation is true whenever E_b/N_J is large enough, and it extends to the $n = 3$ and 4 cases as well.

For the special case $n = M$, (2.110) reduces to

$$F = \min_{\lambda \geq 0} \left[\frac{m\alpha e^{-\lambda} I_0\left(\frac{2\lambda}{\sqrt{\alpha}}\right)}{KE_b/N_J} \right] \tag{2.112}$$

which implies that for $n = M$, β is inversely proportional to K in (2.111); consequently, the performance improves with K for $n = M$ as shown in Figure 2.54. Also, (2.112) reveals that the worst case jamming parameter α_0 is independent of K under this condition, an observation that is reinforced in Table 2.8.

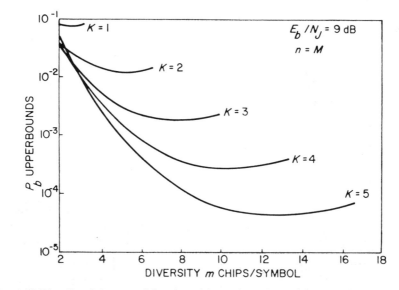

Figure 2.54. Special case of band multitone jamming where $n = M$, i.e., every symbol of a jammed M-ary band contains a jamming tone. For the record, the performance improves with K; however, this is a relatively ineffective jamming strategy which would probably not be used in practice.

Based on (2.111), we can define the optimum diversity condition:

$$P_b \leq \frac{M}{2} e^{-\delta E_b/N_J} \quad \text{where } \delta \equiv \frac{1}{\beta e}$$

$$m_{opt} = \delta E_b/N_J \qquad (2.113)$$

$$\alpha_{wc} = \alpha_0$$

provided

$$E_b/N_J \geq 1/\delta \equiv \gamma_0$$

so that $m_{opt} \geq 1$. We must also verify that we are not in the saturation region: that is,

$$m_{opt} \leq \frac{1}{\zeta} (E_b/N_J)$$

$$\Downarrow \qquad\qquad (2.114)$$

$$\delta \zeta \leq 1.$$

The validity of this condition is demonstrated in Table 2.8, which also lists values of δ and γ_0.

We observed in connection with Figure 2.52 that the performance for arbitrary m and fixed K improves with n, since Table 2.8 shows that β decreases with n. Because δ is inversely proportional to β, the performance

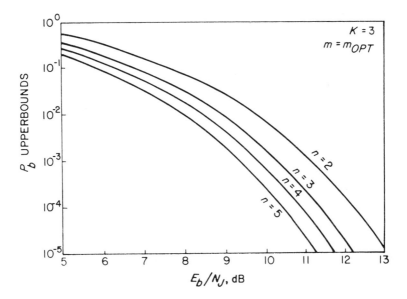

Figure 2.55. Variation in performance of FH/8-ary FSK signals with optimum diversity m_{opt} in worst case band multitone jamming as a function of the number n of jamming tones per jammed band. Jamming effectiveness deteriorates monotonically with n.

with optimum diversity also improves with n for a given K, as shown in Figure 2.55. Recall also that Figure 2.53 suggested that with optimum diversity and worst case band multitone jamming, the best performance for $n = 2$ is achieved when $K = 3$ over some range of E_b/N_J; Figure 2.56 supports this observation and shows that it is valid for $E_b/N_J \gtrsim 11$ dB. (2.113) says that the exponential component in the P_b bound, which determines the asymptotic performance for large E_b/N_J, decreases with δ for a given SNR. Table 2.8 shows that δ is in fact maximized at $K = 3$ when $n = 2$, as well as when $n = 3$ or 4 (but not for $n = 5$). So 8-ary FSK does have an advantage against the more effective band multitone jammers (except for the $n = 1$ case where 4-ary FSK is better) for low bit error rates (i.e., large E_b/N_J). Finally, for the special case $n = M$, Table 2.8 shows that δ increases with K so that the performance improves with K as illustrated in Figure 2.57.

2.3.1.3 Independent Multitone Jamming

As effective as band multitone jamming is against FH/MFSK signals, especially the $n = 1$ case, this strategy loses most of its potency if the M-ary band structure is removed. For example, a sophisticated anti-jam system might elect to use M frequency synthesizers to independently hop each transmitted symbol as a way of defeating band multitone or repeat-back

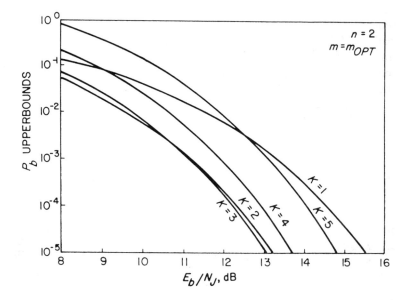

Figure 2.56. Superiority of FH/8-ary FSK signals with optimum diversity in worst case $n = 2$ band multitone jamming, for $E_b/N_J \lesssim 11$ dB.

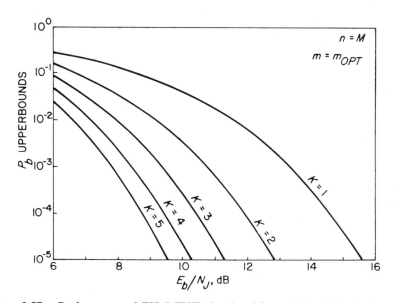

Figure 2.57. Performance of FH/MFSK signals with optimum diversity in worst case $n = M$ band multitone jamming improves with $K = \log_2 M$.

jamming [12]. Recognizing this possibility, a jammer may turn to the independent multitone jamming format as the best remaining ECM option.

Recall from Section 2.2.2.3 that the assumption implicit in the independent multitone strategy is that each FH tone frequency is independently jammed with a CW signal of power S/α with probability ρ; with m-diversity, the expression of (2.51) is modified to

$$\rho = \frac{\alpha m}{K E_b / N_J} \equiv \zeta \alpha. \tag{2.115}$$

With perfect side information, an error can only occur if at least one of the $M - 1$ untransmitted symbols is hit by a jamming tone on each diversity hop. The likelihood of this occurring on the j-th hop is

$$\Pr\{H_j\} = 1 - (1 - \rho)^{M-1} \equiv \varepsilon. \tag{2.116}$$

Independent of H_j, the transmitted symbol on the j-th hop is hit with probability ρ; however, conditioned on H_j, a particular untransmitted symbol on that hop is jammed with probability ρ/ε. Again referring to Table 2.2 and assuming that symbol 1 is sent on the j-th hop, we can write

$$\Pr\left\{e_{2j} - e_{1j} = \beta | H_j, \phi \right\}$$

$$= \begin{cases} \dfrac{\rho^2}{\varepsilon}; & \beta = -1 - \dfrac{2}{\sqrt{\alpha}} \cos \phi \\[2mm] \rho\left(1 - \dfrac{\rho}{\varepsilon}\right); & \beta = -1 - \dfrac{1}{\alpha} - \dfrac{2}{\sqrt{\alpha}} \cos \phi \\[2mm] \dfrac{\rho}{\varepsilon}(1 - \rho); & \beta = \dfrac{1}{\alpha} - 1 \\[2mm] (1 - \rho)\left(1 - \dfrac{\rho}{\varepsilon}\right); & \beta = -1. \end{cases} \tag{2.117}$$

As in the band multitone jamming case of (2.107), an error can only occur if $\beta > 0$ with a non-zero probability, so α is restricted to the range $(0, 4]$. Note that from (2.116),

$$\varepsilon - \rho = (1 - \rho)\left[1 - (1 - \rho)^{M-2}\right]. \tag{2.118}$$

Applying the identity of (2.108), and expressing ρ as a function of α defined in terms of the parameter ζ in (2.115), the union/Chernoff bound on P_b is given by (2.109) with

$$F \equiv \min_{\lambda \geq 0} \left[\overline{\varepsilon e^{\lambda(e_{2j} - e_{1j})}}^{|H_j} \right]$$

$$= \min_{\lambda \geq 0} \left(e^{-\lambda} \left[\zeta \alpha I_0 \left(\frac{2\lambda}{\sqrt{\alpha}} \right) + (1 - \zeta \alpha) e^{\lambda/\alpha} \right] \right.$$

$$\left. \times \left\{ \zeta \alpha + (1 - \zeta \alpha)\left[1 - (1 - \zeta \alpha)^{M-2}\right] e^{-\lambda/\alpha} \right\} \right). \tag{2.119}$$

As in previous jammer cases analyzed in this section on diversity, we want to determine optimum values of α and m for arbitrary but fixed values of K and E_b/N_J. From (2.115), we can replace m by the normalized diversity parameter ζ:

$$m = \zeta K E_b/N_J \qquad (2.120)$$

so that (2.109) becomes

$$P_b \leq \frac{M}{2} e^{-\zeta K \ln(1/F) E_b/N_J}. \qquad (2.121)$$

Now we want to *minimize* the (positive) exponential coefficient $\zeta K \ln(1/F)$ over $\alpha \in (0,4)$ for given values of K, E_b/N_J, and ζ; then we want to *maximize* the result over ζ.

As always, we must verify that all of the required extrema are unique. From the form of (2.119), it can be easily argued that the minimization over λ is always unique. However, it can be shown numerically that *the subsequent optimization over α and ζ is only unique for $K \geq 2$*. In the $K = 3$ case, for example, Figure 2.58 shows that the P_b bound has a well defined *saddlepoint* at the approximate values $\zeta = .28$ and $\alpha = .537$. The existence of a saddlepoint implies that it does not matter whether the optimization is performed first over α or ζ. Recall that this was not the case for partial-band noise [11], although it is in fact true for band multitone jamming. When

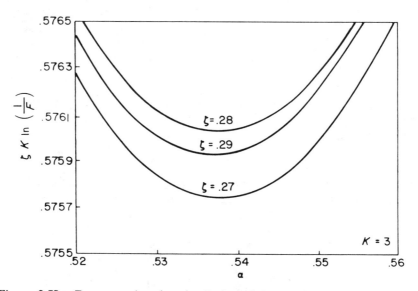

Figure 2.58. Demonstration that the P_b bound for FH/8-ary FSK signals with diversity in independent multitone jamming has a saddlepoint with respect to the normalized diversity ζ and the jammer power coefficient α. Similar results are obtained whenever $K \geq 2$.

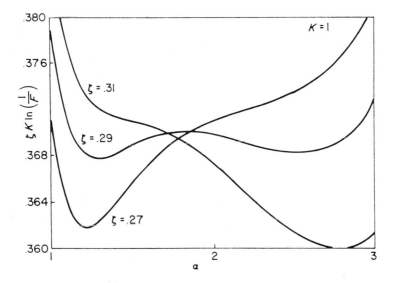

Figure 2.59. Unlike Figure 2.58, there is no saddlepoint when $K = 1$. Instead, the worst case α that minimizes the coefficient $\zeta K \ln(1/F)$ in the P_b bound of (2.121) has an abrupt discontinuity near $\zeta = .29$. In this case, the order in which the P_b bound is optimized over α and ζ affects the performance, although not as drastically as in the partial-band noise scenario.

$K = 1$, Figure 2.59 demonstrates that the optimum α has an abrupt discontinuity at $\zeta \cong .29$: to be more precise,

$$\alpha_{\text{wc}} = \begin{cases} 1.283; & \zeta = .2911_- \\ 2.552; & \zeta = .2911_+. \end{cases} \qquad (2.122)$$

That is, the P_b bound coefficient $\zeta K \ln(1/F)$ has two *equal-valued* local minima at the disjoint values of α given in (2.122) when $\zeta = .2911$. And it is this value of ζ, denoted by ζ_{opt}, that subsequently determines the optimum diversity condition by maximizing $\zeta K \ln(1/F)$ for any α_{wc} when $K = 1$:

$$\delta \equiv \max_{\zeta \geq (KE_b/N_J)^{-1}} \min_{0 < \alpha \leq \min(4, 1/\zeta)} \left[\zeta K \ln(1/F) \right]$$

$$= .3679; \qquad K = 1 \qquad (2.123)$$

provided

$$E_b/N_J \geq \left(K\zeta_{\text{opt}} \right)^{-1} \equiv \gamma_0 = 5.36 \text{ dB}; \qquad K = 1.$$

Then the P_b bound of (2.121), jointly optimized over α and ζ, has the compact form

$$P_b \leq \frac{M}{2} e^{-\delta E_b/N_J}; \qquad E_b/N_J \geq \gamma_0 \qquad (2.124)$$

Table 2.9

Parameters associated with performance of FH/MFSK signals with optimum diversity in worst case independent multitone jamming, as defined in (2.124).

K	α_{wc}	ρ_{opt}	δ	γ_0, dB
1	1.283 or 2.552	.291	.3679	5.36
2	.793	.354	.5495	1.50
3	.537	.282	.5760	0.73
4	.395	.213	.5243	0.70
5	.298	.158	.4379	1.02

where

$$m_{opt} = \zeta_{opt}KE_b/N_J = \frac{E_b/N_J}{\gamma_0}$$

from (2.120). In (2.123), the constraint on α ensures that $\rho \leq 1$ in (2.115), while the limit on ζ (which translates into the restriction that $E_b/N_J \geq \gamma_0$) guarantees that $m_{opt} \geq 1$. Interestingly, $e^{-1} = .3679$, but we have not been able to prove explicitly that $\delta = e^{-1}$ for $K = 1$. Since we do not have a saddlepoint for $K = 1$, it is illuminating to reverse the order of optimization in (2.123): we find that $\zeta_{opt} = .2911$ again, but

$$\delta' \equiv \min_{0 < \alpha \leq 4} \max_{\alpha \leq 1/\zeta \leq KE_b/N_J} [\zeta K \ln(1/F)]$$

$$\cong .370 \quad \text{and} \quad \alpha_{wc} \cong 1.8; \quad K = 1. \tag{2.125}$$

The performance improvement is only $\delta'/\delta = .025$ dB for $K = 1$, which is insignificant so that the absence of a saddlepoint here is only of academic interest, unlike the partial-band noise case.

(2.123) and (2.124) are valid for any K. The computed values of α_{wc}, ζ_{opt}, δ, and γ_0 are shown in Table 2.9 for various K. The effectiveness of diversity in combatting independent multitone jamming is illustrated in Figure 2.60 for 8-ary FSK. As a benchmark, at $P_b = 10^{-5}$, the improvement with optimum diversity relative to no diversity is at least 37.8 dB, and is closer than 4.4 dB to the broadband noise performance, since the m_{opt} curve is an upperbound. The best asymptotic performance for small P_b with optimum diversity is achieved at $K = 3$, which has the largest exponential coefficient δ; this is underscored graphically in Figure 2.61. (Recall that $K = 3$ was also asymptotically optimum for $n = 2$, 3, and 4 band multitone jamming.)

2.3.1.4 Time Diversity Overview

At this stage, we are in a position to assess the relative effectiveness of the different kinds of jammers we have analyzed against FH/MFSK systems with simple repetition coding. For sufficiently large E_b/N_J (how large

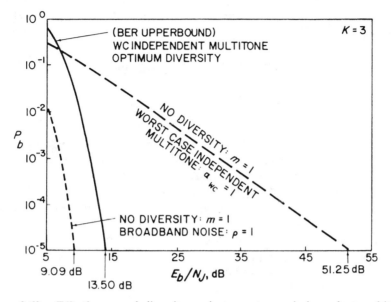

Figure 2.60. Effectiveness of diversity against worst case independent multitone jamming for FH/8-ary FSK signals. Improvement with optimum diversity relative to no diversity exceeds 37.8 dB, and performance is degraded less than 4.4 dB relative to broadband noise, at $P_b = 10^{-5}$.

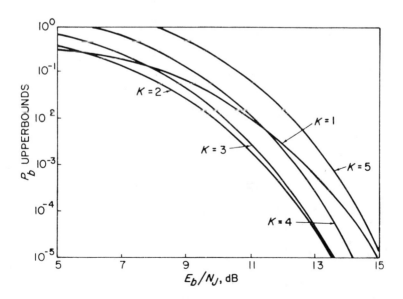

Figure 2.61. Comparison of performance of FH/MFSK systems with optimum diversity in worst case independent multitone jamming as a function of $K = \log_2 M$. Best performance for small P_b (large E_b/N_J) is achieved for 8-ary signalling.

depends on the type of jamming), the P_b upperbounds for worst case jamming with optimum diversity have the form

$$
P_b \leq
\begin{cases}
\dfrac{M}{4} e^{-\delta E_b/N_J}; & \text{noise jamming} \\[2ex]
\dfrac{M}{2} e^{-\delta E_b/N_J}; & \text{multitone jamming}
\end{cases}
\tag{2.126}
$$

where δ is enumerated in Table 2.10. With the caveat that our comparisons are based on exponentially tight upperbounds, it would appear that *the worst case $n = 1$ band multitone jammer is the most asymptotically effective of those considered against FH/MFSK signals with optimum diversity for $K \geq 2$.* Although Table 2.10 suggests that partial-band noise is more effective than band multitone jamming for $K = 1$, it must be stressed that δ is pessimistically low for the partial-band noise case since it is based on a P_b upperbound, while the $n = 1$ band multitone δ is exact; so, it is conceivable that even in this case $n = 1$ band multitone jamming is the winner.

When we analyzed the performance of FH/MFSK signals without diversity in worst case jamming, we saw that for small enough P_b's, the $n = 1$ band and independent multitone jammers were equally effective (see Figure 2.21). We argued that the reason for this equivalence is that the worst case independent multitone jammer sets the probability ρ that a given FH slot is jammed so small that a jammed M-ary band is likely to contain a single jamming tone, just as in the $n = 1$ band multitone structure. Table 2.10 shows that with optimum diversity, $n = 1$ band multitone jamming is more

Table 2.10

P_b upperbound exponential coefficients (see (2.126)) for FH/MFSK signalling with optimum diversity in worst case jamming of various classes. The smaller δ is, the more effective the jammer is for a given E_b/N_J at low P_b. The upper bound for $n = 1$ band multitone jamming and $K = 1$ is identical with the exact performance.

Type of Jammer	P_b Bound Exponential Coefficient δ				
	$K = 1$	$K = 2$	$K = 3$	$K = 4$	$K = 5$
Broadband Noise	.5000	1.0000	1.5000	2.0000	2.5000
Partial-Band Noise	.2500	.5000	.7500	1.0000	1.2500
Independent Multitone	.3679	.5495	.5760	.5242	.4379
$n = 1$ Band Multitone	.3679 (Exact)	.4631	.4493	.3839	.3014
$n = 2$ Band Multitone	.3232	.5835	.6379	.5790	.4734
$n = M$ Band Multitone	.3232	.6465	.9697	1.2929	1.6162

effective than independent multitone jamming. The difference now is that the use of a detection metric that can make an error only if all m diversity transmissions are jammed (perfect side information assumption) forces the jammer to use a much larger value of ρ; these values, shown in Table 2.11, are based on the optimum values of α and ζ shown in Table 2.9 and (2.115). Note that the probability that k slots in an M-ary band are jammed is simply $\binom{M}{k}\rho^k(1 - \rho)^{M-k}$. A relevant parameter is the likelihood that one slot in an M-ary band is jammed relative to the probability that two or more slots are jammed:

$$P_1 \equiv \frac{\Pr\{k = 1\}}{\Pr\{2 \le k \le M\}} = \frac{M\rho(1 - \rho)^{M-1}}{1 - (1 - \rho)^M - M\rho(1 - \rho)^{M-1}}.$$

(2.127)

In the absence of diversity, $P_1 \gg 1$; with optimum diversity, Table 2.11 shows that $P_1 \sim 1$. This means that a jammed band will often contain two or more jamming tones, which is different from, and apparently a less efficient use of available jamming power than the $n = 1$ band multitone structure, for which P_1 is by definition infinite.

Table 2.10 also reiterates the relative impotence of $n = M$ band multitone jamming.

All of these observations are illustrated graphically in Figures 2.62–2.64. Note that as indicated in Figure 2.64, the optimum diversity m_{opt} varies with the type of jamming for a given K and E_b/N_J. Which raises an interesting point: with the exception of independent multitone jamming (refer to (2.124)), all of the remaining jammers have

$$m_{\text{opt}} = \delta E_b/N_J.$$ (2.128)

Now we know from (2.126) that a larger δ ensures better performance for a given E_b/N_J. But (2.128) reminds us that this benefit is balanced by increased system complexity due to the need for more diversity.

In designing an FH/MFSK system with diversity, it is prudent to select parameters that will allow it to achieve the desired performance under the

Table 2.11
Parameters associated with performance of FH/MFSK signals with optimum diversity in worst case independent multitone jamming, as defined in (2.115) and (2.127).

K	ρ_{wc}	P_1
1	.37 or .74	3.4 or .69
2	.28	1.3
3	.15	1.1
4	.084	.91
5	.047	.75

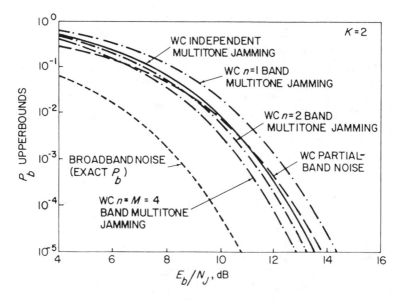

Figure 2.62. Relative performance of FH/4-ary FSK signalling with optimum diversity in different worst case jamming environments. Note that the optimum diversity varies with the type of jammer for a given E_b/N_J.

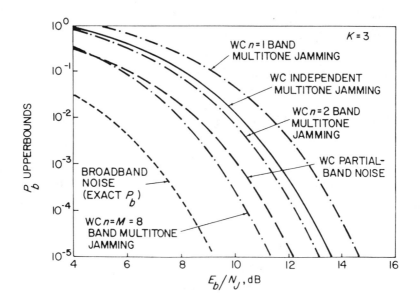

Figure 2.63. Same as Figure 2.62, but for 8-ary FSK.

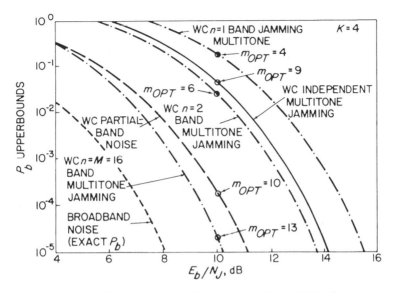

Figure 2.64. Same as Figures 2.62 and 2.63, but for 16-ary FSK. To emphasize the variation in optimum diversity m_{opt} with jammer type, values of m_{opt} are indicated for $E_b/N_J = 10$ dB.

worst conceivable jamming conditions. With the possible exception of $K = 1$ as discussed above, this implies the assumption of $n = 1$ band multitone jamming. However, since this class of jammer has the lowest δ for a given K, it calls for less diversity than is optimum for the other kinds of jamming. So we must verify that the suboptimum, degraded performance for these other jammers does not exceed the optimized P_b for $n = 1$ band multitone jamming. This kind of information is not available from curves of the form of Figures 2.62 2.64. As an example, using Table 2.10, we see that

$$\min_{K} \quad \max_{\text{type of jammer}} \quad \delta = .4493; \qquad n = 1 \text{ band multitone jamming}, \quad K = 3.$$

$$(2.129)$$

On this basis, it is reasonable to use $K = 3$, and if we assume a desired P_b of 10^{-5}, (2.126) and (2.128) suggest that we use $E_b/N_J = 14.58$ dB and $m = m_{opt} = 13$. Using previously derived bounds on P_b for worst case jamming with arbitrary diversity, curves of performance versus diversity are drawn in Figures 2.65 for $n = 1$ band multitone, independent multitone, and partial-band noise jamming with the selected K and E_b/N_J. Although $m_{opt} = 24$ and 22 respectively for the last two jammers, we see that the bound on P_b for these jammers at $m = 13$ does not exceed the desired bit error rate, so we do indeed have a viable system design.

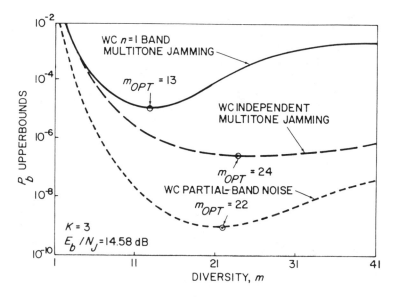

Figure 2.65. Variation in performance of FH/8-ary FSK systems with diversity for several kinds of worst case jamming when $E_b/N_J = 14.58$ dB, which was selected to achieve a P_b of 10^{-5} or better under any jamming conditions with $m = 13$.

2.3.2 Coding without Diversity

We have seen in the previous sections that time diversity techniques can significantly reduce the effectiveness of worst case noise and multitone jammers against FH/MFSK signals. In particular, optimum diversity restores the exponential relationship between P_b and E_b/N_J from the inverse linear dependence characterizing worst case jamming without any coding redundancy.

Yet, diversity transmission corresponds to a simple repetition code. And there are many block and convolutional codes that are much more powerful than repetition codes, based on experience with classical additive Gaussian noise channels. It remains to be seen how effective these codes are in a non-stationary, non-Gaussian jamming environment. In fact, it would probably be valuable to derive codes and detection metrics specifically matched to the more insidious pulsed, partial-band, or tone jamming channels characteristic of the electronic warfare era.

Several papers in the open literature have investigated the effectiveness of various block, convolutional, and concatenated codes for FH/MFSK systems in partial-band noise [2], [14]–[16]. Of these, the paper by Ma and Poole [16] is perhaps the most comprehensive. In this section, we will consider multitone as well as partial-band noise jamming. This is important since we have previously seen that $n = 1$ band multitone jamming is the most effective type of interference against uncoded FH/MFSK signals, with or without diversity, for $K \geq 2$. It will also provide some insights into the effect of the jamming channel characteristics on the performance of some of the more popular codes in use today.

Our intention is not to exhaustively analyze the performance of every known code in all types of jamming. The emphasis will be on providing the analytical tools for evaluating the effectiveness of coding against worst case FH/MFSK jammers. Specifically, we will consider several binary and M-ary convolutional codes, Reed-Solomon codes (because their burst error correction capabilities make them attractive for non-stationary interference), and concatenated schemes employing Reed-Solomon outer codes and convolutional inner codes. With regard to jamming categories, we will restrict our analysis to partial-band noise and $n = 1$ band multitone jamming. The detection metric remains the unweighted linear combination of received chip energies, with perfect jamming state side information. We will be concerned with the relative performance of these codes in worst case jamming, and how they compare with the unsophisticated repetition codes with optimum diversity previously analyzed.

2.3.2.1 Convolutional Codes

Our analytical approach mirrors that of Chapter 4, Volume I, and [8]. Simulation techniques aside, most coding performance analyses are based on bounds of the form (see (4.24), Volume I)

$$P_b \leq G(D) \tag{2.130}$$

where the structure of $G(D)$ depends only on the coder and decoder characteristics, while the parameter D is *separably* dependent only on the type of jamming and the detection metric.

In particular, for the special case of partial-band noise with spectral fill factor ρ, (2.130) is modified by the addition of the familiar factor of $1/2$ (refer to [10] and Appendix 2A):

$$P_b \leq \tfrac{1}{2} G(D); \text{ noise jamming} \tag{2.131}$$

and, for the soft decision linear combining metric with perfect side information (see (4.92), Volume I),

$$D = \left(\frac{\rho}{1 - \lambda^2} \right) e^{-(\lambda/1+\lambda)(\rho E_c/N_J)} \tag{2.132}$$

where

$$\lambda = \tfrac{1}{2}\left[\sqrt{1 + 6\beta + \beta^2} - \beta - 1 \right]$$

$$\beta \equiv \frac{\rho E_c}{2 N_J}.$$

The parameter E_c/N_J is the MFSK chip energy-to-jamming noise spectral power ratio (see (4.2), Volume I):

$$\frac{E_c}{N_J} = R\left(\frac{E_b}{N_J}\right) \qquad (2.133)$$

where R is the code rate expressed in information bits/chip. Recall that the transmission of a chip represents a single use of the M-ary channel; a chip is an M-ary symbol from the innermost code. For example, in an uncoded system with diversity m (we could also regard this as a simple repetition code as discussed previously), each diversity transmission is a single M-ary chip and $R = K/m$.

For worst case partial-band noise [5], we want to maximize D over $\rho \in (0, 1]$:

$$D = \frac{4e^{-1}}{E_c/N_J} \quad \text{and} \quad \rho_{wc} = \frac{3}{E_c/N_J}; \qquad \frac{E_c}{N_J} \geq 3 \qquad (2.134)$$

while for $E_c/N_J \leq 3$, D is given by (2.132) with $\rho = \rho_{wc} = 1$.

For multitone jamming, the P_b bound of (2.130) must be used. For $n = 1$ band multitone jamming, the performance bounds have the form (refer to (2.75), (2.86), and (2.87))

$$P_b \leq \frac{M}{2} F^m. \qquad (2.135)$$

Referring to the definition of D in (4.20) in Volume I, we see that D is simply equal to F, with KE_b/mN_J replaced by E_c/N_J. In terms of the jammer power parameter α, which is restricted to the range (see (2.89))

$$0 < \alpha < \min\left(1, \frac{E_c}{MN_J}\right) \qquad (2.136)$$

we have

$$D = \begin{cases} \dfrac{\alpha}{E_c/N_J}; & K = 1 \\[3mm] \dfrac{1}{E_c/N_J}\left[\dfrac{\alpha(M-2)}{1-\alpha}\right]^{1-\alpha}; & K \geq 2. \end{cases} \qquad (2.137)$$

Maximizing D over α for worst case $n = 1$ band multitone jamming, for the special case $K = 1$,

$$D = \begin{cases} \dfrac{1}{E_c/N_J} \text{ and } \alpha_{wc} = 1_-; & \dfrac{E_c}{N_J} \geq 2 \\[3mm] \dfrac{1}{2} \text{ and } \alpha_{wc} = \dfrac{E_c}{2N_J}; & \dfrac{E_c}{N_J} < 2. \end{cases} \qquad (2.138)$$

Similarly, for $K \geq 2$,

$$D = \begin{cases} \dfrac{\beta K}{E_c/N_J} \text{ and } \alpha_{\mathrm{wc}} = \alpha_0; & \dfrac{E_c}{N_J} \geq \alpha_0 M \\[3mm] \dfrac{1}{E_c/N_J} \left[\dfrac{\alpha_{\mathrm{wc}}(M-2)}{1 - \alpha_{\mathrm{wc}}} \right]^{1 - \alpha_{\mathrm{wc}}} & \\[3mm] \text{and } \alpha_{\mathrm{wc}} = \dfrac{E_c}{MN_J}; & \dfrac{E_c}{N_J} \leq \alpha_0 M \end{cases} \qquad (2.139)$$

where β and α_0 are given in Table 2.7 for $2 \leq K \leq 5$.

We now want to specify $G(D)$ for the codes of interest in this section. As a reference point, for the m-diversity repetition code (see (2.62)),

$$G(D) = \frac{M}{2} D^m \quad \text{and} \quad R = \frac{K}{m}. \qquad (2.140)$$

In worst case partial-band noise, the diversity that minimizes D^m (the so-called quasi-optimum diversity that minimizes the bound on P_b in [2]) is given by (see (2.70))

$$m_{\mathrm{opt}} = \frac{KE_b}{4N_J}; \qquad \frac{E_b}{N_J} \geq \frac{4}{K}. \qquad (2.141)$$

Combining (2.131), (2.133), (2.134), (2.140), and (2.141), we have

$$P_b \leq \frac{M}{4} e^{-KE_b/4N_J}; \qquad \frac{E_b}{N_J} \geq \frac{4}{K}. \qquad (2.142)$$

For the special case of $n = 1$ band multitone jamming with $K = 1$, we could use (see (2.77))

$$m_{\mathrm{opt}} = e^{-1}\left(\frac{E_b}{N_J}\right); \qquad \frac{E_b}{N_J} \geq e \qquad (2.143)$$

although we saw in Figure 2.36 that we could achieve arbitrarily small error rates by using large enough values of m. Here, we combine (2.130), (2.133), (2.138), (2.140), and (2.143) to yield the performance bound

$$P_b \leq e^{-e^{-1}(E_b/N_J)}; \qquad \frac{E_b}{N_J} \geq e. \qquad (2.144)$$

For $n = 1$ band multitone jamming with $K \geq 2$, we want to use (see (2.104))

$$\left. \begin{array}{c} m_{\mathrm{opt}} = \delta\left(\dfrac{E_b}{N_J}\right) \\[3mm] \Downarrow \\[3mm] P_b \leq \dfrac{M}{2} e^{-\delta(E_b/N_J)} \end{array} \right\} \quad \begin{array}{c} \dfrac{E_b}{N_J} \geq \dfrac{1}{\delta} \end{array} \qquad (2.145)$$

where δ is specified in Table 2.7 for $2 \leq K \leq 5$.

Perhaps the most commonly used convolutional codes for *binary* ($K = 1$) channels are the constraint length 7, rate $1/2$ and $1/3$ structures determined by Odenwalder [17]; assuming soft decision Viterbi decoding, the performance is given by

$$G(D) = 36D^{10} + 211D^{12} + 1404D^{14} + 11{,}633D^{16}$$

$$+ 77{,}433D^{18} + 502{,}690D^{20} + 3{,}322{,}763D^{22}$$

$$+ 21{,}292{,}910D^{24} + 134{,}365{,}911D^{26} + \cdots$$

$$\text{and } R = \tfrac{1}{2}; \left(7, \tfrac{1}{2}\right) \text{ code} \qquad (2.146)$$

$$G(D) = D^{14} + 20D^{16} + 53D^{18} + 184D^{20} + \cdots$$

$$\text{and } R = \tfrac{1}{3}; \left(7, \tfrac{1}{3}\right) \text{ code.} \qquad (2.147)$$

((2.147) is not accurate for $D \gtrsim 10^{-1}$.)

For larger channel alphabets, Trumpis found the optimum constraint length 7, rate $1/2$ and $1/3$ (over these larger alphabets) convolutional codes for 4-ary and 8-ary orthogonal signal sets, respectively [18]:

$$G(D) = 7D^7 + 39D^8 + 104D^9$$

$$+ 352D^{10} + 1187D^{11} + \cdots; \left(7, \tfrac{1}{2}\right) \text{ code} \qquad (2.148)$$

$$G(D) = D^7 + 4D^8 + 8D^9 + 49D^{10}$$

$$+ 92D^{11} + \cdots; \left(7, \tfrac{1}{3}\right) \text{ code.} \qquad (2.149)$$

Since each information bit produces an M-ary code symbol above, $R = 1$ for both cases.

For $K \geq 2$, the M-ary orthogonal convolutional codes [19] have

$$G(D) = \frac{D^K(1 - D)^2}{(1 - 2D - D^K)^2} \quad \text{and} \quad R = 1. \qquad (2.150)$$

Finally, for all values of K, we have the powerful class of dual-K convolutional codes with code rate $1/\nu$ over GF(2^K) [2], [20]. That is, for every M-ary (K-bit) input word, ν M-ary code symbols are generated, corresponding to $R = K/\nu$ bits/chip. The performance is defined by

$$G(D) = \frac{MD^{2\nu}}{2\left[1 - \nu D^{\nu-1} - (M - \nu - 1)D^\nu\right]^2}. \qquad (2.151)$$

We now want to investigate how effective these convolutional codes are at combatting worst case partial-band noise and $n = 1$ band multitone jamming. The performance of these codes with FH/MFSK signals is illustrated in Figures 2.66–2.75 for $1 \leq K \leq 5$ and both kinds of jamming. As a benchmark, each graph plots the upperbound on P_b for "uncoded" signalling with optimum diversity. In one sense, this comparison is unfavorable to the convolutional codes. The optimum diversity structure corresponds to the block repetition code with the code rate K/m that provides the best

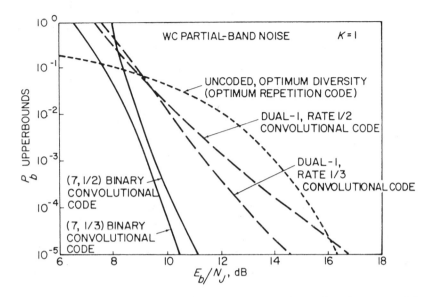

Figure 2.66. Performance of several convolutional codes and optimum repetition code with FII binary FSK signals in worst case partial-band noise. Odenwalder's binary convolutional codes are particularly effective (e.g., about 5–6 dB better than uncoded signalling with optimum diversity at $P_b = 10^{-5}$). Assumes soft decision energy detection, linear combining, and perfect side information.

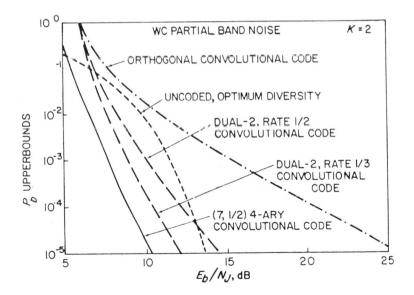

Figure 2.67. Same as Figure 2.66, except $K = 2$. The 4-ary orthogonal convolutional code performs poorly, with an inverse linear characteristic for small P_b's. Trumpis' optimum constraint length 7 convolutional code for 4-ary orthogonal signalling is about 3.5 dB better than optimum diversity at $P_b = 10^{-5}$.

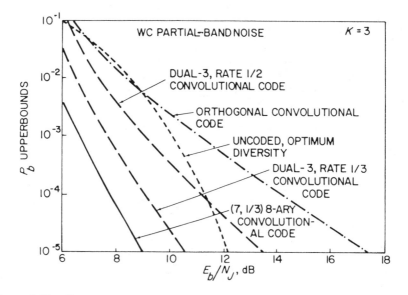

Figure 2.68. Same as Figures 2.66 and 2.67, except $K = 3$. The orthogonal convolutional code continues to be ineffective, while the dual-3, rate $1/3$ code outperforms the optimum diversity system over P_b's of interest. The best performance is again provided by Trumpis' optimum $(7, \frac{1}{3})$ code for 8-ary orthogonal channels.

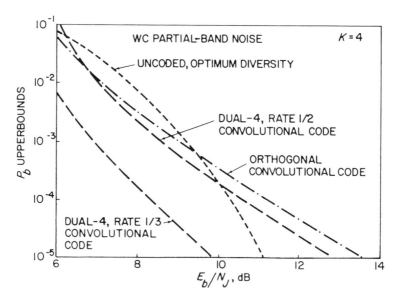

Figure 2.69. Same as previous figures with $K = 4$. The dual-4, rate $1/2$ convolutional code performance is almost as poor as orthogonal convolutional coding. The dual-4, rate $1/3$ performance is marginally better than optimum diversity (e.g., about 1 dB better at $P_b = 10^{-5}$).

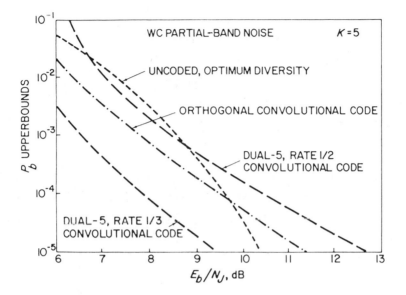

Figure 2.70. Same as previous figures, except $K = 5$. The 32-ary orthogonal convolutional code is now better than the dual-5, rate $1/2$ code, although both codes perform poorly at low P_b's. The dual-5, rate $1/3$ code is again about 1 dB better than optimum diversity at $P_b - 10^{-5}$.

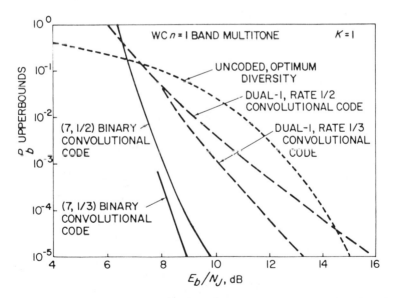

Figure 2.71. Same as Figure 2.66, but for worst case $n = 1$ band multitone jamming. Remember that this entire series of graphs is for the soft decision energy detection metric with linear combining and perfect side information. Odenwalder's codes again provide the best performance, with a 1.5 dB improvement over the noise jamming case at $P_b = 10^{-5}$. Some of the curves are truncated because $G(D)$ is invalid for small E_b/N_J.

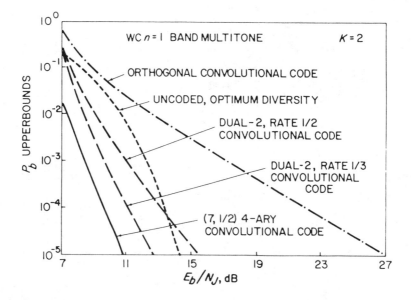

Figure 2.72. Same as Figure 2.67 for multitone jamming. The orthogonal convolution codes continue to be vulnerable to worst case jamming. Trumpis' $(7, \frac{1}{2})$ code is about 3.5 dB better than optimum diversity as in the noise jamming case, but the multitone jammer degrades its performance about .75 dB relative to the noise jammer.

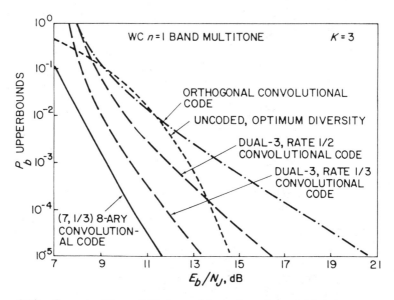

Figure 2.73. Same as Figure 2.68 for multitone jamming, which is more effective than partial-band noise. However, relative performance of codes is essentially unchanged, with Trumpis' $(7, \frac{1}{3})$ code on top.

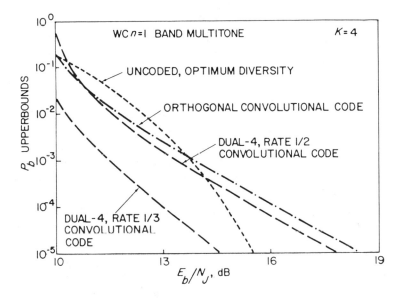

Figure 2.74. Same as Figure 2.69, with worst case multitone jamming proving more effective than partial-band noise again (this is generally true for smaller P_b's and $K \geq 2$).

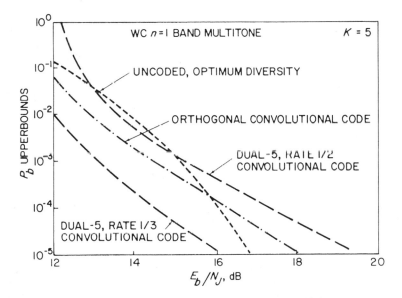

Figure 2.75. Same as Figure 2.70, except jamming is worst case $n = 1$ band multitone. Performance of dual-5, rate 1/3 code is about 6.5 dB worse than in partial-band noise, and 1.5 dB worse than for $K = 4$ in multitone jamming at $P_b = 10^{-5}$.

performance for a given combination of FH/MFSK signals and jamming. No attempt has been made to similarly optimize the convolutional code rates. We will examine this issue later in connection with block codes, particularly Reed-Solomon codes.

Some specific observations about the relative effectiveness of the various convolutional codes in worst case jamming are included in the commentaries with Figures 2.66–2.75. Overall, the Odenwalder optimum binary convolutional codes and the Trumpis optimum 4-ary and 8-ary convolutional codes are a significant improvement over uncoded FH/MFSK with optimum diversity. At the other extreme, orthogonal convolutional coding is quite vulnerable to worst case jamming, especially for smaller values of K; in fact, for lower P_b's the performance curves exhibit the inverse linear dependence characteristic of uncoded signalling. In between are the dual-K convolutional codes, which can be fairly effective against worst case jamming, with improved performance for lower code rates $1/\nu$.

For a given class of codes, there are observations to be made with regard to the impact of the channel dimensionality K and the type of jamming on system performance. The variation in performance with K is illustrated in Figures 2.76–2.81. Recall that for uncoded FH/MFSK signalling with optimum diversity, the performance in worst case partial-band noise improved monotonically with K for moderate and lower P_b's (Figure 2.31); this behavior is adhered to by the orthogonal and dual-K convolutional codes. On the other hand, for worst case $n = 1$ band multitone jamming, Figure 2.51 showed that the best performance for uncoded signalling with optimum diversity is achieved for $K = 2$, with increasingly degraded performance for larger values of K. (This should be qualified by the reminder that the $K = 1$ performance could be made arbitrarily good for large enough amounts of diversity.) The dual-K performance in worst case multitone jamming conforms to this behavior (without the $K = 1$ qualifier above), while the orthogonal convolutional codes in Figure 2.77 do not.

Focussing on the dual-K convolutional code behavior, since performance improves with K for worst case partial-band noise, and degrades with $K \geq 2$ for worst case $n = 1$ band multitone jamming, the impact of the type of jamming becomes more significant for larger values of K. This effect is illustrated in Figure 2.82 for the rate $1/3$ codes. In particular, at $P_b = 10^{-5}$, noise jamming is about 1 dB more effective than multitone jamming for $K = 1$; the jamming effectiveness is reversed for $K \geq 2$, with multitone jamming exhibiting a 3 dB advantage at $K = 3$, and a 7 dB advantage at $K = 5$ for this P_b.

The relative performance of all of these codes in noise and multitone jamming is summarized in Table 2.12 at the benchmark $P_b = 10^{-5}$. From a min/max perspective, if we wanted to choose the coded system that required the smallest SNR to provide a guaranteed P_b of 10^{-5} *in any kind of non-adaptive jamming*, we would use the Odenwalder $(7, \frac{1}{3})$ binary convolutional code: this requires $E_b/N_J = 10.38$ dB for worst case partial-band noise, with better performance for any other jammer. By comparison, if we

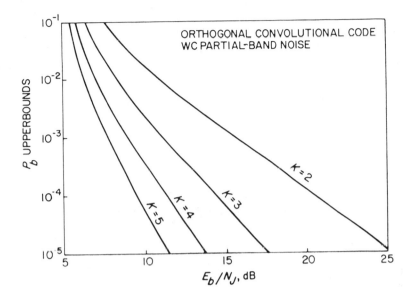

Figure 2.76. Variation with K in performance of orthogonal convolutional codes with FH/MFSK signals in worst case partial-band noise, assuming soft decision energy detection with linear combining metric and perfect jamming state side information. Performance is generally poor, but improves with K.

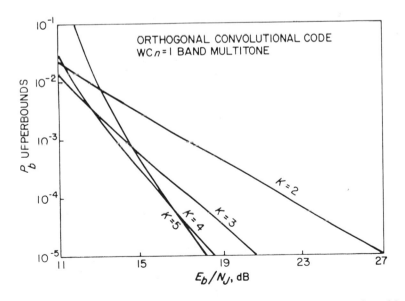

Figure 2.77. Same as Figure 2.76, but for worst case $n = 1$ band multitone jamming. For sufficiently small P_b's, performance improves with K.

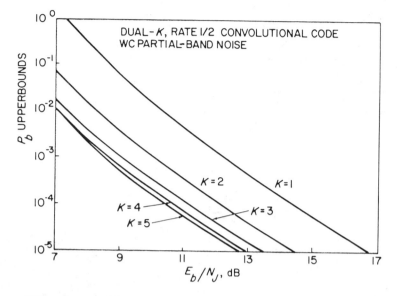

Figure 2.78. Same as Figure 2.76, but for dual-K, rate 1/2 convolutional coding. Performance improves with K for moderate to small P_b's, but improvement is minor for $K \gtrsim 4$.

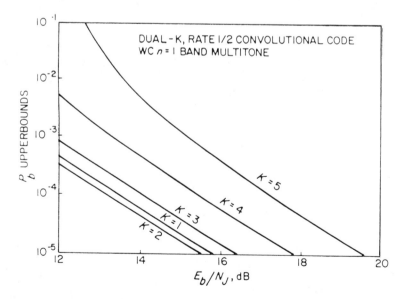

Figure 2.79. Same as Figure 2.78, but for multitone jamming. Performance is optimized over range of interest of P_b at $K = 2$. For $K \geq 2$, performance degrades with K (similar effect was observed for uncoded FH/MFSK signals with optimum diversity in Figure 1.51).

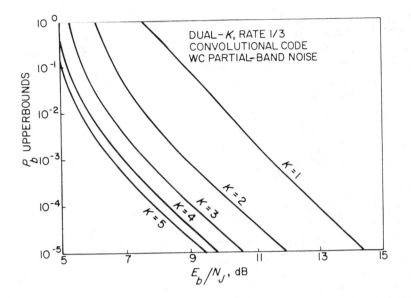

Figure 2.80. Same as Figure 2.78, but for rate $1/3$, dual-K convolutional coding. Performance improves with K.

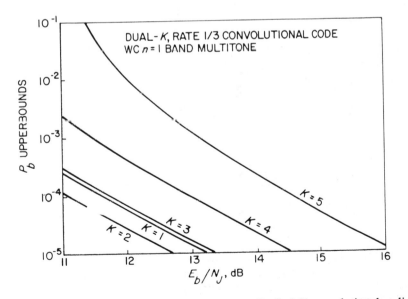

Figure 2.81. Same as Figure 2.79, but for rate $1/3$, dual-K convolutional coding. Variation with K is qualitatively identical to rate $1/2$ code in Figure 2.79.

adopted the same system design philosophy for uncoded signalling with optimum diversity, we would need $E_b/N_J = 14.21$ dB with $K = 2$, and the worst case jammer would be $n = 1$ band multitone. In this sense, the $(7, \frac{1}{3})$ binary convolutional code is about 4 dB more powerful than the best repetition code against worst case jamming of FH/MFSK systems.

Figure 2.83 examines the relative sensitivity of several coded FH/MFSK systems to variations in the partial-band noise fill factor ρ, for the specific case of $K = 4$ and $E_b/N_J = 10$ dB. The ρ-axis uses a logarithmic scale so that equal length intervals anywhere correspond to identical relative (e.g., percent) variations in ρ. From the communicator's viewpoint, a highly peaked curve is desirable because this implies that small deviations in ρ away from ρ_{wc} produce large reductions in jamming effectiveness. However, all three systems exhibit approximately the same sensitivity to changes in ρ. Note though that $\rho_{wc} = .75$ for uncoded signalling with optimum diversity, $\rho_{wc} = .30$ for 16-ary orthogonal convolutional coding, and $\rho_{wc} = .15$ for the dual-4, rate 1/2 convolutional code; furthermore, all three systems achieve comparable performance. This observation suggests a more sophisticated hybrid coding scheme for defeating partial-band noise jamming. Consider a two-way FH/16-FSK system which can select any of the coding schemes of Figure 2.83. The receivers in this link must be able to monitor the jamming parameter ρ (e.g., by performing a fast Fourier transform over the entire

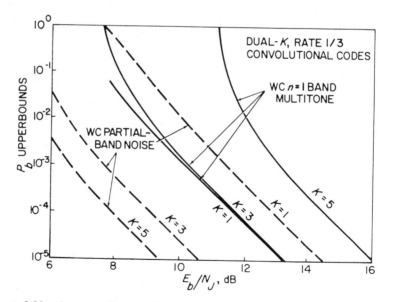

Figure 2.82. Impact of type of jamming on performance of dual-K, rate 1/3 convolutional codes with FH/MFSK signalling as a function of K. Assumes soft decision energy detection, linear combining metric, and perfect jamming state side information.

Table 2.12

Required E_b/N_J in dB to achieve $P_b = 10^{-5}$ for coded FH/MFSK ($M = 2^K$) system in worst case partial-band noise (upper number) and $n = 1$ band multitone jamming (lower number), assuming soft decision energy detection with perfect jamming state side information.

Type of Code	K				
	1	2	3	4	5
Optimum Diversity (Repetition)	16.36	13.62	12.12	11.11	10.36
	14.95	14.21	14.58	15.49	16.76
Optimum Binary	11.11	n/a	n/a	n/a	n/a
$(7, \frac{1}{2})$ Convolutional	9.71	n/a	n/a	n/a	n/a
Optimum Binary	10.38	n/a	n/a	n/a	n/a
$(7, \frac{1}{3})$ Convolutional	8.89	n/a	n/a	n/a	n/a
Optimum 4-ary	n/a	10.12	n/a	n/a	n/a
$(7, \frac{1}{2})$ Convolutional	n/a	10.84	n/a	n/a	n/a
Optimum 8-ary	n/a	n/a	8.95	n/a	n/a
$(7, \frac{1}{3})$ Convolutional	n/a	n/a	11.55	n/a	n/a
Orthogonal	n/a	25.19	17.42	13.58	11.30
Convolutional	n/a	27.03	20.63	18.46	18.05
Dual-K, Rate $\frac{1}{2}$	16.72	14.43	13.40	12.89	12.62
Convolutional	15.75	15.47	16.34	17.76	19.56
Dual-K, Rate $\frac{1}{3}$	14.38	11.86	10.60	9.85	9.39
Convolutional	13.19	12.68	13.31	14.49	16.04

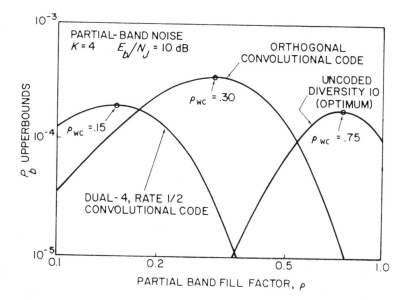

Figure 2.83. Sensitivity of three coded FH/MFSK systems to variations in partial-band noise fill factor ρ, for $K = 4$ and $E_b/N_J = 10$ dB. Although all three cases exhibit comparable sensitivity to ρ, differences in ρ_{wc} suggest a hybrid coding scheme for defeating partial-band noise jamming (see text for details).

spread-spectrum bandwidth). As an example, if one of the receivers determines that it is being jammed with partial-band noise for which $\rho = .75$, it could request that the transmitter at the other end of the link switch to the dual-4, rate 1/2 convolutional code. If the jammer can somehow detect this change and reduce ρ to .15, or if it simply varies ρ pseudorandomly, the FH system can react accordingly. If the link can monitor ρ and adjust its code quickly enough, it can maintain a mismatch between ρ and ρ_{wc} most of the time. The net effect of this mismatch is that the effectiveness of the partial-band noise jammer will be significantly reduced, to the extent that the jammer may in fact be better off emitting broadband noise.

2.3.2.2 Reed-Solomon Codes

We now consider Reed-Solomon (RS) codes (refer to Section 4.10.3, Volume I) in place of convolutional codes, with FH/MFSK signals in worst case partial-band noise and multitone jamming. These are block codes, with input and output characters (symbols) over $GF(2^Q)$. The output block length of a conventional RS code is $2^Q - 1$ Q-bit characters, although extended RS codes can have block lengths of 2^Q or $2^Q + 1$.

As an example, the popular (255, 191) RS code uses $Q = 8$ bit characters. The 64-character redundancy of this code allows it to correct up to 32 character errors (or, equivalently, 256 bit errors) within each received block of length 255. The code rate over $GF(2^8)$ is $191/255$. If an FH/256-ary FSK channel is available, each RS character can be ideally transmitted as a single tone. Other less desirable implementations are the transmission of each RS character as two consecutive tones over a 16-ary channel, or four tones over a 4-ary channel, or even eight binary tones. In general, a Q-bit RS character can be sent as L tones over an $M = 2^K$-ary channel, where

$$M^L = 2^{KL} = 2^Q \Rightarrow L = \frac{Q}{K}. \tag{2.152}$$

Because of complexity considerations, RS decoders in use today typically make hard decisions on each received code symbol, and use this reduced observable space to recreate the transmitted data. Berlekamp has generated a table of 2^Q-ary RS character error rates P_Q required for a hard-decision decoder to achieve some selected P_b's for codes with power of 2 redundancies up to 64 [21]. If L uses of an M-ary channel are needed to send each RS character, and the channel K-bit symbol error rate is P_K, then

$$P_Q = 1 - (1 - P_K)^L. \tag{2.153}$$

With no diversity, the bit error rate for FH/MFSK signals in worst case partial-band noise and $n = 1$ band multitone jamming is specified by (2.25) and (2.41), respectively, except that E_b/N_J *must be reduced by the RS code rate*. The bit error rate over the M-ary channel then determines the symbol error rate P_K according to the familiar relation

$$P_b = \frac{M}{2(M-1)} P_K. \tag{2.154}$$

Using this procedure, we were able to measure the relative performance of various RS codes, without additional levels of coding or diversity, over an FH/MFSK channel in worst case jamming. The results are presented in Table 2.13 in the form of the E_b/N_J required to achieve $P_b = 10^{-5}$ out of the RS decoder. These are *exact* SNR's based on the tabulated P_Q's in [21]. For the sake of comparison, the convolutional code performance in Table 2.12 is generally much better, particularly in the multitone case, even though the tabulated values of E_b/N_J are upperbounds. As a single benchmark, the minimum E_b/N_J required to ensure $P_b = 10^{-5}$ for any non-adaptive jammer in Table 2.12 is less than 10.38 dB; for Table 2.13, the corresponding requirement is 19.70 dB (for $K = 2$, the $(63, 31)$ RS code, with worst case $n = 1$ band multitone jamming). The performance might be improved for some of the higher order RS codes if redundancies greater than 64 were used; however, Berlekamp's tabulated values of P_Q do not extend to this region, and the complexity of high-speed RS coder/decoder implementations increases with redundancy [21, p. 571]. There are some expanded tables of P_Q covering a larger range of redundancies, and we will use these later to explore the issue of optimum RS code rates in the context of combined coding and diversity.

Table 2.13

Required E_b/N_J to achieve $P_b = 10^{-5}$ for Reed-Solomon coding with FH/MFSK signals in worst case partial-band noise and $n = 1$ band multitone jamming, assuming hard decisions on received code symbols. Terminology: Reed-Solomon codes over $GF(2^Q)$; $L = Q/K$ = number of $M = 2^K$-ary channel uses/code symbol; P_Q and P_K are code and channel symbol error rates, respectively.

K	RS Code	L	P_Q	P_K	E_b/N_J, dB for $P_b = 10^{-5}$	
					Noise	Multitone
1	(31, 15)	5	5.88×10^{-2}	1.20×10^{-2}	18.02	22.36
	(63, 31)	6	9.50×10^{-2}	1.65×10^{-2}	16.56	20.91
	(127, 63)	7	1.30×10^{-1}	1.97×10^{-2}	15.76	20.10
	(255, 191)	8	6.51×10^{-2}	8.38×10^{-3}	17.68	22.02
2	(15, 7)	2	2.81×10^{-2}	1.42×10^{-2}	17.22	23.55
	(63, 31)	3	9.50×10^{-2}	3.27×10^{-2}	13.37	19.70
	(255, 191)	4	6.51×10^{-2}	1.67×10^{-2}	14.46	20.79
	(1023, 959)	5	1.72×10^{-2}	3.46×10^{-3}	20.32	26.65
3	(7, 3)	1	9.51×10^{-3}	9.51×10^{-3}	19.24	27.58
	(63, 31)	2	9.50×10^{-2}	4.87×10^{-2}	11.54	19.88
	(511, 447)	3	3.33×10^{-2}	1.12×10^{-2}	15.43	23.77
4	(15, 7)	1	2.81×10^{-2}	2.81×10^{-2}	14.11	24.56
	(255, 191)	2	6.51×10^{-2}	3.31×10^{-2}	11.35	21.80
5	(31, 15)	1	5.88×10^{-2}	5.88×10^{-2}	10.78	23.38
	(1023, 959)	2	1.72×10^{-2}	8.64×10^{-3}	16.27	28.86

2.3.2.3 Concatenated Codes

Since we have determined that Reed-Solomon (RS) codes alone do not provide sufficient redundancy to produce acceptable FH/MFSK performance in worst case jamming, the next recourse is to concatenate an RS outer code with a suitable inner code (usually a convolutional code) as illustrated in Figure 4.21, Volume I.

In particular, the RS outer code generates Q-bit code symbols (or characters). These are decomposed into L smaller K-bit subsymbols, which are then scrambled by an interleaver. Suppose these subsymbols are then fed to an inner dual-K, rate $1/v$ convolutional encoder, producing K-bit chips that are transmitted over an FH/MFSK channel. In the receiver, the dual-K Viterbi decoder makes *hard K-bit subsymbol* decisions based on the *soft* decision *chip* energy, linear combining metric with perfect jammer state side information. These subsymbols are then deinterleaved to break up any burst error patterns, and reconstituted into Q-bit RS characters which are input to the outer RS decoder.

Suppose we want to achieve a particular P_b out of the RS decoder: this maps into a character error rate P_Q [22], which determines the subsymbol error rate P_K using (2.153). Even though the inner decoder makes hard subsymbol rather than bit decisions, its performance is characterized by the P_b of (2.154). Depending on whether we have worst case partial-band noise or multitone jamming, (2.130) or (2.131) is combined with (2.151) to compute the required D. If we have an (n, k) RS outer code (i.e., each block of k Q-bit data symbols produces $n > k$ code symbols), the concatenated code rate is

$$R = \left(\frac{k}{n}\right)\left(\frac{K}{v}\right) \text{ bits/chip} \tag{2.155}$$

which relates E_c/N_J and E_b/N_J via (2.133). Now the required E_b/N_J can be determined from D using (2.132) and (2.134), or (2.138) and (2.139), depending on the type of jamming.

As shown in Table 2.14 for selected combinations of (n, k) and K, the concatenated scheme above appears to be quite promising, with significant performance advantages over the dual-K or RS codes alone (Tables 2.12 and 2.13). Not too surprisingly, the concatenated systems with the rate $1/3$ inner code outperform those with the rate $1/2$ code by anywhere from about .5 to 2 dB. At $P_b = 10^{-5}$, the best system in Table 2.14 is the $(1023, 959)$ RS outer code concatenated with the dual-2, rate $1/3$ convolutional inner code: this system requires an E_b/N_J less than 9.38 dB (upperbound) for any non-adaptive jammer, versus 10.38 dB for the best convolutional code alone in Table 2.12.

In the terminology of Section 4.10, Volume I, or Figure 4.21, Volume I, the inner encoder and decoder, interleaver and deinterleaver, and FH/MFSK channel can be regarded as a super channel. When the inner

Table 2.14

Relative performance of Reed-Solomon (n,k) outer codes concatenated with dual-K, rate $1/\nu$ convolutional inner codes for FH/MFSK signals in worst case non-adaptive jamming. P_K denotes inner code symbol error rate.

K	(n,k)	P_K	Inner Code P_b	E_b/N_J, dB for $P_b = 10^{-5}$			
				Noise		Multitone	
				$\nu=2$	$\nu=3$	$\nu=2$	$\nu=3$
1	$(255, 191)$	8.38×10^{-3}	Same	11.72	11.39	10.61	10.09
	$(511, 447)$	3.76×10^{-3}	as	11.70	9.95	10.61	9.88
	$(1023, 959)$	1.73×10^{-3}	P_K	12.06	10.03	11.00	10.05
2	$(15, 7)$	1.42×10^{-2}	9.47×10^{-3}	11.52	10.95	12.37	11.63
	$(63, 31)$	3.27×10^{-2}	2.18×10^{-2}	10.74	10.34	11.53	10.09
	$(255, 191)$	1.67×10^{-2}	1.11×10^{-2}	9.36	8.82	10.19	9.49
	$(1023, 959)$	3.46×10^{-3}	2.31×10^{-3}	9.57	8.65	10.49	9.38
3	$(7, 3)$	9.51×10^{-3}	5.43×10^{-3}	11.39	10.41	14.12	12.98
	$(63, 31)$	4.87×10^{-2}	2.78×10^{-2}	9.81	9.15	12.42	11.63
	$(511, 447)$	1.12×10^{-2}	6.40×10^{-3}	8.18	7.24	10.90	9.79
4	$(15, 7)$	2.81×10^{-2}	1.50×10^{-2}	10.11	9.02	14.64	13.43
	$(255, 191)$	3.31×10^{-2}	1.77×10^{-2}	7.98	6.91	12.49	11.31
5	$(31, 15)$	5.88×10^{-2}	3.03×10^{-2}	9.77	8.43	16.21	14.78
	$(1023, 959)$	8.64×10^{-3}	4.46×10^{-3}	7.76	6.14	14.36	12.61

code is a dual-K convolutional code, the input and output symbols for this super channel lie in $GF(2^K)$. Suppose now that we use an Odenwalder or Trumpis convolutional inner code: although the FH/MFSK channel proper operates over $GF(2^\nu)$ for the rate $1/\nu$ Trumpis codes, the super channel digests and regurgitates *binary* data. Consequently, the interleaver between the outer and inner encoders has to scramble data at the binary level. In our earlier notation, each RS code symbol is transmitted over the super channel as Q binary subsymbols. Now the RS character error rate P_Q needed to produce a given overall (outer code) P_b is achieved by establishing a super channel (inner code) $P_b = P_K$ in (2.153) with $K = 1$ and $L = Q$:

$$\text{inner code } P_b = 1 - \left(1 - P_Q\right)^{1/Q}. \tag{2.156}$$

We will consider three particular inner codes: the Odenwalder $(7, \frac{1}{2})$ code for $K = 1$ (we would use the rate $1/3$ code except that the $G(D)$ expression of (2.147) does not contain enough terms to be accurate at the higher P_b's defined by (2.156)), and the Trumpis $(7, \frac{1}{2})$ and $(7, \frac{1}{3})$ codes for $K = 2$ and 3, respectively (this K refers to the MFSK data modulation rather than the parameter in (2.152) and (2.153); with the dual-K inner code, the two parameters were synonymous). Analogous to (2.155), the overall code rate to be used in (2.133) is

$$R = \frac{k}{n} \times \text{inner code rate.} \tag{2.157}$$

We should note that the inner Viterbi decoder makes hard binary subsymbol decisions based on the soft decision/side information metric applied to the K-bit chips out of the FH/MFSK channel, so D is still given by (2.132), (2.134), (2.138), or (2.139).

The SNR's needed to achieve a system P_b of 10^{-5} with selected RS outer codes for worst case partial-band noise and multitone jamming are shown in Table 2.15. The concatenated systems employing the Trumpis inner codes are particularly attractive: in fact, the $(511, 447)$ RS outer code combined with the Trumpis $(7, \frac{1}{2})$ convolutional inner code over the 4-ary channel achieves $P_b = 10^{-5}$ in any non-adaptive jamming with $E_b/N_J = 8.31$ dB, representing a 1 dB improvement over the best RS/Dual-K system in Table 2.14. Table 2.15 also adheres to the previously observed behavior that the worst case jammer is partial-band noise for $K = 1$ and multitone for $K \geq 2$. Also, whereas the performance improves with K in partial-band noise, it peaks at $K = 2$ in multitone jamming for small P_b's.

2.3.3 Coding with Diversity

We demonstrated in the previous section that convolutional codes and concatenated Reed-Solomon/convolutional codes can provide improved performance for FH/MFSK signals in worst case jamming over simple diversity. The next question to be answered is whether these codes are powerful enough for this type of non-stationary, non-Gaussian channel.

Table 2.15

Comparison of several Reed-Solomon outer codes concatenated with binary convolutional inner codes and FH/MFSK signalling in worst case partial-band noise and multitone jamming. Inner codes consist of $(7,\frac{1}{2})$ Odenwalder code for $K = \log_2 M = 1$, and $(7,\frac{1}{2})$ and $(7,\frac{1}{3})$ Trumpis codes for $K = 2$ and $K = 3$, respectively.

RS Code n,k	RS Code P_Q	Inner Code P_b	E_b/N_J, dB for $P_b = 10^{-5}$					
			Noise			Multitone		
			$K=1$	$K=2$	$K=3$	$K=1$	$K=2$	$K=3$
127,63	1.30×10^{-1}	1.970×10^{-2}	11.74	9.32	8.27	10.23	9.98	10.81
255,191	6.51×10^{-2}	8.379×10^{-3}	10.16	7.93	6.87	8.66	8.60	9.42
511,447	3.33×10^{-2}	3.756×10^{-3}	9.69	7.64	6.57	8.20	8.31	9.12
1023,959	1.72×10^{-2}	1.733×10^{-3}	9.60	7.72	6.63	8.13	8.39	9.19

Suppose we add diversity to the codes analyzed above; we will employ the usual scheme of transmitting each M-ary symbol as m chips on separate hops. Effectively, we will have a concatenated coding system with an m-fold repetition code as the innermost code. We will then determine the optimum diversity m_{opt} for a given coded FH/MFSK structure in worst case partial-band noise or multitone jamming. If the outer code has sufficient redundancy and is well matched to the jamming environment, we should find that $m_{opt} = 1$. If instead we discover that m_{opt} is significantly larger, the implication is that diversity is required to boost the capabilities of the given code.

As in Table 4.3, Volume I, an m-fold repetition code has

$$G(D) = \frac{M}{2}D^m \qquad (2.158)$$

for an M-ary orthogonal channel. As a result, *all of the performance expressions in the previous section remain valid, except that D is replaced by D^m for m-diversity.* That is, (2.130) and (2.131) become

$$P_b \leq \begin{cases} \frac{1}{2}G(D^m); & \text{noise jamming} \\ G(D^m); & \text{multitone jamming.} \end{cases} \qquad (2.159)$$

Aside from this change of arguments, the functional form of $G(\cdot)$ for the various codes in (2.146)–(2.151) remains the same. Depending on the type of jamming, D is computed from (2.132), (2.134), (2.138), or (2.139), with E_c/N_J related to E_b/N_J by the overall code rate R. Since the diversity innermost code rate is $1/m$, it is convenient to use

$$\frac{E_c}{N_J} = \frac{R'}{m}\left(\frac{E_b}{N_J}\right) \qquad (2.160)$$

in the expressions for D, where R' is the outer code rate in bits per M-ary symbol. This explicit dependence on m will be useful in determining the optimum diversity later.

As a check on (2.158)–(2.160), consider an uncoded FH/MFSK system with diversity m in worst case partial-band noise. Using (2.134) and (2.160), with $R' = K$ for uncoded M-ary signalling, we compute

$$D^m = \left(\frac{4mN_J}{eKE_b}\right)^m; \qquad \frac{E_b}{N_J} > \frac{3m}{K}. \qquad (2.161)$$

This combines with (2.158) and (2.159) to produce the same P_b expression derived earlier for this system (i.e., (2.64)).

These m-diversity formulas can now be used to determine the variation in performance with m of any uncoded or coded FH/MFSK system in worst case jamming. As an example, consider the uncoded 4-ary system with diversity: Table 2.12 says that we require $E_b/N_J = 14.2$ dB to guarantee $P_b \leq 10^{-5}$ in any non-adaptive jamming environment if we use the opti-

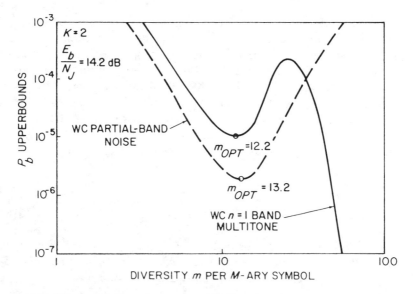

Figure 2.84. Variation in performance with diversity of uncoded FH/4-ary FSK signals in worst case jamming, for $E_b/N_J = 14.2$ dB; this SNR was selected to guarantee that $P_b \leq 10^{-5}$ at $m = m_{opt} = 12.2$ for any non-adaptive jamming.

mum amount of diversity. For this value of E_b/N_J, Figure 2.84 illustrates the dependence of the system performance on diversity for worst case partial-band noise and $n = 1$ band multitone jamming. (We have seen similar curves in Figures 2.30 and 2.41.) The multitone jamming curve reminds us that although there is a local minimum at $m = 12.2$, we can force P_b to be arbitrarily small for large enough m. However, as we previously argued in Section 2.3.1.2, for practical implementation considerations, we want to keep m small; consequently, we will continue to refer to the diversity at the local minimum P_b for worst case $n = 1$ band multitone jamming as m_{opt}, defined in (2.79) and (2.104).

The variation in performance of several coded systems in worst case noise and multitone jamming as a function of diversity is shown in Figures 2.85 and 2.86. Each curve is characterized by a different value of E_b/N_J, selected to achieve a P_b upperbound of 10^{-5} at the corresponding m_{opt}. In each graph, the curves are approximately lateral translations of each other, implying that all four systems are essentially equally sensitive to *proportional* deviations from m_{opt} (since the m axis is logarithmic). However, on an *absolute* basis, the curves with smaller values of m_{opt} are less robust with respect to m: for example, the performance of the $(7, \frac{1}{2})$ convolutionally coded system in Figure 2.85 is significantly degraded at $m = m_{opt} \pm 1$. (Recall that although it is analytically convenient to treat m as if it were continuously valued, only integer values are physically meaningful.) In the

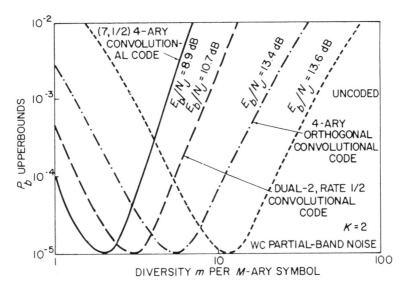

Figure 2.85. Comparison of sensitivities of various FH/MFSK systems to variations in diversity for $K = 2$ and worst case partial-band noise. More powerful codes (requiring smaller E_b/N_J to achieve $P_b = 10^{-5}$ at $m = m_{opt}$) need less diversity to combat jamming.

absence of diversity, we saw in Figures 2.67 and 2.72 and Table 2.12 that for FH/4-ary FSK signalling in worst case jamming, the most powerful of the convolutional codes considered was the $(7, \frac{1}{2})$ Trumpis code, followed by the dual-2 and 4-ary orthogonal codes. When these codes are combined with optimum diversity, this anti-jam effectiveness hierarchy is preserved as indicated by the required values of E_b/N_J in Figures 2.85 and 2.86. Furthermore, the more powerful codes need less additional diversity redundancy to cope with worst case jamming, although they all need some improvement since $m_{opt} > 1$ in all cases. The orthogonal convolutional code is particularly weak, providing a meager 0.2 dB advantage over uncoded signalling, both with optimum diversity. In the multitone jamming case of Figure 2.86, each curve has a local minimum, which we still use to define the "optimum" diversity m_{opt}, as well as a monotonically decreasing P_b for sufficiently large values of m. In the uncoded case, we argued that the values of m needed to provide better performance than at m_{opt} were too large for practical implementations; for example, in Figure 2.86, $E_b/N_J = 14.2$ dB will provide the uncoded system with $P_b = 10^{-5}$ at $m = m_{opt} \cong 12$, whereas lower bit error rates require diversities in excess of 40. However, the $(7, \frac{1}{2})$ Trumpis convolutional code has $m_{opt} = 2$ with extremely low P_b's for diversities of the order of 10. We will continue to use our previous definitions of m_{opt} (refer to (2.79) and (2.104)); however, we should remember that some of the more capable codes can achieve arbitrarily good performance in worst case $n = 1$ band multitone jamming for moderate amounts of diversity.

We will now mathematically define the performance of coded FH/MFSK systems concatenated with optimum diversity. For arbitrary diversity m, P_b for the concatenated system is proportional to $G(D^m)$, where D itself depends on m through (2.160). Since $G(\cdot)$ is monotonic for small P_b's, the optimum diversity is the value of m that minimizes D^m. Recall from (2.158) that the performance of uncoded FH/MFSK signals with diversity is proportional to D^m, and we computed m_{opt} for such systems in Section 2.3.1. The only difference here is that R' is arbitrary, whereas $R' = K$ in the previous analysis. Generalizing the earlier results for arbitrary outer code rates, we conclude that for values of E_b/N_J large enough to yield $m_{\text{opt}} > 1$,

$$D^m = e^{-m_{\text{opt}}} \tag{2.162}$$

as in (2.70), (2.77), and (2.104), where, for worst case noise or tone jamming, the optimum diversity has the form

$$m_{\text{opt}} = \frac{R'}{\gamma}\left(\frac{E_b}{N_J}\right) \tag{2.163}$$

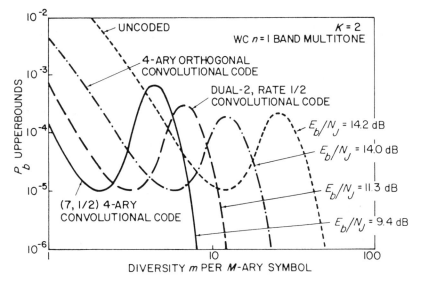

Figure 2.86. Same as Figure 2.85, but for worst case $n = 1$ band multitone jamming. Note that although we consider the local minimum to be m_{opt} in each curve, arbitrarily low P_b's can be achieved for sufficiently large values of m; *for the more powerful codes, these larger amounts of diversity are small enough to be of practical interest.*

and the factor γ varies with the type of jamming according to

$$\gamma = \begin{cases} 4; & WC \text{ partial-band noise} \\ \beta K e; & WC\ n = 1 \text{ band multitone} \end{cases} \qquad (2.164)$$

with the definition of β in Table 2.7 extended to include $\beta = 1$ for $K = 1$. Of course, for smaller values of E_b/N_J, $m_{opt} = 1$ and the performance is based on formulas in Section 2.3.2.

In Figures 2.66–2.75, we compared the performance of convolutionally coded FH/MFSK systems in worst case noise and tone jamming for $1 \le K \le 5$. We are now in a position to extend the scope of that comparison to include optimum diversity inner codes; the results are illustrated in Figures 2.87–2.96. These graphs contain no major surprises: the more powerful codes without diversity still provide better performance with optimum diversity. Orthogonal convolutional codes continue to be relatively inadequate, although they manage to slightly outperform uncoded systems with optimum diversity. In all of the graphs, the performance of the rate 1/2 and 1/3 dual-K convolutional codes with optimum diversity converges for low P_b's. Furthermore, these systems are consistently 3 dB better than uncoded MFSK signals with optimum diversity for small enough P_b's. These observations are explained analytically later in this section.

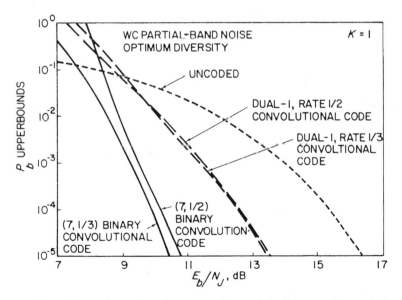

Figure 2.87. Comparison of performance of uncoded and several convolutionally coded FH/binary FSK systems with optimum diversity in worst case partial-band noise. With the addition of optimum diversity, the dual-1 system performance is almost identical for both code rates considered.

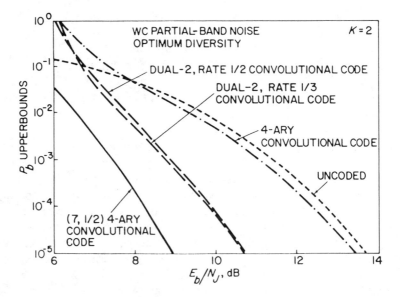

Figure 2.88. Same as Figure 2.87, but for $K = 2$. Again, with optimum diversity, we see that the dual-K system performance is negligibly affected by the code rate; in fact, we show in the text that this statement is asymptotically true for small P_b's for any K and all code rates $1/\nu$.

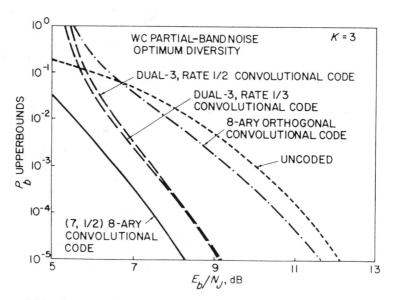

Figure 2.89. Same as Figure 2.87, but for $K = 3$. With the addition of optimum diversity, the M-ary orthogonal code performs moderately better than uncoded signalling, but it is clearly not a very strong code for worst case partial-band noise. The $(7, \frac{1}{3})$ Trumpis code with diversity is clearly superior (about 4 dB better than the uncoded case at $P_b = 10^{-5}$).

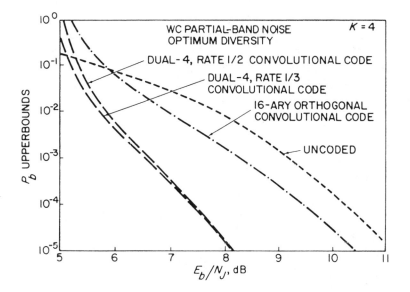

Figure 2.90. Same as Figure 2.87, but for $K = 4$. The dual-4 convolutional codes (both rates) perform about 3 dB better than uncoded signalling with optimum diversity at $P_b = 10^{-5}$.

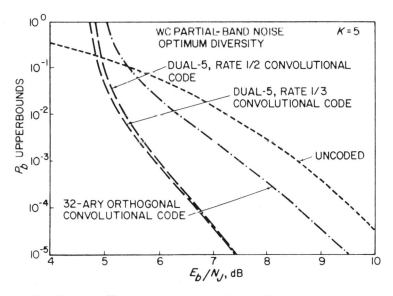

Figure 2.91. Same as Figure 2.87, but for $K = 5$. With optimum diversity, the orthogonal convolutional code is about 1 dB better than uncoded signalling at $P_b - 10^{-5}$, while the dual-K codes provide another 2 dB improvement.

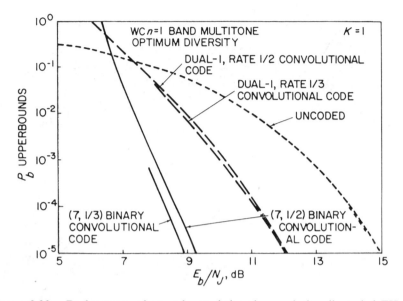

Figure 2.92. Performance of several uncoded and convolutionally coded FH/binary FSK systems with optimum diversity in worst case $n = 1$ band multitone jamming. At $P_b = 10^{-5}$, the dual-1 codes are about 3 dB better than uncoded signalling, and about 3 dB worse than the Odenwalder convolutional codes.

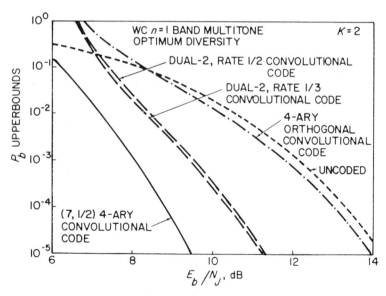

Figure 2.93. Same as Figure 2.92, but for $K = 2$. With diversity, we see that the uncoded and orthogonally coded systems have comparable performance, the dual-2 codes are about 3 dB better at $P_b = 10^{-5}$, and the Trumpis $(7, \frac{1}{2})$ code provides another 2 dB improvement.

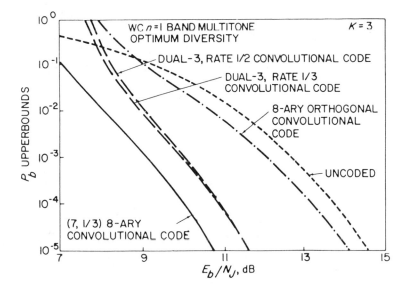

Figure 2.94. Same as Figure 2.92, but for $K = 3$. As in other cases considered, the dual-K convolutional codes provide a significant improvement over uncoded signalling with optimum diversity, while the Trumpis code is better still.

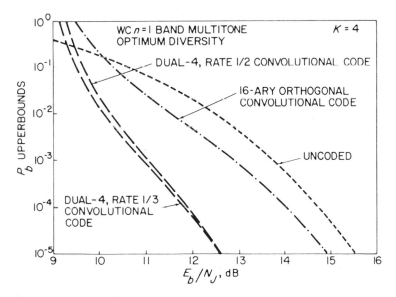

Figure 2.95. Same as Figure 2.92, but for $K = 4$. The dual-K code with optimum diversity is about 3 dB better than the uncoded system for small P_b's.

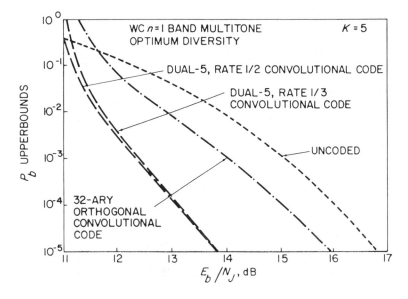

Figure 2.96. Same as Figure 2.92, but for $K = 5$. The 3 dB performance advantage of the dual-K code over the uncoded system with optimum diversity continues for small P_b's at $K = 5$.

The variation in performance of a given coded system concatenated with optimum diversity as a function of the channel dimensionality K and the type of jamming is exemplified by Figures 2.97 and 2.98. For both the M-ary orthogonal and dual-K convolutional codes in worst case partial-band noise, the performance improves with K, just as it did in Figures 2.76, 2.78, and 2.80 in the absence of diversity. However, unlike Figure 2.77 for orthogonal convolutionally coded systems in worst case $n = 1$ band multi-tone jamming, the addition of optimum diversity causes the performance in Figure 2.97 to degrade with increasing K. By comparison, for the dual-K convolutional codes in worst case $n = 1$ band multitone jamming, with or without diversity, Figures 2.79, 2.81, and 2.98 show that the best performance is obtained when $K = 2$ for low P_b's. In general, because the performance of coded FH/MFSK systems with optimum diversity tends to improve with K in worst case partial-band noise, and degrade with $K \geq 2$ for low bit error rates in worst case tone jamming, the impact of the type of jamming on the AJ capability becomes more significant for larger K. For example, at $P_b = 10^{-5}$, Figure 2.98 shows that worst case tone jamming is about .6 dB more effective against dual-K, rate 1/2 convolutional codes with optimum diversity than worst case noise jamming at $K = 2$; this difference rises to 6.4 dB at $K = 5$.

The performance improvements in various coded FH/MFSK systems provided by the addition of optimum diversity is illustrated in Figures

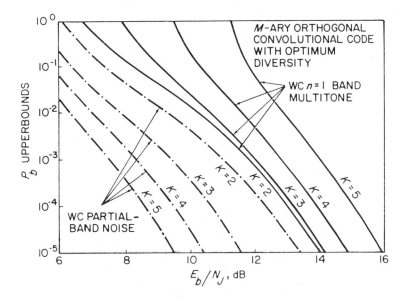

Figure 2.97. Variation in performance of FH/MFSK systems with orthogonal convolutional coding and optimum diversity versus $K = \log_2 M$ and type of jammer. By comparison with no diversity in multitone jamming (Figure 2.77), performance degrades with increasing K.

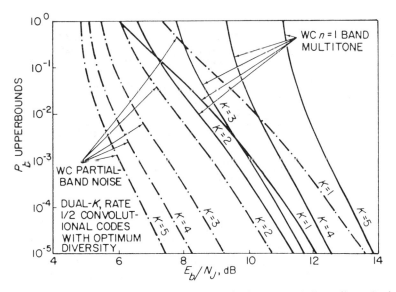

Figure 2.98. Same as Figure 2.97, but for dual-K, rate 1/2 coding. As in no diversity case with multitone jamming (Figure 2.79), best performance at low P_b's occurs when $K = 2$. Recall that all of these results are for soft decision energy detection, linear combining, and perfect side information.

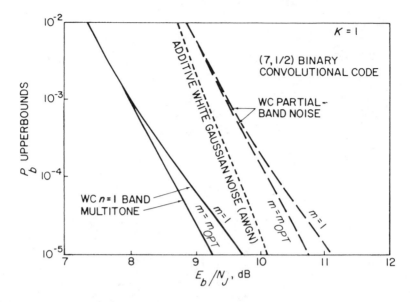

Figure 2.99. Performance improvement available from diversity inner code for FH/BFSK system employing Odenwalder $(7, \frac{1}{2})$ binary convolutional code in worst case noise and multitone jamming. In this case, diversity provides only a small performance gain. Note that broadband noise jamming (AWGN) is more effective than multitone jamming here.

2.99–2.102. Some codes are powerful enough over jamming channels that diversity does not produce significant performance gains: the Odenwalder binary convolutional codes fall into this category, and, in particular, the $(7, \frac{1}{3})$ code has $m_{opt} = 1$ for both noise and tone jamming implying no benefits from diversity. At the other extreme, the dual-K (Figure 2.102) and M-ary orthogonal (not illustrated) convolutional codes show major improvements with the addition of optimum diversity, indicating that these codes by themselves are not well-matched to the jamming channel and require additional redundancy to provide acceptable AJ performance. We also see in Figures 2.99 and 2.100 that *for the binary convolutional codes with FH/BFSK modulation, broadband noise jamming is more effective than worst case tone jamming*. This is not a general result for other $K = 1$ systems: for example, (2.20b) tells us that an uncoded FH/BFSK system in broadband noise requires $E_b/N_J = 13.35$ dB to produce $P_b = 10^{-5}$, whereas Table 2.12 shows that the same system with optimum diversity in worst case tone jamming needs $E_b/N_J = 14.95$ dB to achieve this performance.

As in previous sections, it is useful to compare the performance of all of these systems in worst case noise and tone jamming by tabulating the required E_b/N_J to achieve $P_b = 10^{-5}$: Table 2.16 does this for all of the FH/MFSK systems considered with convolutional outer codes and opti-

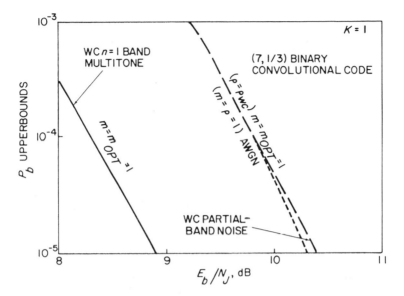

Figure 2.100. Same as Figure 2.99, except for $(7, \frac{1}{3})$ Odenwalder code; this is the only code of those considered in this section which is powerful enough by itself ($m_{\text{opt}} = 1$) to combat worst case non-adaptive jamming. In the noise case, the code is insensitive to ρ (ρ_{wc} is not much worse than $\rho = 1$), but performance is about 1.5 dB better in multitone jamming.

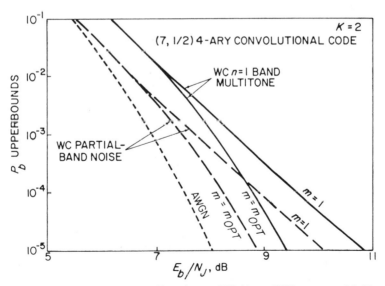

Figure 2.101. Effect of optimum diversity on FH/4-ary FSK system with Trumpis $(7, \frac{1}{2})$ code in worst case noise and multitone jamming. Optimum diversity provides a moderate improvement: e.g., it cuts the degradation due to worst case partial-band noise approximately in half.

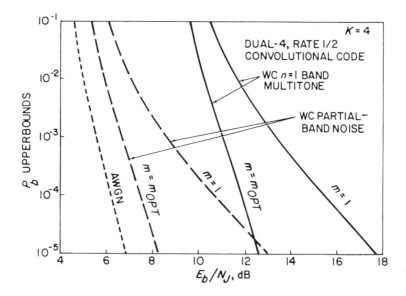

Figure 2.102. For FH/16-ary FSK system with dual-4, rate 1/2 convolutional code, diversity provides a major performance improvement; for example, optimum diversity recovers most of the degradation inflicted by worst case partial-band noise.

mum diversity inner codes. An overall benchmark is the system that requires the minimum E_b/N_J to achieve this P_b in any kind of non-adaptive jamming. Recall from Table 2.12 that without diversity, the Odenwalder $(7, \frac{1}{3})$ binary convolutional code gained this distinction with $E_b/N_J = 10.38$ dB; with the addition of optimum diversity, the Trumpis $(7, \frac{1}{2})$ 4-ary convolutional code is the overall winner with $E_b/N_J = 9.43$ dB, an improvement of almost 1 dB over the best convolutional code alone and nearly 5 dB better than optimum diversity alone ($E_b/N_J = 14.21$ dB at $K = 2$). Not surprisingly, m_{opt} is much smaller for the concatenated convolutionally coded systems (we already noted that the $(7, \frac{1}{3})$ binary convolutional code has $m_{opt} = 1$). In fact, m_{opt} is small enough that the overall code rate $R = R'/m_{opt}$ for the concatenated systems with the exception of the M-ary orthogonal codes is approximately double that of the uncoded systems with optimum diversity for a given K. Table 2.16 reiterates the close performance of the rate 1/2 and 1/3 dual-K convolutional codes with optimum diversity for low bit error rates; and, although m_{opt} is smaller for the lower rate (R') code, the overall code rate R is essentially independent of R' for any K.

We can illuminate this observation and provide further analytical insights by deriving asymptotic expressions for the performance of these concatenated systems in the region of large E_b/N_J. From (2.159), (2.162), and (2.163), we know that the performance of coded FH/MFSK systems with optimum diversity is given by

$$P_b \leq \eta G [e^{-R'/\gamma(E_b/N_J)}] \tag{2.165}$$

Table 2.16

Required E_b/N_J in dB (left columns) to achieve $P_b = 10^{-5}$ for coded FH/MFSK systems with optimum diversity m_{opt} (right columns) in worst case partial-band noise (upper rows) and $n = 1$ band multitone jamming (lower rows).

Type of Outer Code	$K = \log_2 M$									
	1		2		3		4		5	
Uncoded	16.36	10.8	13.62	11.5	12.12	12.2	11.11	12.9	10.36	13.6
	14.95	11.5	14.21	12.2	14.58	12.9	15.49	13.6	16.76	14.3
$(7, \frac{1}{2})$ Binary	10.73	1.5	n/a	n/a	n/a	n/a	n/a	n/a	n/a	n/a
Convolutional	9.24	1.5	n/a	n/a	n/a	n/a	n/a	n/a	n/a	n/a
$(7, \frac{1}{3})$ Binary	10.38	1.0	n/a	n/a	n/a	n/a	n/a	n/a	n/a	n/a
Convolutional	8.89	1.0	n/a	n/a	n/a	n/a	n/a	n/a	n/a	n/a
$(7, \frac{1}{2})$ 4-ary	n/a	n/a	8.91	1.9	n/a	n/a	n/a	n/a	n/a	n/a
Convolutional	n/a	n/a	9.43	2.0	n/a	n/a	n/a	n/a	n/a	n/a
$(7, \frac{1}{3})$ 8-ary	n/a	n/a	n/a	n/a	8.26	1.7	n/a	n/a	n/a	n/a
Convolutional	n/a	n/a	n/a	n/a	10.70	1.8	n/a	n/a	n/a	n/a
M-ary Orthogonal	n/a	n/a	13.36	5.4	11.61	3.6	10.40	2.7	9.48	2.2
Convolutional	n/a	n/a	13.96	5.8	14.10	3.8	14.81	2.9	15.90	2.3
Dual-K, Rate $\frac{1}{2}$	13.46	2.8	10.70	2.9	9.18	3.1	8.16	3.3	7.42	3.5
Convolutional	12.03	2.9	11.27	3.1	11.63	3.3	12.53	3.4	13.79	3.6
Dual-K, Rate $\frac{1}{3}$	13.41	1.8	10.66	1.9	9.15	2.1	8.13	2.2	7.39	2.3
Convolutional	11.99	1.9	11.24	2.1	11.60	2.2	12.51	2.3	13.78	2.4

where $\eta = \frac{1}{2}$ in worst case partial-band noise, $\eta = 1$ in worst case $n = 1$ band multitone jamming, and γ varies with the type of jamming according to (2.164). If E_b/N_J is large enough, the argument of $G(\cdot)$ in (2.165) is sufficiently small to approximate the expressions of (2.146)–(2.151) by their dominant terms, which all have the form

$$G(D) \cong CD^k; \qquad D \ll 1 \qquad (2.166)$$

where C and k are listed in Table 2.17 for the systems under consideration. Therefore, the anti-jam performance of these systems for sufficiently large E_b/N_J is of the order of

$$P_b \lesssim \eta C e^{-(kR'/\gamma)(E_b/N_J)}. \qquad (2.167)$$

This is a very succinct and useful gauge of the relative performance of all of these systems in worst case jamming for small bit error rates. Since the exponential term dominates the behavior of (2.167) in this region, a larger coefficient kR'/γ usually indicates a more capable anti-jam system; however, if two systems have comparable values of kR'/γ, the one with the smaller coefficient ηC will provide better performance. If we compare systems restricted to worst case partial-band noise ($\gamma = 4$), or worst case tone jamming with a fixed K, γ is fixed and the exponential coefficient kR' (see Table 2.17) is often a sufficient performance indicator, with a minor impact from C (η is fixed for a given jammer).

Table 2.17 shows us why the performance of the dual-K systems are asymptotically independent of the code rate $1/\nu$ for small P_b's: C and kR' are not functions of ν. And since $kR' = K$ for both the uncoded and M-ary orthogonal convolutionally coded systems, their performance is asymptotically similar, with the orthogonal convolutional code achieving a slight advantage because its factor C is smaller for $K \geq 2$. Another observation made earlier based on Figures 2.87–2.96 and Table 2.16 is that the dual-K systems provide 3 dB better AJ performance than uncoded systems with optimum diversity at low error rates; Table 2.17 shows that this behavior occurs because both systems have the same factor C, while the dual-K scheme's exponential coefficient kR' is double that of the uncoded implementation.

Table 2.17 also reveals some other heretofore unnoted asymptotic performance comparisons. In particular, the $(7, \frac{1}{3})$ Odenwalder (binary) and Trumpis (M-ary) convolutional codes are modestly superior to their $(7, \frac{1}{2})$ counterparts primarily because of their smaller values of C rather than due to any exponential advantage: this is underscored by the fact that kR' is actually smaller for the rate $1/3$ binary convolutional code.

If we temporarily restrict our attention to the noise jamming case where $\gamma = 4$ independent of K, we can use Table 2.17 to make some interesting comparisons of systems with different channel alphabets. For example, based on the exponential coefficient kR', we see that the uncoded system with optimum diversity begins to exceed the asymptotic performance of the

Table 2.17

For given outer code with rate R', and optimum diversity, performance of FH/MFSK system is approximated by $P_b \leq \eta C e^{-(kR'/\gamma)(E_b/N_J)}$ (see (2.167)), where η and γ vary with the type of jamming.

Type of Outer Code	R'	C	k	kR'	Range of $K = \log_2 M$
Uncoded	K	$\dfrac{M}{2}$	1	K	≥ 1
$(7, \tfrac{1}{2})$ Binary Convolutional	$\tfrac{1}{2}$	36	10	5	1
$(7, \tfrac{1}{3})$ Binary Convolutional	$\tfrac{1}{3}$	1	14	$\tfrac{14}{3} = 4.7$	1
$(7, \tfrac{1}{2})$ 4-ary Convolutional	1	7	7	7	2
$(7, \tfrac{1}{3})$ 8-ary Convolutional	1	1	7	7	3
M-ary Orthogonal Convolutional	1	1	K	K	≥ 2
Dual-K, Rate $\tfrac{1}{\nu}$ Convolutional	$\dfrac{K}{\nu}$	$\dfrac{M}{2}$	2ν	$2K$	≥ 1

binary convolutional codes when the former operates with $K \geq 5$: this is supported by the results in Table 2.16. Similarly, we conclude that the dual-K systems with $K \geq 4$ asymptotically outperform the Trumpis codes in worst case partial-band noise, an observation that is again corroborated by Table 2.16.

Recall that at the beginning of this chapter we considered an unsophisticated AJ communication system using uncoded FH/MFSK signals confronted by equally simplistic broadband noise (i.e., additive white Gaussian noise or AWGN) jamming. The asymptotic performance of this system is approximated by the upperbound of (2.20c), repeated here for convenience:

$$P_b \leq \frac{M}{4} e^{-KE_b/2N_J}.$$

Interestingly, this is *identical* to the asymptotic upperbound of (2.167) for the dual-K system with optimum diversity operating in worst case partial-band noise. Imagine then an electronic warfare (EW) scenario in which both adversaries begin with innocuous equipment and then feel compelled to alternately escalate the EW ante. Initially, there is an uncoded FH/MFSK system opposed by a broadband noise jammer. Then the ECM opponent introduces a worst case partial-band noise jammer, with the dramatic result that the communication link performance degrades severely to the inverse linear relationship of (2.21). To combat this increased threat, the communication system counters with a series of increasingly more sophisticated coding techniques, beginning with simple diversity. Each improved coding scheme brings the AJ performance closer to the initial AWGN situation, and with the concatenated dual-K/optimum diversity system, a balance is achieved between the ECM and ECCM upgrade effectiveness. Of course, the jammer can still resort to the increased capability of worst case multitone jamming if $K \geq 2$, but the communicator has an arsenal of even more powerful codes (such as the RS/convolutional code/optimum diversity scheme analyzed below) to neutralize this ploy.

We can also use the asymptotic results to explain some of the multitone jamming behavior that we have previously observed graphically and in Table 2.16. From (2.164) and Table 2.17, we see that the exponential performance coefficient kR'/γ in (2.167) is proportional to $(\beta e)^{-1}$ for uncoded, M-ary orthogonal, and dual-K convolutionally coded signals with optimum diversity in worst case multitone jamming; the same coefficient is proportional to $K/4$ in worst case partial-band noise. These quantities are shown in Table 2.18. Since a larger coefficient kR'/γ contributes to a smaller P_b, we can understand why the noise jamming performance of these systems improves monotonically with K, while the tone jamming performance is best at $K = 2$, as seen earlier in Figures 2.97 and 2.98 and Table 2.16. In the uncoded and dual-K cases, we also see why worst case noise jamming is more effective than worst case tone jamming for $K = 1$, whereas the reverse is true for $K \geq 2$.

Table 2.18

Relative performance of coded FH/MFSK signals with optimum diversity
in worst case multitone and partial-band noise jamming based
on exponential coefficient kR'/γ in (2.167).

K	$(\beta e)^{-1}$	$K/4$
1	.368	.250
2	.463	.500
3	.449	.750
4	.384	1.000
5	.301	1.250

Now let us briefly consider the performance of FH/MFSK systems employing RS outer codes concatenated with optimum diversity. Using the notation introduced in Section 2.3.2.2, each RS code symbol over GF(2^Q) is sent by using the M-ary channel L times. Although the soft decision/side information metric is used with linear combining of the m diversity chip energies, hard decisions are made on each M-ary symbol, and these are combined in groups of L to recreate the received RS characters. Using Berlekamp's table of RS symbol error rates P_Q required to achieve a given P_b for a particular (n, k) RS code [21], (2.153) converts it to an M-ary channel symbol error rate P_K, which implies an effective *channel* P_b defined by (2.154). For worst case noise and tone jamming, the channel P_h is related to E_b/N_J by (2.159) and (2.162)–(2.164), where

$$G(D) = \frac{M}{2}D \tag{2.168}$$

for uncoded M-ary signalling, and

$$R' = \frac{k}{n}K \tag{2.169}$$

is the overall code rate exclusive of diversity. These equations reduce to

$$\frac{E_b}{N_J} = \left(\frac{\gamma n}{Kk}\right) m_{\text{opt}}$$

$$m_{\text{opt}} \leq \begin{cases} \ln\left(\dfrac{M-1}{2P_K}\right); & WC \text{ noise} \\[2ex] \ln\left(\dfrac{M-1}{P_K}\right); & WC \text{ multitone.} \end{cases} \tag{2.170}$$

Table 2.19 compares the performance of various combinations of RS codes, M-ary channels, and worst case jamming for the system P_b of 10^{-5}. The results are not as good as those in Table 2.16 for convolutional outer codes: for example, the best system in Table 2.16 required $E_b/N_J = 9.43$ dB for any kind of non-adaptive jamming, whereas we need $E_b/N_J = 11.75$ dB

Table 2.19

Required SNR to achieve $P_b = 10^{-5}$ for FH/MFSK systems with (n, k) Reed-Solomon outer codes and optimum diversity m_{opt}, in worst case partial-band noise and $n = 1$ band multitone jamming; P_K is M-ary ($M = 2^K$) channel symbol error rate.

K	n, k	P_K	Noise E_b/N_J, dB	Noise m_{opt}	Multitone E_b/N_J, dB	Multitone m_{opt}
1	31, 15	1.20×10^{-2}	14.89	3.7	13.95	4.4
	63, 31	1.65×10^{-2}	14.43	3.4	13.56	4.1
	127, 63	1.97×10^{-2}	14.16	3.2	13.33	3.9
	255, 191	8.38×10^{-3}	13.39	4.1	12.39	4.8
	511, 447	3.76×10^{-3}	13.49	4.9	12.39	5.6
	1023, 959	1.73×10^{-3}	13.83	5.7	12.66	6.4
2	15, 7	1.42×10^{-2}	13.00	4.7	13.94	5.4
	63, 31	3.27×10^{-2}	11.92	3.8	12.97	4.5
	255, 191	1.67×10^{-2}	10.80	4.5	11.75	5.2
	1023, 959	3.46×10^{-3}	11.12	6.1	11.93	6.8
3	7, 3	9.51×10^{-3}	12.64	5.9	15.35	6.6
	63, 31	4.87×10^{-2}	10.64	4.3	13.52	5.0
	511, 447	1.12×10^{-2}	9.42	5.7	12.14	6.4
4	15, 7	2.81×10^{-2}	10.78	5.6	15.45	6.3
	255, 191	3.31×10^{-2}	8.60	5.4	13.28	6.1
5	31, 15	5.88×10^{-2}	9.65	5.6	16.33	6.3
	1023, 959	8.64×10^{-3}	8.06	7.5	14.62	8.2

in Table 2.19, and the diversity required for the RS systems is also larger by more than a factor of 2.

As in the section on coding without diversity, we again conclude that RS codes need to be concatenated with other good codes to achieve satisfactory performance. In particular, we want to look at systems that combine RS outer codes with convolutional inner codes over FH/MFSK channels with optimum diversity.

First assume that the inner code is a dual-K, rate 1/2 convolutional code. Each RS code symbol over $GF(2^Q)$ is composed of L inner code symbols over $GF(2^K)$. The M-ary channel symbols are transmitted as m_{opt} diversity chips which are energy detected in the receiver. These chip energies are linearly combined using the soft decision metric with perfect side information. The convolutional decoder uses these observables to make hard M-ary decisions. These M-ary symbols are unscrambled (assuming interleaving between the inner and outer codes) and reformatted into Q-bit RS characters, which are then fed to the RS decoder.

As before, a given outer code P_b maps into an RS symbol error rate P_Q [21], which translates into an M-ary symbol error rate P_K, corresponding to an inner code P_b: combining (2.153) and (2.154), we have

$$\text{inner code } P_b = \frac{M}{2(M-1)}\left[1 - (1 - P_Q)^{1/L}\right]. \qquad (2.171)$$

Finally, this inner code P_b is converted into a required SNR for a dual-K, rate 1/2 convolutional code with optimum diversity using (2.151), (2.159), and (2.162)–(2.164), with overall code rate

$$R' = \frac{k}{n}\left(\frac{K}{2}\right) \qquad (2.172)$$

exclusive of diversity. That is,

$$\text{inner code } P_b \leq \begin{cases} \frac{1}{2}G(e^{-m_{opt}}); & \text{noise} \\ G(e^{-m_{opt}}); & \text{multitone} \end{cases} \qquad (2.173)$$

and

$$m_{opt} = \frac{kK}{2n\gamma}\left(\frac{E_b}{N_J}\right) \qquad (2.174)$$

where $G(\cdot)$ and γ are defined in (2.151) and (2.164), respectively.

The performance of these systems is illustrated in Table 2.20 for an overall bit error rate of 10^{-5} and worst case noise and tone jamming. The best of these systems requires $E_b/N_J = 9.39$ dB to achieve the desired P_b in any non-adaptive jamming environment. Recall that the best concatenated FH/MFSK system with an RS outer code, a dual-K inner code, and *no diversity* required $E_b/N_J = 9.38$ dB under the same design constraints (Table 2.14); however, this was with a rate 1/3 inner code, whereas Table 2.20 is restricted to rate 1/2 inner codes. (If we extended Table 2.20 to

Table 2.20

Required SNR and optimum diversity m_{opt} to achieve $P_b = 10^{-5}$ for FH/MFSK systems with RS outer codes, dual-K, rate 1/2 inner codes, and optimum diversity for worst case partial-band noise and $n = 1$ band multitone jamming. Note that $m_{opt} \sim 2$ in almost all cases, implying concatenated RS/dual-K coding is quite powerful even without diversity.

K	RS Code n, k	Inner Code P_b	Noise		Multitone	
			$\frac{E_b}{N_J}$, dB	m_{opt}	$\frac{E_b}{N_J}$, dB	m_{opt}
1	255, 191	8.38×10^{-3}	11.52	1.3	10.25	1.5
	511, 447	3.76×10^{-3}	11.32	1.5	10.02	1.6
	1023, 959	1.73×10^{-3}	11.44	1.6	10.13	1.8
2	15, 7	9.47×10^{-3}	11.11	1.5	11.77	1.6
	63, 31	2.18×10^{-2}	10.49	1.4	11.14	1.5
	255, 191	1.11×10^{-2}	8.98	1.5	9.64	1.6
	1023, 959	2.31×10^{-3}	8.74	1.8	9.39	1.9
3	7, 3	5.43×10^{-3}	10.48	1.8	12.98	1.9
	63, 31	2.78×10^{-2}	9.29	1.6	11.76	1.7
	511, 447	6.40×10^{-3}	7.32	1.8	9.81	1.9
4	15, 7	1.50×10^{-2}	9.05	1.9	13.40	2.0
	255, 191	1.77×10^{-2}	6.95	1.9	11.29	1.9
5	31, 15	3.03×10^{-2}	8.32	2.1	14.62	2.1
	1023, 959	4.46×10^{-3}	5.85	2.3	12.21	2.4

include rate 1/3 inner codes, the performance with optimum diversity would not improve much as argued earlier.) The 9.39 dB SNR benchmark in Table 2.20 should more fairly be compared with the best rate 1/2 inner code in Table 2.14, which requires $E_b/N_J = 10.19$ dB, to measure the improvement due to diversity. And the amount of diversity in Table 2.20 is small (of the order of 2). Still, to place this result in perspective, the best RS/dual-K/optimum diversity system is only slightly improved over the $E_b/N_J = 9.43$ dB requirement of Table 2.16 for the best binary convolutional code with optimum diversity.

So, as a final exercise, we will now determine how much additional improvement is afforded by combining an RS outer code, a binary convolutional inner code, and optimum diversity. As in Section 2.3.2.3 when we considered the same systems without diversity, because the inner code operates on binary data, the RS outer code sees a binary rather than an M-ary super channel. Analytically, this requires that we use (2.171) with $M = 2$ and $L = Q$, which reduces it to (2.156). As before, we will use the Odenwalder $(7, \frac{1}{2})$ convolutional inner code for the $K = 1$ FH/MFSK system, and the Trumpis $(7, \frac{1}{2})$ and $(7, \frac{1}{3})$ codes for $K = 2$ and $K = 3$, respectively. The inner code P_b translates into a required E_b/N_J and m_{opt} through (2.163–2.164) and (2.173), with $G(\cdot)$ specified by (2.146), (2.148), or (2.149) depending on the particular inner code, and

$$R' = \frac{k}{n} \times \text{inner code rate} . \tag{2.175}$$

These systems are compared in Table 2.21 for an overall $P_b = 10^{-5}$ in worst case noise and tone jamming. In all cases, $m_{opt} \sim 1$ so that the concatenated RS/binary convolutional coding structure is capable of coping with the jamming without the need for additional coding redundancy through diversity; this is underscored by the similarity between the results in Tables 2.15 and 2.21. In fact, the addition of optimum diversity only provides an improvement of .2 dB in the sense that the best system in Table 2.21 (the (1023, 959) RS outer code with the $(7, \frac{1}{2})$ Trumpis convolutional inner code over a 4-ary channel) requires $E_b/N_J = 8.11$ dB to achieve the benchmark bit error rate of 10^{-5} in worst case non-adaptive jamming, whereas $E_b/N_J = 8.31$ dB for the best system in Table 2.15. Still, this particular coding scheme provides the best performance of all implementations considered so far.

Lest these results be taken out of context without regard for the associated design constraints, we should note that there are realistic scenarios which reduce the E_b/N_J requirement for a given P_b. For example, band multitone jamming is only possible if the M-ary bands are distinct; that is, the MFSK tones on a given hop occupy M adjacent FH slots as shown in Figure 2.3(a) or (b). If this band structure is destroyed, for example, by hopping each MFSK tone independently as suggested in [22] (a relatively expensive implementation requiring not one but M separate frequency

Table 2.21

Performance of several concatenated FH/MFSK systems employing RS outer codes, binary convolutional inner codes, and optimum diversity on the M-ary symbols. The inner codes are the $(7, \frac{1}{2})$ Odenwalder code for $K = \log_2 M = 1$, and the $(7, \frac{1}{2})$ and $(7, \frac{1}{3})$ Trumpis codes for $K = 2$ and $K = 3$, respectively. Hard decisions are made on inner code M-ary symbols based on soft decision diversity chip metric with perfect jamming state side information and linear chip energy combining.

RS Code n, k	Convolutional Code P_b	E_b/N_J, dB and m_{opt} (lower parameter) for $P_b = 10^{-5}$					
		Noise			Multitone		
		$K = 1$	$K = 2$	$K = 3$	$K = 1$	$K = 2$	$K = 3$
255, 191	8.379×10^{-3}	10.16	7.89	6.87	8.66	8.50	9.42
		1.0	1.2	1.0	1.0	1.2	1.0
511, 447	3.756×10^{-3}	9.69	7.53	6.57	8.19	8.13	9.11
		1.0	1.2	1.0	1.1	1.3	1.1
1023, 959	1.733×10^{-3}	9.59	7.53	6.62	8.10	8.11	9.14
		1.1	1.3	1.1	1.1	1.4	1.2

synthesizers), the worst case multitone jammer must use the independent tone placement strategy. And we have already seen that independent multitone jamming is less effective than $n = 1$ band multitone jamming. If we assume only worst case partial-band noise jamming, the best system considered is the $(1023, 959)$ RS outer code with the dual-5, rate $1/2$ convolutional inner code and optimum diversity (~ 2), which achieves $P_b = 10^{-5}$ with $E_b/N_J = 5.85$ dB; furthermore, the performance improves as K and the RS block length increase.

2.3.3.1 Optimum Code Rates

Many different combinations of coding and optimum diversity have been considered in this section. Using a system $P_b = 10^{-5}$ as a benchmark, in most cases with worst case noise or tone jamming, the optimum diversity was greater than unity, implying that the coding without diversity did not provide enough (or the appropriate kind of) redundancy to effectively counter the assumed threat. Some notable exceptions where diversity was not needed (i.e., $m_{opt} = 1$) were the $(7, \frac{1}{3})$ Odenwalder convolutional code (see Table 2.16), and several of the concatenated Reed-Solomon/binary convolutional codes (Table 2.21).

The RS codes by themselves were particularly ineffective, as evidenced by Table 2.19. This poor performance could in part be attributed to the use of a hard decision decoder. However, it has been shown that for a given signalling scheme and jamming environment, there is a unique code rate which optimizes the system performance [23]. So far, we have only considered a single (n, k) RS code (rate k/n) over GF(2^Q) for each block length $n - 2^Q - 1$.

To investigate the significance of optimizing the RS code rate for FH/MFSK systems in worst case jamming, we will examine the values of m_{opt} and E_b/N_J required to achieve $P_b = 10^{-5}$ for the class of $(255, k)$ RS codes with optimum diversity used over a 16-ary channel. Recall from Table 2.19 that the $(255, 191)$ code over this channel required $m_{opt} \sim 5$ or 6 depending on whether the jamming was noise or multitone. It is certainly plausible that a lower rate $n - 255$ RS code might have sufficient redundancy to provide better system performance with less or no diversity.

In general, the performance is specified by (2.153) and (2.170): for the case of interest (i.e., $M = 16$, $K = 4$, $Q = 8$, $L = \frac{1}{2}$, $n = 255$), with γ defined by (2.164) and Table 2.7, these equations reduce to

$$m_{opt} \leq \begin{cases} \ln\left[\dfrac{7.5}{1 - \sqrt{1 - P_Q}}\right]; & \text{noise} \\[4ex] \ln\left[\dfrac{15}{1 - \sqrt{1 - P_Q}}\right]; & \text{multitone} \end{cases}$$

where

$$\frac{E_b}{N_J} = \begin{cases} \left(\dfrac{255}{k}\right) m_{\text{opt}}; & \text{noise} \\[3mm] \left(\dfrac{255 \times .9583 \times e}{k}\right) m_{\text{opt}}; & \text{multitone} \end{cases} \qquad (2.176)$$

The RS code rate is k/n over $GF(2^Q)$; however, the entire FH/MFSK system including the RS outer code and the optimum diversity has an overall code rate

$$r = \frac{R}{K} = \frac{k}{nm_{\text{opt}}} \qquad (2.177)$$

over $GF(2^K)$. Recall that R was defined in Section 2.3.2 (e.g., see (2.133)) as the system code rate in information *bits* per *M*-ary channel use: r is the same parameter expressed in *M-ary* information *symbols* per channel use, so that $r \le 1$. Therefore, $E_b/N_J \propto r^{-1}$ for both kinds of jamming, and *the optimum RS code rate that minimizes the required E_b/N_J is the one that maximizes the overall system code rate.*

Berlekamp's table of RS character error rates P_Q for a given block length n and P_b with hard decision decoding was limited to redundancies $n - k$ that were powers of 2 up to a maximum of 64 [21]. In an unpublished report, L. Deutsch and R. Miller of the Jet Propulsion Laboratory in Pasadena, California, extended Berlekamp's table to include *all* odd values of k; these results were used in conjunction with (2.176) to derive the performance curves of Figure 2.103. We found that for the system under consideration, with worst case partial-band noise jamming, the optimum RS code rate $k/n = .81$, for which $E_b/N_J \le 8.55$ dB, $m_{\text{opt}} \le 5.8$, and $r \ge .14$; for worst case $n = 1$ band multitone jamming, the optimum $k/n = .83$, with $E_b/N_J \le 13.19$ dB, $m_{\text{opt}} \le 6.6$, and $r \ge .12$. These results are only marginally better than the $(255, 191)$ RS code in Table 2.19. But, what is really surprising is that the system performance is optimized by a weak, high rate RS outer code concatenated with a simple repetition code, rather than a more powerful low rate RS code alone. Thus, although the overall system code rate r is quite low, indicating that a significant amount of redundancy is needed to counter the postulated threats, *the RS code by itself with hard decision decoding is apparently poorly matched to the worst case FH/MFSK jamming channel.*

In general, the traditional codes in use today were designed to provide good performance in stationary, additive Gaussian noise. *The worst case jamming environment is often neither stationary nor Gaussian,* so we should not be terribly surprised if these codes are vulnerable to this kind of pathological interference. What is really needed is the development of new codes specifically matched to the various anticipated jamming threats.

Do such codes exist? To provide some insight into this issue, let us apply random coding arguments to the FH/MFSK channel with worst case noise

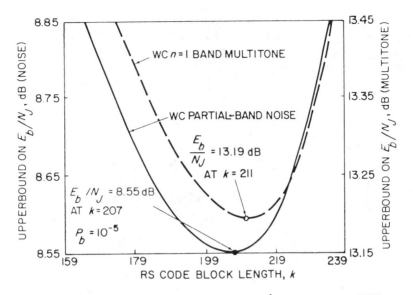

Figure 2.103. Required E_b/N_J to achieve $P_b = 10^{-5}$ for FH/16-ary FSK system with $(255, k)$ Reed-Solomon code and optimum diversity in worst case noise and tone jamming. Minima occur at RS code rates $k/n > .8$, with $m_{\text{opt}} > 6$.

and tone jamming. And since the convolutional codes performed generally better than the one class of block codes considered (i.e., the Reed-Solomon codes), we will restrict this exercise to random (n, k) block codes with characters over $GF(2^K)$. So, for each block of k K-bit information symbols, the encoder generates a block of n K-bit code symbols, according to a randomly generated mapping as in [9, pp. 309–320]; this corresponds to a code rate

$$R' = \frac{kK}{n} \frac{\text{information bits}}{M\text{-ary channel use}} \tag{2.178}$$

in the notation of Section 2.3.3. The random coding scheme selects a set of $M^k = 2^{Kk} = 2^{nR'}$ codewords $\{s_i\} \in \mathrm{GF}(2^K)$, with a mapping of the j-th message, m_j, into codeword s_j. If all sets of codewords and messages are equally probable, the union bound on the word error rate is [9, (5.47)]

$$P_w = \Pr\{\text{word error}|m_j\}$$

$$\leq \sum_{\substack{l=1 \\ (l \neq j)}}^{2^{nR'}} \overline{P_2(s_l, s_j)}^{\{s_i\}}$$

$$\leq (2^{nR'} - 1)\, 2^{-nR_0} \tag{2.179}$$

where $P_2(s_l, s_j)$ is the probability that codeword s_j is sent and the decoder decides message m_l was intended, the expectation is over the codeword set $\{s_i\}$, and R_0 is the M-ary channel cutoff rate in bits/channel use (see Chapter 4, Volume I). We can convert R_0 to the parameter D used in this chapter according to (see (4.27), Volume I)

$$R_0 = \log_2\left[\frac{M}{1+(M-1)D}\right] \tag{2.180}$$

which implies that

$$2^{-nR_0} = \left[\frac{1+(M-1)D}{M}\right]^n. \tag{2.181}$$

By analogy with (2.20a), with $M^k = 2^{nR'}$ replacing M,

$$P_b = \frac{2^{nR'-1}}{2^{nR'}-1}P_w$$

$$\leq 2^{nR'-1}\left[\frac{1+(M-1)D}{M}\right]^n. \tag{2.182}$$

Suppose we concatenate the (n, k) random block code with an m-diversity inner code. That is, the block code operates over an M-ary channel with D replaced by D^m as in Section 2.3.3. When this diversity is optimized, (2.162)–(2.164) apply for worst case noise and tone jamming: substituting these expressions into (2.182), we find that

$$m_{\text{opt}} \leq \ln\left[\frac{M-1}{M2^{-R'}(2P_b)^{1/n}-1}\right]$$

$$\frac{E_b}{N_J} = \left(\frac{\gamma}{R'}\right)m_{\text{opt}}. \tag{2.183}$$

As an example, (2.183) was used to plot an upperbound on E_b/N_J required to achieve $P_b = 10^{-5}$ as a function of R' with parameter n, for $K = \log_2 M = 2$ and worst case partial-band noise jamming. Figure 2.104 shows that for each n, the required E_b/N_J upperbound is minimized at an interior code rate $R' = R_{\text{opt}}$. If we compute m_{opt} at this optimum code rate as a function of n, Figure 2.105 shows that m_{opt} decreases monotonically with n, finally reaching unity at $n = n^* \cong 92$, for which $R_{\text{opt}} = .76$. Remember that (2.183) represents the *average* performance over all possible code sets $\{s_i\}$, with the obvious implication that some *particular* codes must perform even better. Generalizing these results for arbitrary system parameters, we know that *for sufficiently large block lengths and optimized code rates, codes exist that are powerful enough to defeat any kind of non-adaptive jamming without the need for additional diversity redundancy.* It remains for information theorists to determine how to generate these codes.

We can derive closed form solutions for n^* and the corresponding R_{opt}. Independent of the kind of jamming, (2.183) tells us that E_b/N_J is bounded

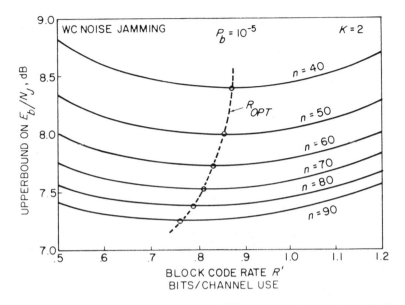

Figure 2.104. Performance of FH/4-ary FSK system employing $(n, nR'/2)$ random block code with optimum diversity in worst case partial-band noise. For a given block length n over the 4-ary field, the required E_b/N_J to achieve $P_b = 10^{-5}$ is minimized at $R' = R_{opt}$.

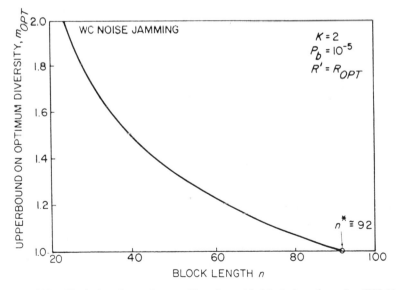

Figure 2.105. Variation in optimum diversity with block length n for FH/4-ary FSK system with (n, k) random block code and optimized code rate, when $P_b = 10^{-5}$ and signals are received in worst case partial-band noise. For $n > n^* \cong 92$, $m_{opt} = 1$, implying that diversity is not required.

by an expression that is proportional to

$$f(R') \equiv \frac{1}{R'} \ln\left(\frac{A}{B2^{-R'} - 1} \right)$$

where, for convenience,

$$A \equiv M - 1$$

$$B \equiv M(2P_b)^{1/n}. \qquad (2.184)$$

Differentiating $f(R')$ with respect to R' and setting this result equal to zero shows that R_{opt} must satisfy the transcendental constraint

$$\ln\left(\frac{A}{B2^{-R_{opt}} - 1} \right) = \frac{R_{opt}B2^{-R_{opt}}\ln 2}{B2^{-R_{opt}} - 1} \qquad (2.185)$$

for a given n (remember that B depends on n). If we further require that the bound on m_{opt} in (2.183) be unity at $n = n^*$, we have the additional constraint

$$B2^{-R_{opt}} = Ae^{-1} + 1. \qquad (2.186)$$

For this particular value of n, (2.185) simplifies to

$$R_{opt} = \frac{Ae^{-1}}{(Ae^{-1} + 1)\ln 2} \qquad (2.187)$$

in bits/channel use. Since $m_{opt} = 1$ at $n = n^*$, R_{opt} in (2.187) is the *overall* system code rate. If we prefer to express this parameter in M-ary information symbols/channel use, as in (2.177), we have

$$r_{opt} \equiv \frac{R_{opt}}{K} = \frac{(M - 1)e^{-1}}{K \ln 2[(M - 1)e^{-1} + 1]}. \qquad (2.188)$$

Finally, from (2.184) and (2.186) and (2.187), we find that

$$n^* = \frac{\ln(2P_b)}{\ln\left[\frac{(M - 1)e^{-1} + 1}{M} \right] - \left[\frac{(M - 1)e^{-1}}{(M - 1)e^{-1} + 1} \right]}. \qquad (2.189)$$

Note that r_{opt} depends only on K, n^* depends only on K and P_b, and neither parameter depends on the type of jamming. From (2.183), the corresponding SNR is

$$\frac{E_b}{N_J} = \frac{\gamma \ln 2[(M - 1)e^{-1} + 1]}{(M - 1)e^{-1}} \qquad (2.190)$$

which does depend on the jamming environment through (2.164), as well as K, *but not explicitly on* P_b. These parameters are shown in Table 2.22 as a function of K for worst case noise and tone jamming when $P_b = 10^{-5}$. The random block coding results perpetuate a pattern that we observed for

Table 2.22

Random (n, k) block coding with FH/MFSK signals in worst case jamming. If $n \geq n^*$, the performance cannot be improved by adding diversity; at n^*, $r_{opt} = k/n^*$ minimizes the required E_b/N_J.

K	n^*	r_{opt}	E_b/N_J, dB for $P_b = 10^{-5}$	
			Noise	Tone
1	97.5	.39	10.13	8.45
2	91.7	.38	7.23	7.56
3	127.1	.35	5.85	8.08
4	210.5	.31	5.15	9.31
5	382.1	.27	4.79	10.97

almost all of the specific codes considered in this chapter. *The performance improves monotonically with K in worst case partial-band noise, but it peaks at K = 2 for worst case multitone jamming. If we must be capable of dealing with either threat, the best performance for a coded FH/MFSK system is usually achieved at K = 2.* Recall that of all of the coded FH/MFSK systems considered, the smallest E_b/N_J that could guarantee $P_b \leq 10^{-5}$ in any non-adaptive jamming environment was 8.11 dB, realized for the $(1023, 959)$ RS outer code combined with the $(7, \frac{1}{2})$ Trumpis 4-ary convolutional inner code and optimum diversity. Table 2.22 says that block codes exist that can perform at least .5 dB better than that system, presumably because they are better matched to the jamming channel; the incentive to find these codes is evident.

2.4 SLOW FADING UNIFORM CHANNELS

We assume a slow fading channel where each hopped FH/MFSK signal experiences an independent fade with the same statistics. In particular, assume that during each hop the signal amplitude A at the receiver is a Rayleigh random variable with probability density function

$$p_A(a) = \frac{a}{\sigma^2} e^{-a^2/2\sigma^2}, \qquad a \geq 0 \tag{2.191}$$

where this fade is independent from hop to hop. Amplitude A is constant during each hop interval. This is called a uniform channel since the probability density function (2.191) does not depend on which part of the spread spectrum frequency band the signal hops to at any time.

Consider a CW signal of T seconds duration at frequency ω_0 with amplitude A and phase ϕ together with white Gaussian noise $n(t)$ of double-sided power spectral density $N_0/2$, which is given by

$$y(t) = A\sin(\omega_0 t + \phi) + n(t)$$

$$0 \leq t \leq T. \tag{2.192}$$

Using the sine expansion

$$y(t) = A\sin\phi\cos\omega_0 t + A\cos\phi\sin\omega_0 t + n(t)$$

$$0 \leq t \leq T. \tag{2.193}$$

In detecting such a signal, there is no loss of generality[3] in basing decisions on the normalized cosine and sine components of $y(t)$ given by

$$y_c = \int_0^T y(t)\sqrt{\frac{2}{T}}\cos\omega_0 t\, dt = \sqrt{\frac{T}{2}}\, A\sin\phi + n_c$$

$$y_s = \int_0^T y(t)\sqrt{\frac{2}{T}}\sin\omega_0 t\, dt = \sqrt{\frac{T}{2}}\, A\cos\phi + n_s \tag{2.194}$$

[3] This is because the signal has only cosine and sine components.

where n_c and n_s are independent zero mean Gaussian random variables with variance $N_0/2$.

Note that if A is a Rayleigh random variable with probability density given by (2.191) and ϕ is independent of A and uniformly distributed over $[0, 2\pi]$, then

$$z_s = A \sin \phi$$

and

$$z_c = A \cos \phi \tag{2.195}$$

are independent zero mean Gaussian random variables with variance σ^2. Thus, y_c and y_s are independent zero mean Gaussian random variables with variance $\sigma^2 T/2 + N_0/2$. Denoting

$$\overline{E} = \sigma^2 T \tag{2.196}$$

as the average signal energy, the joint density for y_c and y_s is given by

$$p_1(y_c, y_s) = \frac{1}{\pi(\overline{E} + N_0)} e^{-(y_c^2 + y_s^2)/(\overline{E} + N_0)}. \tag{2.197}$$

If the received signal was noise alone then

$$y(t) = n(t) \tag{2.198}$$

and the joint probability density of y_c and y_s is

$$p_0(y_c, y_s) = \frac{1}{\pi N_0} e^{-(y_c^2 + y_s^2)/N_0}. \tag{2.199}$$

Thus in a Rayleigh fading channel, at the receiver each FH/MFSK signal is a narrowband Gaussian random process with the resulting sine and cosine components being zero mean independent Gaussian random variables with variance $(\overline{E} + N_0)/2$. At a frequency[4] where there is no FH/MFSK signal, the corresponding cosine and sine components would be zero mean Gaussian random variables with variance $N_0/2$.

Note that when the CW signal is present and A and ϕ are known, then y_c and y_s are nonzero mean Gaussian random variables with variance $N_0/2$. The conditional probability density function is

$$p_1(y_c, y_s | A, \phi) = \frac{1}{\pi N_0} \cdot$$

$$\exp\left\{ -\frac{\left(y_c - \sqrt{\frac{T}{2}} A \sin \phi\right)^2 + \left(y_s - \sqrt{\frac{T}{2}} A \cos \phi\right)^2}{N_0} \right\}.$$

$$\tag{2.200}$$

---

[4] We assume all possible carrier frequencies are spaced so as to result in orthogonal signals.

Averaging this probability density over ϕ, which is uniformly distributed over $[0, 2\pi]$ and independent of A, gives [9]

$$p_1(y_c, y_s|A) = \frac{1}{\pi N_0} e^{-(y_c^2 + y_s^2)/N_0} e^{-A^2 T/2N_0} I_0\left(\sqrt{2(y_c^2 + y_s^2) A^2 T/N_0^2}\right).$$

(2.201)

Averaging this result over the Rayleigh random variable A returns us to the unconditioned probability density function given by (2.197).

2.4.1 Broadband Jamming—No Diversity

The performance of the FH/MFSK system against broadband noise jamming is the same as that of conventional non-coherent MFSK systems in white Gaussian noise with single-sided spectral density $N_J = J/W_{ss}$. Denote

$$y_m = (y_{mc}, y_{ms})$$
$$m = 1, 2, \ldots, M$$

(2.202)

as the cosine and sine components of the received signal at the M possible tones representing an M-ary symbol. For the Rayleigh fading channel described above, the maximum-likelihood (ML) decision rule is based on probability density functions

$$p(y_1, y_2, \ldots, y_M|m \text{ is sent}) = F(y_1, y_2, \ldots, y_M) \frac{p_1(y_m)}{p_0(y_m)} \quad (2.203)$$

where

$$F(y_1, y_2, \ldots, y_M) = \prod_{l=1}^{M} p_0(y_l) \quad (2.204)$$

and $p_1(y)$ and $p_0(y)$ are given by (2.197) and (2.199), respectively. Without any coding or diversity, the ML decision rule is to choose that symbol m that yields the maximum value of

$$\frac{p_1(y_m)}{p_0(y_m)} = \frac{N_J}{E + N_J} \exp\left\{\frac{\overline{E}/N_J}{\overline{E} + N_J}(y_{mc}^2 + y_{ms}^2)\right\} \quad (2.205)$$

or equivalently the maximum value of

$$e_m = y_{mc}^2 + y_{ms}^2, \quad (2.206)$$

the energy of the m-th frequency component of the received signal. Thus, the optimum decision rule is the same as that for the non-fading channel when there is no coding or diversity.

Since the decision rule is the same whether or not the fading amplitude A is given, consider the conditional bit error probability given A (see Section 2.1)

$$P_b(A) = \frac{\frac{1}{2}M}{M-1} \sum_{l=1}^{M-1} \binom{M-1}{l} \frac{(-1)^{l+1}}{l+1} e^{-(l/l+1)(A^2 T/2N_J)}. \quad (2.207)$$

Averaging this over the Rayleigh random variable with probability density function (2.191) results in the bit error probability expression

$$P_b = \frac{\frac{1}{2}M}{M-1} \sum_{l=1}^{M-1} \binom{M-1}{l} \frac{(-1)^{l+1}}{1 + l[1 + (\overline{E}/N_J)]} \qquad (2.208)$$

where $\overline{E}$ is given by (2.196). In Chapter 4, Section 4.7.1, Volume I, a simple union bound for (2.207) was given by

$$P_b(A) \leq \frac{1}{4} M e^{-A^2 T/4N_J} \qquad (2.209)$$

which, when averaged over the Rayleigh fading random variable, becomes

$$P_b \leq \frac{M}{4 + 2(\overline{E}/N_J)}. \qquad (2.210)$$

In Figure 2.106 we plot P_b and its bound versus $\overline{E}_b/N_J$ where $\overline{E}_b = \overline{E}/K$ is the average bit energy and $M = 2^K$ with $K = 1, 2, 3, 4,$ and 5. Note that

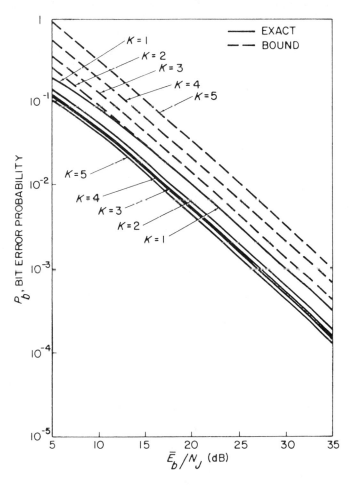

Figure 2.106. Rayleigh fading–uncoded.

the impact of fading is similar to the worst case partial-band and multitone jammer. Also notice that although the union bound is very tight for the non-fading case (see Figure 4.11 in Chapter 4, Volume I), this bound is quite weak in this uncoded Rayleigh fading case. Furthermore, the bound increases as K increases, whereas the true bit error probability decreases with increasing K. With no fading the union bound results in a sum of terms that decrease exponentially with E_b/N_J, whereas in the Rayleigh case these terms decrease at a much slower rate, namely inversely with $\overline{E}_b/N_J$.

We saw earlier that coding and/or diversity can improve the performance of worst case partial-band jamming. These techniques similarly improve the performance in Rayleigh fading channels.

2.4.2 Broadband Jamming—Diversity and Coding

Consider diversity of order m where each MFSK signal of duration T seconds is transmitted as m MFSK chip tones each of duration $T_c = T/m$ seconds. Each chip is independently hopped over the spread-spectrum frequency band. Also assume that each chip tone has independent Rayleigh fading of the same statistics. This is a reasonable assumption for FH/MFSK signals that hop over a wide frequency band W_{ss} where

$$W_{ss} \gg \Delta f_c \qquad (2.211)$$

where Δf_c is the "coherence bandwidth" of the fading channel [24].

If we denote

$$y_{lk} = (y_{lkc}, y_{lks}) \qquad (2.212)$$

as the cosine and sine components of the received signal at the l-th frequency during the k-th chip interval and

$$Y_l = (y_{l1}, y_{l2}, \ldots, y_{lm}) \qquad (2.213)$$

as the sequence of such components for the l-th tone, then the conditional probability of the entire set of mM chip cosine and sine components given the l-th tone was transmitted is

$$p(Y_1, Y_2, \ldots, Y_M | l \text{ is sent}) = F(Y_1, Y_2, \ldots, Y_M) \frac{p_{1m}(Y_l)}{p_{0m}(Y_l)} \qquad (2.214)$$

where

$$F(Y_1, Y_2, \ldots, Y_M) = \prod_{j=1}^{M} p_{0m}(Y_j), \qquad (2.215)$$

$$p_{0m}(Y_j) = \prod_{k=1}^{m} p_0(y_{jk}) \qquad (2.216)$$

and

$$p_{1m}(Y_j) = \prod_{k=1}^{m} p_1(y_{jk}). \qquad (2.217)$$

The ML decision rule chooses l that maximizes

$$\frac{p_{1m}(\mathbf{Y}_l)}{p_{0m}(\mathbf{Y}_l)} = \left(\frac{N_J}{\overline{E} + N_J}\right)^m \exp\left\{\frac{\overline{E}/N_J}{\overline{E} + N_J} \sum_{k=1}^m \left(y_{lkc}^2 + y_{lks}^2\right)\right\} \quad (2.218)$$

or equivalently maximizes

$$e_l = \sum_{k=1}^m e_{lk} \qquad (2.219)$$

where

$$e_{lk} = y_{lkc}^2 + y_{lks}^2 \qquad (2.220)$$

is the energy of the l-th tone in the k-th chip interval. $\overline{E}$ is the average energy per MFSK chip.

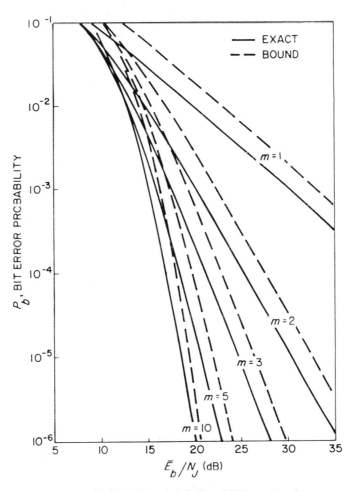

Figure 2.107. Rayleigh fading BFSK—diversity.

Note that here the ML decision rule is based on the sum of the m chip energies for each of the M tones. The energy metric is thus optimum for the Rayleigh fading channel with equivalent white Gaussian noise or broadband noise jamming. This was not the case with coding over a non-fading white Gaussian noise channel. There, however, the energy metric was used since it is convenient.

For m diversity, the bit error probability is given by [24],

$$P_b = \frac{\frac{1}{2}M}{M-1} \sum_{l=1}^{M-1} \frac{\binom{M-1}{l}(-1)^{l+1}}{\left\{1 + l\left[1 + (\bar{E}/N_J)\right]\right\}^m} \cdot$$

$$\sum_{k=0}^{l(m-1)} \beta_{kl} \frac{(m-1+k)!}{(m-1)!} \left(\frac{1 + (\bar{E}/N_J)}{1 + l\left[1 + (\bar{E}/N_J)\right]}\right)^k \qquad (2.221)$$

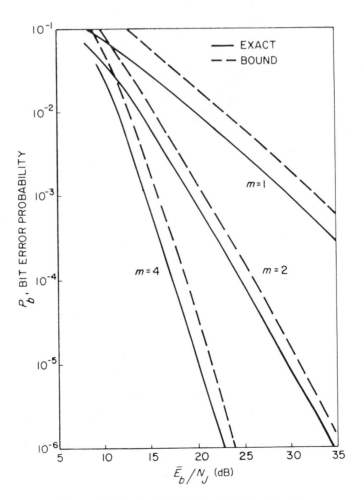

Figure 2.108. Rayleigh fading 4FSK—diversity.

where $\bar{E} = (K/m)\bar{E}_b$, $M = 2^K$, and β_{kl} satisfies

$$\left(\sum_{k=0}^{m-1} \frac{x^k}{k!}\right)^l = \sum_{k=0}^{l(m-1)} \beta_{kl} x^k. \tag{2.222}$$

For the binary case where $M = 2$, the bit error probability is simply

$$P_b = \delta^m \sum_{k=0}^{m-1} \binom{m-1+k}{k}(1-\delta)^k \tag{2.223}$$

where

$$\delta = \frac{m}{2m + (\bar{E}_b/N_J)}. \tag{2.224}$$

Figure 2.107 shows this bit error probability versus $\bar{E}_b/N_J$ for various values of diversity. Note that, for each value of E_b/N_J, there is an optimum

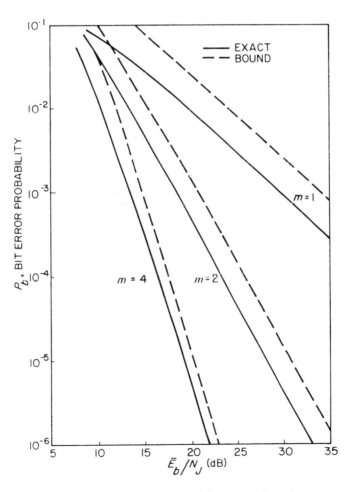

Figure 2.109. Rayleigh fading 8FSK—diversity.

diversity. Excessive diversity results in non-coherent combining losses that begin to cancel the beneficial effects of having independent observations.

Next, a Chernoff bound will be derived for the bit error probability when $M = 2$. Assuming the ML metric, the optimized Chernoff bound is the Bhattacharyya bound given by

$$P_b \le \frac{1}{2} \int \sqrt{p(Y_1, Y_2 | l = 1) \, p(Y_1, Y_2 | l = 2)} \, dY_1 \, dY_2$$

$$= \frac{1}{2} \int F(Y_1, Y_2) \sqrt{\frac{p_{1m}(Y_1) p_{1m}(Y_2)}{p_{0m}(Y_1) p_{0m}(Y_2)}} \, dY_1 \, dY_2$$

$$= \frac{1}{2} \int F(Y_1, Y_2) \prod_{k=1}^{m} \left(\frac{N_J}{E + N_J} \right). \tag{2.225}$$

$$\exp\left\{ \frac{\bar{E}/N_J}{\bar{E} + N_J} \left(\frac{y_{1kc}^2 + y_{1ks}^2 + y_{2kc}^2 + y_{2ks}^2}{2} \right) \right\} \, dY_1 \, dY_2.$$

But $F(Y_1, Y_2)$ is the joint Gaussian density function where all random variables are i.i.d. with zero mean and variance $N_J/2$. Thus

$$P_b \le \frac{1}{2} \left[\frac{4m(m + (\bar{E}_b/N_J))}{(2m + (\bar{E}_b/N_J))^2} \right]^m \tag{2.226}$$

Figure 2.107 shows this bound (dotted lines) together with the exact bit error probability.

For $M > 2$ the union bound can be combined with the above Chernoff bound. The symbol error probability is union bounded by

$$P_s = Pr\{\hat{l} \ne 1 | l = 1\}$$

$$\le \sum_{k=2}^{M} Pr\{\hat{l} = k | l = 1\}$$

$$= \sum_{k=2}^{M} Pr\{1 \rightarrow k\}$$

$$= (M - 1) Pr\{1 \rightarrow 2\} \tag{2.227}$$

where $l = 1$ was assumed to be the transmitted symbol and $Pr\{1 \rightarrow \hat{l}\}$ is the pairwise error bound, which is the same as the $M = 2$ bit error probability. Thus, using the Chernoff bound

$$Pr\{1 \rightarrow 2\} \le \frac{1}{2} \left[\frac{4m(m + (\bar{E}/N_J))}{(2m + (\bar{E}/N_J))^2} \right]^m \tag{2.228}$$

and noting that here $\overline{E} = K\overline{E}_b$, we obtain the bit error bound

$$P_b = \frac{\frac{1}{2}M}{M-1}P_s$$

$$\leq 2^{K-2}\left[\frac{4m(m + K(\overline{E}_b/N_J))}{(2m + K(\overline{E}_b/N_J))^2}\right]^m. \qquad (2.229)$$

Figures 2.108 to 2.111 show these bounds for various values of K and diversity m together with the corresponding exact bit error probabilities given by (2.221) and (2.222).

Next choose the value of diversity m that minimizes the bound in (2.228). This is shown in Figure 2.12. Figure 2.113 shows the same bound with

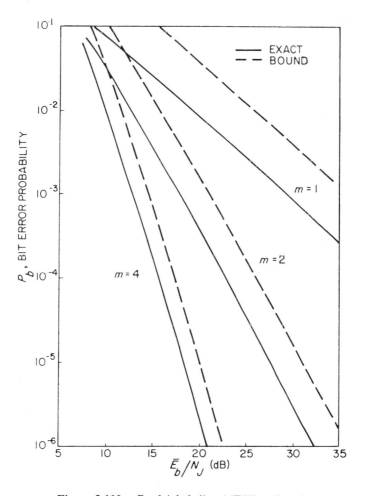

Figure 2.110. Rayleigh fading 16FSK—diversity.

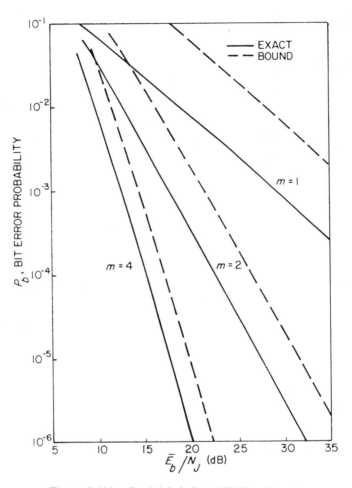

Figure 2.111. Rayleigh fading 32FSK—diversity.

$m = K$ where $M = 2^K$. For fixed data rate, this case results in the bandwidth of each FH/MFSK chip signal being the same for $M = 2, 4, 8, 16$, and 32.

Regarding each FH/MFSK chip as a coded M-ary symbol, a coding channel (see Figure 4.1 in Chapter 4, Volume I) with output

$$y = (y_1, y_2, \ldots, y_M) \qquad (2.230)$$

where y_l is the cosine and sine components of the channel output signal at the l-th chip tone is obtained. In a coded system, a sequence of coded chips is transmitted and the channel outputs are the corresponding vectors of these cosine and sine components. Any coding system uses a metric $m(y, l)$ which assigns a value to each channel output corresponding to each possible

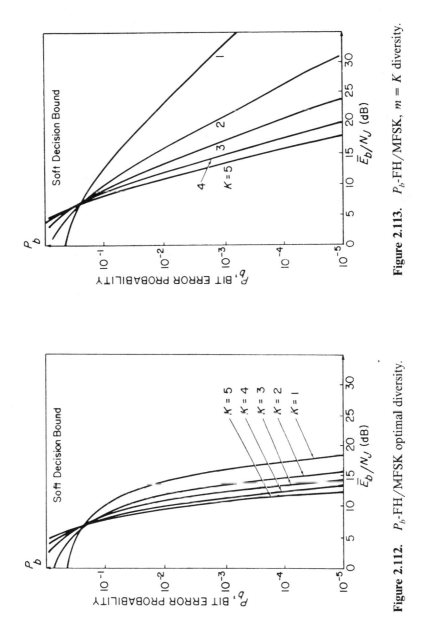

Figure 2.113. P_b-FH/MFSK, $m = K$ diversity.

Figure 2.112. P_b-FH/MFSK optimal diversity.

input. The ML energy metric we have considered previously is

$$m(y, l) = e_l$$
$$= y_{lc}^2 + y_{ls}^2 \tag{2.231}$$
$$l = 1, 2, \ldots, M.$$

For the ML metric and simple diversity, the expression for the bit error probability given by (2.221) and (2.222) is quite complex. Obtaining an exact bit error probability expression is much more difficult with other metrics. Hence we examine bounds on the coded bit error probabilities.

Based on the general approach of Chapter 4, Volume I, for an arbitrary metric $m(y, l)$, the general coded bit error bound is given by

$$P_b \leq G(D) = B(R_0) \tag{2.232}$$

where for FH/MFSK the cutoff rate R_0 is given by (2.180) where

$$D = \min_{\lambda \geq 0} E\{ \exp\lambda[m(y, l') - m(y, l)] \,|\, l \} \tag{2.233}$$

with $l' \neq l$. For the energy metric (2.231), D is given by

$$D = \frac{4(1 + (\bar{E}/N_J))}{(2 + (\bar{E}/N_J))^2}. \tag{2.234}$$

Also, for the special case of an m diversity code, the function $G(\cdot)$ is

$$G(D) = 2^{K-2}D^m \tag{2.235}$$

with code rate $R = K/m$ bits per chip and $\bar{E} = R\bar{E}_b = (K/m)\bar{E}_b$. This is the result given by (2.229).

Rather than the energy metric, if a hard decision is made for each MFSK chip then

$$D = 2\sqrt{\frac{(1 - \varepsilon)\varepsilon}{M - 1}} + \left(\frac{M - 2}{M - 1}\right)\varepsilon \tag{2.236}$$

where

$$\varepsilon = \sum_{l=1}^{M-1} \binom{M-1}{l} \frac{(-1)^{l+1}}{1 + l[1 + (\bar{E}/N_J)]} \tag{2.237}$$

which is the chip error probability given by (2.208). Figures 2.114 and 2.115 show the performance for optimum diversity and diversity $m = K$ where $M = 2^K$ for the hard decision metric.

2.4.3 Partial-Band Jamming

Next consider the case of a Rayleigh fading channel together with a partial-band noise jammer with average power J that jams a fraction ρ of the total spread-spectrum band. Assume that when an FH/MFSK signal

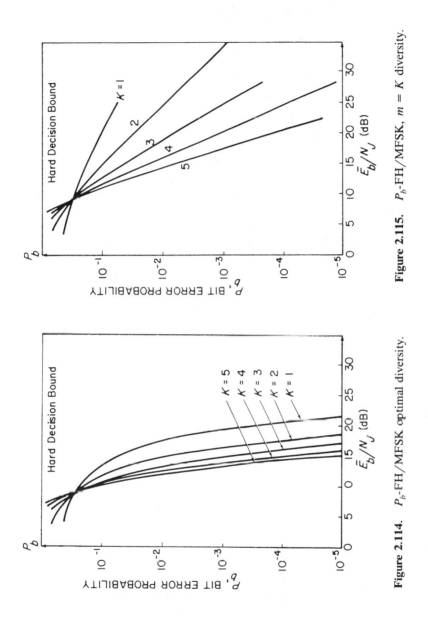

Figure 2.115. P_b-FH/MFSK, $m = K$ diversity.

Figure 2.114. P_b-FH/MFSK optimal diversity.

hops into the part of the band where there is no noise jamming, the bit error probability is negligibly small since the signal-energy-to-background noise is large.[5] When a noise jamming signal is present in the transmitted signal band, the symbol-energy-to-noise ratio is $\rho E/N_J$. Thus, with no coding or diversity, the bit error probability is given by

$$P_b = \rho P_b(\rho \overline{E}/N_J) + (1 - \rho) P_b(\infty)$$

$$= \rho P_b(\rho \overline{E}/N_J) \tag{2.238}$$

where $P_b(x)$ is the uncoded bit error probability when the symbol-energy-to-noise ratio is x. For the binary case, $M = 2$,

$$P_b = \frac{\rho}{2 + \rho(\overline{E}_b/N_J)}. \tag{2.239}$$

The choice of the worst partial-band parameter ρ that maximizes this bit error probability is $\rho = 1$. *Thus, the broadband jammer is the worst noise jammer in a Rayleigh fading channel.* This is primarily due to the fact that Rayleigh fading has already created the same impact as the worst case partial-band jammer with a resulting uncoded bit error probability that changes slowly with increasing values of $\overline{E}_b/N_J$. Changes in the signal-to-noise ratio caused by the signal hopping to a jammed part of the frequency band no longer results in large changes in the uncoded bit error probability and the broadband jammer turns out to be the worst case. Coding in the form of diversity still shows dramatic improvements, but this time to overcome the effects of Rayleigh fading.

In general, for arbitrary fade statistics, if the derivative with respect to ρ of the bit error probability given in (2.238) is positive for a given symbol-energy-to-noise ratio and all $\rho\varepsilon[0, 1]$, then $\rho = 1$ maximizes the bit error probability. Taking the derivative of (2.238) with respect to ρ gives

$$P_b'(\rho\overline{E}/N_J) > -\frac{P_b(\rho\overline{E}/N_J)}{\rho(\overline{E}/N_J)} \tag{2.240}$$

which must be satisfied for all $\rho\varepsilon[0, 1]$ where $P_b'(x)$ is the derivative of $P_b(x)$. This condition defines how slowly the bit error probability must decrease with $\overline{E}/N_J$ in order to have broadband noise jamming as the worst case. It is satisfied by all uncoded FH/MFSK signals in a Rayleigh fading channel. For small values of $\overline{E}/N_J$, condition (2.240) is usually satisfied for non-fading channels as well.

With the use of coding, the worst fraction ρ of the partial-band noise jammer depends on the coding metric used. With Rayleigh fading, and the ML metric, which assumes jammer state information, the broadband jammer ($\rho = 1$) is again the worst case. This is not true, however, for the energy metric with no jammer state information. We show this next with a more general noise jammer power distribution.

---

[5] We assume it is infinite.

2.5 WORST NOISE JAMMER DISTRIBUTION—SLOW FADING UNIFORM CHANNEL

Most of this chapter assumed the worst case noise jammer was a partial band jammer that jammed some fraction, ρ, of the spread-spectrum frequency band with constant noise power spectral density. In this section consider the Rayleigh fading channel with arbitrary power spectral density for the noise jammer and then find the worst noise power spectral density. This presentation sets the stage for our discussion of slowly fading non-uniform channels in Section 2.6, which is based on the work of Avidor [25], [26].

Assume that whenever a hopped MFSK signal is transmitted the average signal energy at the receiver is $\bar{E}$ independent of the part of the spread-spectrum frequency band that the signal hopped into during its transmission. This is what we refer to as the uniform channel and is the channel assumed in Section 2.4.

Divide the spread-spectrum frequency band into Q equal size sub-bands where in each sub-band the jammer places noise with constant spectral density across the sub-band. Let J_q be the jammer noise power in the q-th sub-band of bandwidth $W_q = W_{ss}/Q$. Thus, the q-th sub-band has white Gaussian noise of single-sided spectral density

$$
\begin{aligned}
N_q &= J_q/W_q \\
&= QJ_q/W_{ss} \\
q &= 1, 2, \ldots, Q.
\end{aligned}
\tag{2.241}
$$

This is illustrated in Figure 2.116. Note that for large values of Q this can accurately approximate all noise jammer power spectral densities of interest. With total jammer power J, the total power constraint is

$$
J = \sum_{q=1}^{Q} J_q.
\tag{2.242}
$$

2.5.1 Uncoded

The uncoded (no diversity) FH/MFSK signal that hops with uniform probability across the total spread-spectrum frequency band has a bit error probability given by[6]

$$
\begin{aligned}
P_b &= \sum_{q=1}^{Q} \frac{1}{Q} P_b\left(\frac{\bar{E}}{N_q}\right) \\
&= \sum_{q=1}^{Q} \frac{1}{Q} P_b\left(\frac{\bar{E} W_{ss}}{Q J_q}\right)
\end{aligned}
\tag{2.243}
$$

---

[6] We assume the FH/MFSK signal will hop into each sub-band with equal probability $1/Q$.

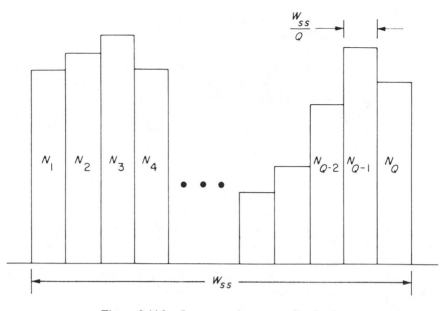

Figure 2.116. Jammer noise power distribution.

where $P_b(x)$ is the Rayleigh fading uncoded non-coherent MFSK bit error probability in white Gaussian noise with average symbol-energy-to-noise ratio x. This is given by (2.208).

As was done in Section 2.5, consider uncoded FH/BFSK in a Rayleigh fading uniform channel where the Rayleigh fading statistics are the same in each sub-band. In this case from (2.208)

$$P_b(x) = \frac{1}{2 + x} \qquad (2.244)$$

and the average bit error probability is

$$P_b = \sum_{q=1}^{Q} \frac{1}{Q} \left(\frac{J_q Q}{2 J_q Q + \overline{E}_b W_{ss}} \right). \qquad (2.245)$$

Now consider the worst distribution of jammer power $J = (J_1, J_2, \ldots, J_Q)$ which satisfies constraint (2.242). This equality constraint can be incorporated into the maximization of P_b given by (2.243) by using a Lagrange multiplier $\lambda \geq 0$ and considering the maximization of

$$C(J) = \sum_{q=1}^{Q} \frac{1}{Q} P_b \left(\frac{\overline{E}_b W_{ss}}{Q J_q} \right) - \lambda \sum_{q=1}^{Q} J_q \qquad (2.246)$$

with respect to $J_1, J_2, \ldots, J_Q$ where

$$J_q \geq 0 \text{ for all } q. \qquad (2.247)$$

Suppose $J^* = (J_1^*, J_2^*, \ldots, J_Q^*)$ maximizes $C(J)$. Then for values of l where

$$J_l^* > 0 \tag{2.248}$$

we have the necessary condition

$$\left. \frac{\partial C(J)}{\partial J_l} \right|_{J_l = J_l^*} = 0 \tag{2.249}$$

or

$$\left. \frac{d}{dJ_l} P_b \left(\frac{\overline{E}_b W_{ss}}{J_l Q} \right) \right|_{J_l = J_l^*} = \lambda Q. \tag{2.250}$$

For those values of l where

$$J_l^* = 0 \tag{2.251}$$

we have the necessary condition

$$\left. \frac{\partial C(J)}{\partial J_l} \right|_{J_l = J_l^*} \leq 0 \tag{2.252}$$

or

$$\left. \frac{d}{dJ_l} P_b \left(\frac{\overline{E}_b W_{ss}}{J_l Q} \right) \right|_{J_l = J_l^*} \leq \lambda Q. \tag{2.253}$$

These necessary conditions become both necessary and sufficient if $P_b(1/z)$ is a concave function of z. (See the Kuhn-Tucker Theorem [19]). In this case, the minimizing jammer noise power distribution is unique.

For the Rayleigh fading FH/BFSK case, $P_b(1/z)$ is concave in z so the necessary and sufficient conditions become

$$\frac{\overline{E}_b W_{ss}}{\left(2J_l^* Q + \overline{E}_b W_{ss} \right)^2} \leq \lambda \tag{2.254}$$

with equality when $J_l^* > 0$. The unique worst jammer noise power distribution is

$$J_l^* = \frac{J}{Q} \tag{2.255}$$

$$l = 1, 2, \ldots, Q$$

which is the broadband jammer with uniform power distribution. This generalizes to arbitrary uncoded FH/MFSK signals including background white Gaussian noise together with the noise jammer.

With no fading, the worst distribution for uncoded FH/MFSK is the usual two-level partial-band noise jammer considered previously in this chapter. To show this for the binary case, suppose that the distinct values of

J_q, $q = 1, 2, \ldots, Q$ are given by $\Delta_1, \Delta_2, \ldots, \Delta L$ where L is the number of these values of the jammer power levels in the sub-bands. Let F_l be the number of sub-bands that have jammer power Δ_l. Then

$$\rho_l = \frac{F_l}{Q} \tag{2.256}$$

is the fraction of sub-bands with jammer power Δ_l for $l = 1, 2, \ldots, L$. The bit error probability thus has the form

$$
\begin{aligned}
P_b &= \sum_{q=1}^{Q} \frac{1}{Q} e^{-E_b W_{ss}/2 J_q Q} \\
&= \sum_{l=1}^{L} \frac{F_l}{Q} e^{-E_b W_{ss}/2 \Delta_l Q} \\
&= \sum_{l=1}^{L} \rho_l e^{-\rho_l E_b W_{ss}/2 F_l \Delta_l}.
\end{aligned}
\tag{2.257}
$$

Using the bound

$$
\begin{aligned}
\rho e^{-\rho \alpha} &\leq \max_{\rho'} \rho' e^{-\rho' \alpha} \\
&= \frac{e^{-1}}{\alpha}
\end{aligned}
\tag{2.258}
$$

which is easily obtained by direct differentiation,

$$
\begin{aligned}
P_b &\leq \sum_{l=1}^{L} F_l \Delta_l \frac{2 e^{-1}}{E_b W_{ss}} \\
&= \frac{2 J e^{-1}}{E_b W_{ss}} \\
&= \frac{2 e^{-1}}{E_b / N_J}
\end{aligned}
\tag{2.259}
$$

using the relationship

$$
\begin{aligned}
J &= \sum_{q=1}^{Q} J_q \\
&= \sum_{l=1}^{L} F_l \Delta_l.
\end{aligned}
\tag{2.260}
$$

The bound in (2.259) can be achieved with two levels ($L = 2$) where

$$\Delta_1 = 0$$

$$\Delta_2 = J/\rho \tag{2.261}$$

with

$$\rho_1 = 1 - \rho$$

$$\rho_2 = \rho = \frac{2}{E_b/N_J} \tag{2.262}$$

provided that

$$E_b/N_J \geq 2. \tag{2.263}$$

If this condition is not satisfied, then the broadband jammer is the worst noise jammer.

2.5.2 Diversity and Coding

Next assume that a coded sequence of FH/MSFK chips is transmitted over the slowly varying Rayleigh fading uniform channel where the noise jammer power is distributed according to J. Each chip is hopped with uniform probability across the spread-spectrum frequency band and assume all chip signals have independent Rayleigh fade envelopes with the same statistics.

If y_c and y_s are the cosine and sine components of the received signal at a carrier frequency in the q-th sub-band, then y_c and y_s are independent Gaussian random variables with joint probability density function

$$p_1(y_c, y_s) = \frac{1}{\pi(\overline{E} + N_q)} e^{-(y_c^2 + y_s^2)/(\overline{E} + N_q)} \tag{2.264}$$

if a chip signal is present at this carrier frequency and

$$p_0(y_c, y_s) = \frac{1}{\pi N_q} e^{-(y_c^2 + y_s^2)/N_q} \tag{2.265}$$

when only jammer noise is present. Here N_q is given by (2.241). The ratio of these probabilities is

$$\frac{p_1(y_c, y_s)}{p_0(y_c, y_s)} = \frac{N_q}{\overline{E} + N_q} \exp\left\{ \frac{\overline{E}/N_q}{\overline{E} + N_q}(y_c^2 + y_s^2) \right\}. \tag{2.266}$$

With diversity of order m where the m chips fall into frequency sub-bands indexed by

$$q = (q_1, q_2, \ldots, q_m), \tag{2.267}$$

and following the same discussion as in Section 2.4.2, the ML decision rule is to choose symbol l that maximizes

$$e_l = \sum_{k=1}^{m} \frac{\overline{E}}{N_{q_k}(\overline{E} + N_{q_k})} e_{lk} \tag{2.268}$$

where e_{lk} is the energy corresponding to the k-th chip of the l-th symbol. Defining the k-th channel output as the vector of M chip energy detector outputs

$$e_k = (e_{1k}, e_{2k}, \ldots, e_{Mk}) \tag{2.269}$$

then the ML metric is

$$m(e_k, l | J_{q_k}) = \frac{\bar{E}}{N_{q_k}(\bar{E} + N_{q_k})} e_{lk}. \tag{2.270}$$

This is the ideal receiver that assumes knowledge of J and corresponds to the case of known jammer state information.

For binary communication ($M = 2$), follow the same discussion leading to (2.226) except now assume the sequence of sub-bands q is given. Thus the Chernoff bit error bound using the ML metric (2.270) is the Bhattacharyya bound given by

$$P_b(q) \leq \frac{1}{2} \prod_{k=1}^{m} \left[\frac{4m(m + (\bar{E}_b/N_{q_k}))}{(2m + (\bar{E}_b/N_{q_k}))^2} \right]. \tag{2.271}$$

Assuming each hop is independent of other hops, then averaging (2.271) over the hop sequence gives the m diversity bit error bound

$$P_b \leq \frac{1}{2} \left\{ \sum_{q=1}^{Q} \frac{1}{Q} \left[\frac{4m(m + (\bar{E}_b/N_q))}{(2m + (\bar{E}_b/N_q))^2} \right] \right\}^m. \tag{2.272}$$

For arbitrary $M = 2^K$ the corresponding bit error bound is

$$P_b \leq 2^{K-2} \left\{ \sum_{q=1}^{Q} \frac{1}{Q} \left[\frac{4m(m + K(\bar{E}_b/N_q))}{(2m + K(\bar{E}_b/N_q))^2} \right] \right\}^m \tag{2.273}$$

For arbitrary coded FH/MFSK signals,

$$P_b \leq G(D(J)) \tag{2.274}$$

where

$$D(J) = \sum_{q=1}^{Q} \frac{1}{Q} \left[\frac{4(1 + (\bar{E}/N_q))}{(2 + (\bar{E}/N_q))^2} \right]$$

$$= \sum_{q=1}^{Q} \frac{1}{Q} \left[\frac{4J_q Q(J_q Q + \bar{E} W_{ss})}{(2J_q Q + \bar{E} W_{ss})^2} \right] \tag{2.275}$$

and $\bar{E}$ is the average chip energy. For a code rate of R bits per chip, $\bar{E} = R E_b$. With diversity m and $M = 2^K$ this is the special case of $R = K/m$ and $G(D) = 2^{K-2} D^m$ given in (2.274). The union Chernoff bound given by (2.274) and (2.275) applies for the ML metric that assume jammer state information J, namely, the jammer noise power distribution.

The jammer noise power distribution that maximizes the coded bit error bound (2.274) is the one that maximizes $D(J)$ given in (2.275). Applying the Kuhn-Tucker Theorem to maximizing $D(J)$ under equality constraint (2.242) results in the necessary and sufficient conditions

$$\frac{4\overline{E}^2 W_{ss}^2}{\left(2J_q^* Q + \overline{E} W_{ss}\right)^3} \leq \lambda \quad \text{for all } q \qquad (2.276)$$

with equality for values of q where $J_q^* > 0$. Here the function $D(J)$ is concave in J so the maximizing $J = J^*$ is unique. Again we see that the worst jammer noise distribution is the broadband noise jammer with

$$J_q^* = \frac{J}{Q} \quad \text{for all } q. \qquad (2.277)$$

For this worst distribution

$$D(J^*) = \frac{4\left(1 + (\overline{E}/N_J)\right)}{\left(2 + (\overline{E}/N_J)\right)^2} \qquad (2.278)$$

which is the result shown earlier in (2.234).

Again for a slowly varying Rayleigh fading uniform channel, the worst jammer noise power distribution is uniform across the band which is the broadband jammer. This is exactly true for uncoded FH/MFSK signals and was shown here for the union Chernoff bound for coded FH/MFSK signals with ideal ML decoders having jammer state information. For coded systems using other metrics, similar union Chernoff bounds can be obtained in which the broadband jammer is not necessarily the worst noise jammer.

Consider the coded FH/MFSK signal using the energy metric

$$m(e, x) = e_x \qquad (2.279)$$

which assumes no knowledge of J, the jammer state. This is known to be the optimum metric when there is broadband noise. Since the broadband noise jammer was the worst case in the uncoded case, we might expect good performance with this metric. The coding parameter obtained using the Chernoff bound has the form

$$D(J) = \min_{\lambda \geq 0} \sum_{q=1}^{Q} \frac{1}{Q} E\left\{ e^{\lambda[e_{l'} - e_l]} | l, q \right\}$$

$$= \min_{\lambda \geq 0} \sum_{q=1}^{Q} \frac{1}{Q} E\left\{ e^{\lambda e_{l'}} | l, q \right\} E\left\{ e^{-\lambda e_l} | l, q \right\} \qquad (2.280)$$

for $l' \neq l$. Here

$$E\left\{ e^{\lambda e_{l'}} | l, q \right\} = \left[\int_{-\infty}^{\infty} e^{\lambda x^2} \frac{1}{\sqrt{\pi N_q}} e^{-x^2/N_q} \, dx \right]^2$$

$$= \frac{1}{1 - \lambda N_q}; \quad 0 \leq \lambda \leq \frac{1}{N_q} \qquad (2.281)$$

and

$$E\{e^{-\lambda e_l}|l,q\} = \left[\int_{-\infty}^{\infty} e^{-\lambda x^2} \frac{1}{\sqrt{\pi(\bar{E}+N_q)}} e^{-x^2/(\bar{E}+N_q)} dx\right]^2$$

$$= \frac{1}{1+\lambda(\bar{E}+N_q)}. \qquad (2.282)$$

Thus,

$$D(J) = \min_{0\leq\lambda\leq\lambda^*} \sum_{q=1}^{Q} \frac{1}{Q}\left[\frac{1}{(1-\lambda N_q)(1+\lambda\bar{E}+\lambda N_q)}\right] \qquad (2.283)$$

where

$$\lambda^* = \min_q \frac{1}{N_q}. \qquad (2.284)$$

Consider the two-level partial-band jammer where

$$N_1 = QJ/W_{ss}$$
$$= QN_J \qquad (2.285)$$

and

$$N_q = 0 \qquad q = 2, 3, \ldots, Q.$$

Then

$$D(J) = \min_{0\leq\lambda\leq\frac{1}{QN_J}} \left\{\frac{1}{Q(1-\lambda QN_J)(1+\lambda\bar{E}+\lambda QN_J)} + \frac{Q-1}{Q(1+\lambda\bar{E})}\right\}.$$

$$(2.286)$$

Note that as $Q \to \infty$, we have $\lambda \to 0$ and thus

$$D(J) \underset{Q\to\infty}{\to} 1. \qquad (2.287)$$

Since in general $D(J) \leq 1$, this shows that the asymptotically worst noise jammer power distribution is the two-level partial-band jammer where the fraction of the sub-band that is jammed approaches zero.

The above example points out the important fact that in coded systems the choice of metric is crucial even in Rayleigh fading channels where the broadband jammer is worst in the uncoded case. Here, without jammer state information, the usual energy metric which is optimum for a broadband noise jammer case gives very poor performance for the worst jammer noise power distribution.

2.6 WORST NOISE JAMMER DISTRIBUTION—SLOW FADING NON-UNIFORM CHANNEL

Based on the work of Avidor [25] a consideration now follows of the non-uniform fading channel where there is Rayleigh fading that can be different for each of the Q sub-bands. In particular, let the average energy of an FH/MFSK symbol transmitted in the q-th sub-band have average energy

$$\overline{E}_q = \alpha_q \overline{E} \tag{2.288}$$

$$q = 1, 2, \ldots, Q.$$

Also assume that the noise jammer's power is non-uniformly attenuated across the band. For the q-th sub-band the jammer's single-sided noise power spectral density is constant at

$$N_q = \beta_q J_q \tag{2.289}$$

where as before the jammer power distribution $J = (J_1, J_2, \ldots, J_Q)$ satisfies the constraint given by (2.242). The channel parameters $\alpha_1, \alpha_2, \ldots, \alpha_Q$ and $\beta_1, \beta_2, \ldots, \beta_Q$ are fixed throughout this section.

Since the channel is no longer uniform, assume the frequency hopping is also non-uniform across the total spread-spectrum frequency band. Certainly if $\overline{E}_q = 0$, it does not make sense to hop into the q-th sub-band. Thus we let p_q be the probability that the FH/MFSK signal hops into the q-th sub-band and define the hopping probability vector

$$\boldsymbol{p} = (p_1, p_2, \ldots, p_Q). \tag{2.290}$$

The non-uniform fading channel described here might model the HF channel where $\alpha_1, \alpha_2, \ldots, \alpha_Q$ describe the propagation conditions across the HF band (3MHz to 30 MHz). These may be known to the transmitter and receiver by using channel sounder signals that periodically measure the channel. The jammer may also be transmitting the jammer signal over the non-uniform channel, thus resulting in propagation parameters $\beta_1, \beta_2, \ldots, \beta_Q$. The jammer has control over the noise power distribution $J = (J_1, J_2, \ldots, J_Q)$ under constraint (2.242) while the FH/MFSK anti-jam system controls the frequency hopping pattern defined by the probability vector $\boldsymbol{p} = (p_1, p_2, \ldots, p_Q)$.

In addition to the generalizations described above, assume there is receiver white Gaussian noise with single-sided spectral density N_0 across the total spread-spectrum band.

2.6.1 Uncoded

The uncoded (no diversity) FH/MFSK signal has conditional symbol error probability

$$P_q(J_q) = \sum_{k=1}^{M-1} \binom{M-1}{k} \frac{(-1)^{k+1}}{1 + k\left(1 + \dfrac{\alpha_q \overline{E}}{N_0 + \beta_q J_q}\right)} \tag{2.291}$$

when the FH/MFSK signal hops into the q-th sub-band. This is the usual symbol error probability for non-coherent Rayleigh fading channels of average signal energy $\alpha_q \overline{E}$ in additive white Gaussian noise of single-sided spectral density $N_0 + \beta_q J_q$. The symbol error probability averaged over the random hopping depends on the jammer noise power distribution $\boldsymbol{J}$ and the hopping probability vector $\boldsymbol{p}$ as follows:

$$P_s(\boldsymbol{p}, \boldsymbol{J}) = \sum_{q=1}^{Q} p_q P_q(J_q)$$

$$= \sum_{q=1}^{Q} p_q \sum_{k=1}^{M-1} \binom{M-1}{k} \frac{(-1)^{k+1}}{1 + k\left(1 + \dfrac{\alpha_q \overline{E}}{N_0 + \beta_q J_q}\right)} \qquad (2.292)$$

which is a concave function of $\boldsymbol{J}$.

Without loss of generality let

$$\alpha_1 \geq \alpha_2 \geq \cdots \geq \alpha_Q. \qquad (2.293)$$

Thus with no jammer the sub-band symbol error probabilities satisfy

$$P_1(0) \leq P_2(0) \leq \cdots \leq P_Q(0). \qquad (2.294)$$

Suppose the jammer is very weak so that even if the jammer used all its power to jam the 1st sub-band, it would still have lower symbol error probability than the other sub-bands. That is, with $J_1 = J$ and $J_q = 0$, $q = 2, 3, \ldots, Q$

$$P_1(J) \leq P_2(0) \leq \cdots \leq P_Q(0). \qquad (2.295)$$

Clearly in this case the optimum choice for the hopping probability vector is $p_1 = 1$ and $p_q = 0$, $q = 2, 3, \ldots, Q$. Using only the 1st sub-band would maximize the bit error probability.

Now define threshold τ_1 that satisfies

$$\frac{\alpha_1 \overline{E}}{N_0 + \beta_1 \tau_1} = \frac{\alpha_2 \overline{E}}{N_0} \qquad (2.296)$$

or

$$\tau_1 = \frac{N_0}{\beta_1}\left[\frac{\alpha_1}{\alpha_2} - 1\right]. \qquad (2.297)$$

Then τ_1 is the jammer power in sub-band 1 that makes it exactly as bad as sub-band 2 without jamming. It is clear that if $J > \tau_1$, then the optimum hopping probability would not have $p_1 = 1$. Now consider the general choice for $\boldsymbol{p}$ under a minimax criterion.

We assume the transmitter has no prior knowledge of the jammer noise power distribution $\boldsymbol{J}$ and so the probability vector $\boldsymbol{p}$ does not depend on $\boldsymbol{J}$. Consider here the choice of $\boldsymbol{p}$ that minimizes the maximum error probability. Specifically we choose $\boldsymbol{p} = \boldsymbol{p}^*$ to minimize

$$\max_{\boldsymbol{J}} P_s(\boldsymbol{p}, \boldsymbol{J}) \qquad (2.298)$$

under the constraint that p is a probability vector. The maximization of $P_s(p, J)$ with respect to J is under the constraints of (2.242) and J have non-negative components. Here the minimax error probability is given by

$$P_s^* = P_s(\mathbf{p}^*, J^*)$$
$$= \min_{\mathbf{p}} \max_{J} P_s(\mathbf{p}, J). \tag{2.299}$$

Necessary and sufficient conditions for (p, J) to be the minimax point (p^*, J^*) have been found and analyzed by Avidor [25]. The minimax solution corresponds to the jammer noise power distribution resulting in the sub-band symbol error probabilities

$$P_1(J_1^*) = P_2(J_2^*) = \cdots = P_L(J_L^*) < P_{L+1}(0) \leq \cdots \leq P_Q(0) \tag{2.300}$$

for some L where

$$J = \sum_{l=1}^{L} J_l^*. \tag{2.301}$$

For the minimax solution the jammer distributes its power so that the strongest sub-bands without jammer noise receive the most jammer power, thus reducing all the better sub-bands to the same level of performance. Here the effective signal-to-noise ratios for the L sub-bands are

$$\delta = \frac{\alpha_q \overline{E}}{N_0 + \beta_q J_q^*} \tag{2.302}$$

$$q = 1, 2, \ldots, L$$

and for the remaining sub-bands

$$\delta > \frac{\alpha_q \overline{E}}{N_0} \tag{2.303}$$

$$q = L + 1, L + 2, \ldots, Q.$$

Note that the minimax jammer noise power distribution is

$$J_q^* = \begin{cases} \dfrac{\alpha_q \overline{E} - \delta N_0}{\beta_q \delta}, & q = 1, 2, \ldots, L \\ 0, & q = L + 1, L + 2, \ldots, Q \end{cases} \tag{2.304}$$

and from the constraint (2.242)

$$J = \sum_{q=1}^{L} \left(\frac{\alpha_q \overline{E} - \delta N_0}{\beta_q \delta} \right)$$

$$= \frac{1}{\delta} \sum_{q=1}^{L} \left(\frac{\alpha_q}{\beta_q} \right) \overline{E} - \sum_{q=1}^{L} \left(\frac{N_0}{\beta_q} \right) \tag{2.305}$$

or

$$\delta = \frac{\sum_{q=1}^{L} \left(\frac{\alpha_q}{\beta_q} \right) \overline{E}}{J + \sum_{q=1}^{L} \left(\frac{N_0}{\beta_q} \right)}. \qquad (2.306)$$

Thus, in terms of the total power J the smallest L is found to satisfy (2.304) and (2.306). Only these L sub-bands are jammed by the noise jammer.

The minimax choice of p^* has the form

$$p_q^* = \begin{cases} \dfrac{\alpha_q/\beta_q}{\sum_{i=1}^{L} (\alpha_i/\beta_i)}, & q = 1, 2, \ldots, L \\[4mm] 0, & q = L+1, \ldots, Q. \end{cases} \qquad (2.307)$$

with resulting minimax bit error probability

$$P_b^* = \frac{\frac{1}{2}M}{M-1} P_s(p^*, J^*)$$

$$= \frac{\frac{1}{2}M}{M-1} \sum_{k=1}^{M-1} \binom{M-1}{k} \frac{(-1)^{k+1}}{1 + k(1+\delta)} \qquad (2.308)$$

where δ and J^* satisfies (2.302) and (2.304) and L is the smallest integer satisfying (2.306).

2.6.2 Diversity and Coding

Following the discussion leading to (2.270), with coding and diversity the soft decision ML metric is

$$m(e, x|J_q) = \frac{\alpha_q \overline{E} e_x}{(N_0 + \beta_q J_q)(\alpha_q \overline{E} + N_0 + \beta_q J_q)} \qquad (2.309)$$

when the chip hops into the q-th sub-band. This ideal receiver assumes knowledge of $\alpha_1, \alpha_2, \ldots, \alpha_Q$; $\beta_1, \beta_2, \ldots, \beta_Q$; and J, the jammer noise power distribution. Here $\alpha_q \overline{E}$ is the average chip energy at the receiver when the signal hops into the q-th sub-band. In this case the coding parameter is

$$D(p, J) = \sum_{q=1}^{Q} p_q \left[\frac{4 \left(1 + \dfrac{\alpha_q \overline{E}}{N_0 + \beta_q J_q} \right)}{\left(2 + \dfrac{\alpha_q \overline{E}}{N_0 + \beta_q J_q} \right)^2} \right] \qquad (2.310)$$

where any coded bit error bound has the form

$$P_b \le G(D(p, J)) \qquad (2.311)$$

and $G(\cdot)$ is determined by the code used. This follows from generalizing the discussion leading to (2.275).

For the coded FH/MFSK system with the ML metric, the minimax point (p^*, J^*) satisfies

$$D^* = D(p^*, J^*)$$
$$= \min_{p} \max_{J} D(p, J). \tag{2.312}$$

This is similar to the uncoded case and in fact, the minimax point (p^*, J^*) is exactly the same. This is because $D(p, J)$ is a concave function of J and the unique minimax point gives the condition

$$D^* = \frac{4\left(1 + \dfrac{\alpha_q \overline{E}}{N_0 + \beta_q J_q^*}\right)}{\left(2 + \dfrac{\alpha_q \overline{E}}{N_0 + \beta_q J_q^*}\right)^2} \tag{2.313}$$

$$q = 1, 2, \ldots, L$$

and

$$D^* < \frac{4\left(1 + \dfrac{\alpha_q \overline{E}}{N_0}\right)}{\left(2 + \dfrac{\alpha_q \overline{E}}{N_0}\right)^2} \tag{2.314}$$

$$q = L + 1, L + 2, \ldots, Q.$$

This implies J^* satisfies (2.304) where δ is given by (2.302) and L is the smallest integer where (2.306) is true.

For the hard decision case assume a hard M-ary decision is made for each FH/MFSK chip. If the FH/MFSK signal is hopped into the q-th sub-band, then the coding channel has M inputs and M outputs with conditional channel probabilities

$$p(y|x, q) = \begin{cases} 1 - P_q(J_q), & y = x \\ P_q(J_q)/(M - 1), & y \neq x \end{cases} \tag{2.315}$$

where $P_q(J_q)$ is given by (2.291) and $x, y \in \{1, 2, \ldots, M\}$. For a sequence of symbols x and y and hop sequence q we have the conditional probability

$$p(y|x, q) = \prod_k p(y_k|x_k, q_k)$$

$$= \prod_k \left[1 - P_{q_k}(J_{q_k})\right]^{1 - w(y_k, x_k)} \left[P_{q_k}(J_{q_k})/(M - 1)\right]^{w(y_k, x_k)}$$

$$= \prod_k \left[\frac{P_{q_k}(J_{q_k})}{(M - 1)\left(1 - P_{q_k}(J_{q_k})\right)}\right]^{w(y_k, x_k)} \left(1 - P_{q_k}(J_{q_k})\right). \tag{2.316}$$

Here the ML hard decision metric is

$$m(y, x|J_q) = w(y, x)\log\left[\frac{P_q(J_q)}{(M-1)(1 - P_q(J_q))}\right]. \quad (2.317)$$

With this metric the coding parameter is

$$D(p, J) = \sum_{q=1}^{Q} p_q\left[2\sqrt{\frac{(1 - P_q(J_q))P_q(J_q)}{M-1}} + \frac{M-2}{M-1}P_q(J_q)\right].$$

$$(2.318)$$

Again $D(p, J)$ is a concave function of J and thus there is a unique minimax solution where the minimax point (p^*, J^*) satisfies (2.302), (2.303), and (2.304) where L is the minimum integer where (2.306) is satisfied.

For the special case of the uniform channel where

$$\alpha_q = 1 \quad (2.319)$$

and

$$\beta_q = 1$$
$$q = 1, 2, \ldots, Q$$

for the soft decision ML metric, broadband jamming is the worst case when jammer state information, J, is assumed. This was shown in Section 2.5.2 and also follows directly from the general results here. Also for the hard decision ML metric, the broadband jammer is the worst case noise jammer. In Section 2.5.2 we had shown that for the energy metric without jammer state information, the broadband jammer was not the worst case for the uniform channel. We shall examine the hard decision channel with no jammer state information after deriving the general coding parameter expression for arbitrary metrics.

For more general metrics the general Chernoff bound can be used to compute the coding parameter D. To examine metrics that assume no knowledge of J, consider the jammer state information or metrics with various quantizations on the chip energy out of each detector. Here assuming the most general metric form, $m(y, x|J_q)$, the usual pairwise error Chernoff bound is

$$P(x \to \hat{x}|q) = \Pr\left\{ \sum_k m(y_k, \hat{x}_k|J_{q_k}) \geq \sum_k m(y_k, x_k|J_{q_k}) \Big| x, q \right\}$$

$$\leq E\left\{ \exp\lambda \left(\sum_k \left[m(y_k, \hat{x}_k|J_{q_k}) - m(y_k, x_k|J_{q_k}) \right] \right) \Big| x, q \right\}$$

$$= \prod_k E\left\{ \exp\lambda \left[m(y_k, \hat{x}_k|J_{q_k}) - m(y_k, x_k|J_{q_k}) \right] \Big| x_k, q_k \right\}$$

$$(2.320)$$

where $E\{\cdot\}$ is the expectation over all channel statistics including jammer noise when x is sent and q is the hopping sequence. Averaging this over the hopping sequence according to probability p and minimizing over the Chernoff bound parameter, λ, gives the final pairwise error bound

$$P(x \to \hat{x}) \leq [D(p, J)]^{w(x, \hat{x})} \tag{2.321}$$

where

$$D(p, J) = \min_{\lambda \geq 0} \sum_{q=1}^{Q} p_q E\left\{\exp \lambda \left[m(y, \hat{x}|J_q) - m(y, x|J_q)\right]\middle| x, q\right\} \tag{2.322}$$

when $\hat{x} \neq x$. The minimax bound is

$$P(x \to \hat{x}) \leq [D^*]^{w(x, \hat{x})} \tag{2.323}$$

where

$$D^* = \min_{p} \max_{J} D(p, J). \tag{2.324}$$

The parameter D^* can be computed for non-fading channels as well as fading channels with any fading statistics. These can all be included in the general analysis discussed in Chapter 4, Volume I.

As an example consider the hard decision metric

$$m(y, x) = -w(y, x) \tag{2.325}$$

which assumes no knowledge of the jammer state J. Here when $\hat{x} \neq x$

$$\begin{aligned}
E\{\exp \lambda \left[m(y, \hat{x}) - m(y, x)\right]|x, q\} \\
= E\{\exp \lambda \left[w(y, x) - w(y, \hat{x})\right]|x, q\} \\
= \left(1 - P_q(J_q)\right)e^{-\lambda} + \frac{P_q(J_q)}{M-1}e^{\lambda} + \left(\frac{M-2}{M-1}\right)P_q(J_q) \tag{2.326}
\end{aligned}$$

and

$$\begin{aligned}
D(p, J) &= \min_{\lambda \geq 0} \sum_{q=1}^{Q} p_q\left|\left(1 - P_q(J_q)\right)e^{-\lambda} + \frac{P_q(J_q)}{M-1}e^{\lambda} + \left(\frac{M-2}{M-1}\right)P_q(J_q)\right| \\
&= \min_{\lambda \geq 0}\left[(1 - \bar{\varepsilon})e^{-\lambda} + \frac{\bar{\varepsilon}}{M-1}e^{\lambda} + \left(\frac{M-2}{M-1}\right)\bar{\varepsilon}\right] \\
&= 2\sqrt{\frac{(1 - \bar{\varepsilon})\bar{\varepsilon}}{M-1}} + \left(\frac{M-2}{M-1}\right)\bar{\varepsilon} \tag{2.327}
\end{aligned}$$

where

$$\bar{\varepsilon} = \sum_{q=1}^{Q} p_q P_q(J_q). \tag{2.328}$$

Note that for the hard decision metric (2.325) where no jammer state information is used, the minimax point (p^*, J^*) for $D(p, J)$ is also the minimax point for $\bar{\varepsilon}$ given by (2.328). But $\bar{\varepsilon}$ is exactly the uncoded symbol error probability with the same minimax point as specified by (2.302), (2.304), and (2.306). For the uniform channel specified by (2.319) the worst noise jammer is again the broadband jammer. Thus with the hard decision uniform channel with no jammer state information the worst noise jammer is again the broadband jammer.

2.7 OTHER CODING METRICS

In this section we continue with the general slow fading non-uniform channel where there are Q sub-bands, jammer noise power distribution given by J, and hopping probability vector p where the channel parameters are $\alpha = (\alpha_1, \alpha_2, \ldots, \alpha_Q)$ and $\beta = (\beta_1, \beta_2, \ldots, \beta_Q)$. Each coded FH/MFSK chip signal has a Rayleigh fading envelope which is independent from hop to hop and there is white Gaussian receiver noise with parameter N_0. Up to this point the two extreme types of coding metrics, soft decision and hard decision, were examined. In this section other coding metrics are now considered. The basic approach illustrated here and the various metrics considered can be applied to other types of channels and jammers such as non-fading channels with multitone jamming.

Recall that each time an M-ary coded symbol $x \in \{1, 2, \ldots, M\}$ is modulated as an FH/MFSK chip signal the receiver dehops the received signal and samples the outputs of M energy detectors. Denote the M-sampled energy detector outputs as $e = (e_1, e_2, \ldots, e_M)$. If the signal hopped into the q-th sub-band, then ε has conditional probability density,

$$p(e|x, q) = \prod_{m=1}^{M} p(e_m|x, q) \qquad (2.329)$$

where

$$p(e_m|x, q) = \begin{cases} \dfrac{1}{\alpha_q \bar{E} + N_0 + \beta_q J_q} e^{-e_m/(\alpha_q \bar{E} + N_0 + \beta_q J_q)}, & m = x \\[3mm] \dfrac{1}{N_0 + \beta_q J_q} e^{-e_m/(N_0 + \beta_q J_q)}, & m \neq x \end{cases}$$

$$(2.330)$$

Thus the energy detector outputs are independent central chi-square random variables of order 2 and their square root values are Rayleigh random variables.

When the M energy detector outputs are taken as the channel output, then the maximum-likelihood metric is given as $m(e, x|J_q)$ defined by

(2.309). This is the soft decision ML metric which assumes knowledge of jammer state information represented by J, the jammer noise power distribution. The suboptimum soft decision energy metric without jammer state information is denoted $m(e, x)$ and is given by (2.279).

In practical applications coded systems require a receiver that uses the coding metric in a digital processing algorithm or decoding process. This means the coding metric must be some quantized function of the M-sampled energy detector outputs. Generally, smaller number of bits used to represent metric values means faster decoding speeds with less memory requirements.[7] Consider the channel output in the form

$$y = f(e) \qquad (2.331)$$

where $f(\cdot)$ is some function of the sampled M energy detector outputs. The simplest form of this function is the hard decision output defined by

$$y = m \qquad (2.332)$$

if

$$e_m \geq e_k \quad \text{all } k \neq m.$$

For this case the channel conditional probability $p(y|x, q)$ is given by (2.315) where $P_q(J_q)$ is the symbol error probability given by (2.291).

For the hard decision channel the ML metric $m(y, x|J_q)$ is given by (2.317). This assumes jammer state information J is known. The suboptimum hard decision metric without jammer station information is given by (2.325), which can be represented by a single binary digit.

The above soft decision and hard decision channels are special cases of the demodulator-to-decoder interface function $f(\cdot)$. This is illustrated in Figure 2.117, where the dotted line shows the equivalent coding channel with input x, output y, and conditional probability $p(y|x, q)$. For a given interface function $y = f(e)$, the conditional probabilities can be computed.

Once a coding channel is defined by the input symbols, output symbols, and conditional probabilities, a coding metric $m(y, x|J_q)$ is chosen. For the given channel and chosen coding metric, the coding parameter is

$$D(p, J) = \min_{\lambda \geq 0} \sum_{q=1}^{Q} p_q E\left\{ e^{\lambda[m(y, \hat{x}|J_q) - m(y, x|J_q)]} \big| x, q \right\} \qquad (2.333)$$

for $\hat{x} \neq x$. Here the expectation $E\{\cdot\}$ is over the random variable y given x, $\hat{x}$, and q. In terms of a given J and p the cutoff rate is then

$$R_0(p, J) = \log_2 M - \log_2\left[1 + (M - 1)D(p, J)\right] \qquad (2.334)$$

in bits per FH/MFSK chip. When an ML metric of the form[8]

$$m(y, x|J_q) = a \log p(y|x, q) + b \qquad (2.335)$$

---

[7] Memory requirements in both the decoder and the deinterleaver.

[8] This is a maximum likelihood metric for the given function $y = f(e)$.

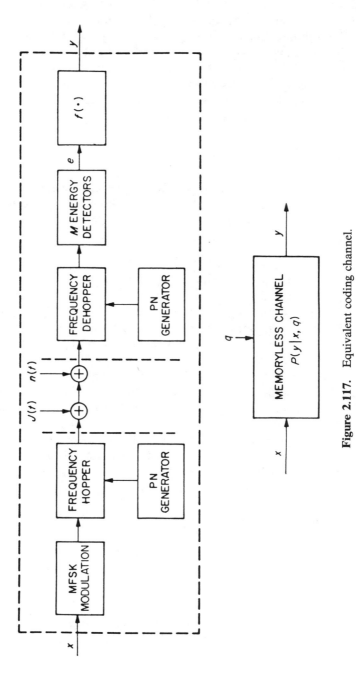

Figure 2.117. Equivalent coding channel.

for any $a > 0$ and b is used, the expression for $D(p, J)$ becomes

$$D(p, J) = \sum_{q=1}^{Q} p_q \int \sqrt{p(y|\hat{x}, q) p(y|x, q)} \, dy \qquad (2.336)$$

for $\hat{x} \neq x$ and where the integral may be a summation when y is a discrete random variable.

We now investigate various quantized channels defined by (2.331) and consider examples of coding metrics associated with these equivalent coding channels.

2.7.1 Energy Quantizer

Consider the function

$$\begin{aligned} y &= f(e) \\ &= (y_1, y_2, \ldots, y_M) \end{aligned} \qquad (2.337)$$

where

$$y_m = \begin{cases} 0, & 0 \le \sqrt{e_m} < v \\ 1, & v \le \sqrt{e_m} < 2v \\ 2, & 2v \le \sqrt{e_m} < 3v \\ 3, & 3v \le \sqrt{e_m} \end{cases} \qquad (2.338)$$

$$m = 1, 2, \ldots, M.$$

This function just quantizes each sampled energy detector output into one of four levels (two-bit quantizer) where the quantization interval is of length v.

Suppose x is the transmitted symbol. Then define probabilities

$$p_0(k|v, q) = \Pr\left\{ kv \le \sqrt{e_m} < (k+1)v \big| m \neq x, q \right\}$$

$$p_1(k|v, q) = \Pr\left\{ kv \le \sqrt{e_m} < (k+1)v \big| m = x, q \right\} \qquad (2.339)$$

$$k = 0, 1, 2$$

and

$$p_0(3|v, q) = \Pr\left\{ 3v \le \sqrt{e_m} \big| m \neq x, q \right\}$$

$$p_1(3|v, q) = \Pr\left\{ 3v \le \sqrt{e_m} \big| m = x, q \right\}. \qquad (2.340)$$

Using the probability density for e_m given by (2.330)

$$p_i(k|v, q) = e^{-k^2 v^2 / 2\sigma_i^2} - e^{-(k+1)^2 v^2 / 2\sigma_i^2}$$

$$p_i(3|v, q) = e^{-9v^2 / 2\sigma_i^2} \qquad (2.341)$$

$$i = 0, 1; \; k = 0, 1, 2$$

where

$$2\sigma_i^2 = \begin{cases} N_0 + \beta_q J_q, & i = 0 \\ \alpha_q \overline{E} + N_0 + \beta_q J_q, & i = 1. \end{cases} \qquad (2.342)$$

The coding channel conditional probability density is then

$$\begin{aligned} p(y|v,q) &= \prod_{m=1}^{M} p(y_m|v,q) \\ &= \prod_{m=1}^{M} p_0(y_m|v,q) \cdot \frac{p_1(y_x|v,q)}{p_0(y_x|v,q)} \\ &= F(y|v,q) \frac{p_1(y_x|v,q)}{p_0(y_x|v,q)} \qquad (2.343) \end{aligned}$$

where

$$F(y|v,q) = \prod_{m=1}^{M} p_0(y_m|v,q) \qquad (2.344)$$

is independent of the channel input symbol x.

For this channel the maximum-likelihood metric is

$$m(y, x|J_q) = \log p_1(y_x|v,q) - \log p_0(y_x|v,q) \qquad (2.345)$$

which will give a coding parameter

$$D(p, J) = \sum_{q=1}^{Q} p_q \sum_{k=0}^{3} \sum_{j=0}^{3} \sqrt{p_0(k|v,q) p_0(j|v,q) p_1(k|v,q) p_1(j|v,q)} . \qquad (2.346)$$

For the uniform channel with no receiver noise where parameters α and β are given by (2.319),

$$J_q = \frac{J}{Q}$$

and

$$p_q = \frac{1}{Q}$$

$$q = 1, 2, \ldots, Q \qquad (2.347)$$

the choice of v that minimizes this coding parameter can be found numerically. Figure 2.118 shows the minimizing value of $\bar{v} = v/N_J$ versus $\overline{E}/N_J$ along with corresponding values of the cutoff rate R_0 for $M = 2$ defined by (2.334).

Instead of the ML metric (2.335) for this coding channel, a simpler metric is to approximate the soft decision energy metric with no jammer state information by choosing metric

$$m(y, x) = y_x. \qquad (2.348)$$

This uses no jammer state information and has coding parameter given by (2.333), which becomes

$$D(p, J) = \min_{\lambda \geq 0} \sum_{q=1}^{Q} p_q E\{ e^{\lambda[y_{\hat{x}} - y_x]} | x, q \}$$

$$= \min_{\lambda \geq 0} \sum_{q=1}^{Q} p_q \sum_{k=0}^{3} \sum_{j=0}^{3} e^{k\lambda} e^{-j\lambda} p_0(k|v, q) p_1(j|v, q)$$

$$= \min_{\lambda \geq 0} \sum_{q=1}^{Q} p_q \left(\sum_{k=0}^{3} e^{k\lambda} p_0(k|v, q) \right) \left(\sum_{j=0}^{3} e^{-j\lambda} p_1(j|v, q) \right).$$

$$(2.349)$$

This, too, can be minimized over $v \geq 0$ and examined for special cases such as the uniform channel. For the uniform channel with large values of $\bar{E}/N_J$ we expect similar results to the ML metric since the two metrics are approximately equal. This follows from the approximation

$$p_i(k|v, q) \cong e^{-k^2 v^2 / 2\sigma_i^2} \qquad (2.350)$$

for large values of signal-to-noise ratios and the optimized choice of $\bar{v} = v/N_J$.

The difficulty with the energy quantization presented here is that a good choice of $\bar{v}$ depends on channel measurements which may be difficult to accurately obtain in jamming and fading channels.

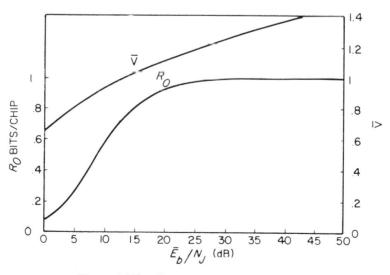

Figure 2.118. Energy quantizer parameters.

2.7.2 Hard Decision with One Bit Quality Measure

The demodulator-to-decoder interface is used to reduce the number of bits representing each channel output. The fewest bits correspond to the hard decision interface function (2.332). A practical compromise is to pass over this interface, a binary vector of low dimension. Viterbi [27] noted that the first (most significant) bits of this vector will correspond to the demodulator's hard decision while subsequent bits provide information on the quality of the channel which require some modification in the conventional demodulator design. He concentrated on adding one quality bit to the modulator's hard decision. This metric is presented here.

Define a modified hard decision demodulator-to-decoder interface function as follows:

$$y = \begin{cases} (m, 1) & \text{if } e_m \geq \gamma \max_{k \neq m} e_k \\ (m, 0), & \text{if } \gamma \max_{k \neq m} e_k > e_m \geq \max_{k \neq m} e_k \end{cases} \quad (2.351)$$

for some parameter $\gamma \geq 1$. This is the hard decision function with a quality bit where "1" indicates a "good" channel while a "0" indicates a "poor" channel. For FH/MFSK this results in a coding channel with M inputs and $2M$ outputs. Figure 2.119 illustrates this for $M = 4$. Now what remains is to determine the coding channel conditional probabilities.

Note that output $y = (1, 1)$ is a correct decision if $x = 1$. This occurs when

$$e_1 \geq \gamma e_m \qquad m = 2, 3, \dots, M. \quad (2.352)$$

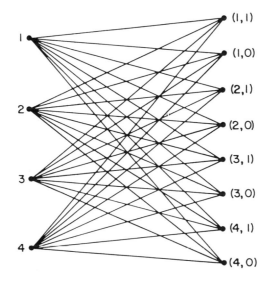

Figure 2.119. FH/4FSK coding channel.

The channel conditional probability is

$$p((1,1)|1) = \Pr\left\{ e_1 \geq \gamma \max_{m \neq 1} e_m \Big| x = 1 \right\}$$

$$= \int_0^\infty \Pr\left\{ \max_{m \neq 1} e_m \leq \frac{z}{\gamma} \Big| x = 1 \right\} p_1(z|q) \, dz \qquad (2.353)$$

where

$$p_i(z|q) = \frac{1}{2\sigma_i^2} e^{-z/2\sigma_i^2}, \qquad z \geq 0$$

$$i = 0, 1 \qquad (2.354)$$

and

$$2\sigma_i^2 = \begin{cases} \alpha_q \overline{E} + N_0 + \beta_q J_q, & i = 1 \\ N_0 + \beta_q J_q, & i = 0. \end{cases} \qquad (2.355)$$

Since

$$\Pr\left\{ \max_{m \neq 1} e_m \leq \frac{z}{\gamma} \Big| x = 1 \right\}$$

$$= \prod_{m=2}^{M} \Pr\left\{ e_m \leq \frac{z}{\gamma} \Big| x = 1 \right\}$$

$$= \prod_{m=2}^{M} \int_0^{z/\gamma} p_0(w|q) \, dw$$

$$= \left[1 - e^{-z/2\gamma\sigma_0^2} \right]^{M-1}$$

$$= \sum_{k=0}^{M-1} \binom{M-1}{k} (-1)^k e^{-kz/2\gamma\sigma_0^2}, \qquad (2.356)$$

the conditional probability is

$$p((1,1)|1) = \sum_{k=0}^{M-1} \binom{M-1}{k} (-1)^k \int_0^\infty e^{-kz/2\gamma\sigma_0^2} \frac{1}{2\sigma_1^2} e^{-z/2\sigma_1^2} \, dz$$

$$= \sum_{k=0}^{M-1} \binom{M-1}{k} \frac{(-1)^k}{1 + k\dfrac{\sigma_1^2}{\gamma\sigma_0^2}}. \qquad (2.357)$$

Defining the parameterized probabilities

$$P(a, S, M) = \begin{cases} \displaystyle\sum_{k=1}^{M-1} \binom{M-1}{k} \frac{(-1)^{k+1}}{a + kS}, & M \geq 2 \\ 0, & \text{otherwise} \end{cases} \qquad (2.358)$$

where, recall from (2.291) the hard decision symbol error probability is

$$\varepsilon = P\left(1, \frac{\sigma_1^2}{\sigma_0^2}, M \right), \qquad (2.359)$$

we have

$$p((1,1)|1) = 1 - P\left(1, \frac{\sigma_1^2}{\gamma\sigma_0^2}, M\right). \qquad (2.360)$$

Next note that

$$p((1,0)|1) = \Pr\left\{ \max_{m \neq 1} e_m \leq e_1 \leq \gamma \max_{m \neq 1} e_m \middle| x = 1 \right\}$$

$$= \Pr\left\{ e_1 \geq \max_{m \neq 1} e_m \middle| x = 1 \right\} - \Pr\left\{ e_1 \geq \gamma \max_{m \neq 1} e_m \middle| x = 1 \right\}$$

$$= P\left(1, \frac{\sigma_1^2}{\gamma\sigma_0^2}, M\right) - P\left(1, \frac{\sigma_1^2}{\sigma_0^2}, M\right). \qquad (2.361)$$

Similarly when errors occur

$$p((2,1)|1) = \Pr\left\{ e_2 \geq \gamma \max_{m \neq 2} e_m \middle| x = 1 \right\}$$

$$= \int_0^\infty \Pr\left\{ \max_{m \neq 2} e_m \leq \frac{z}{\gamma} \middle| x = 1 \right\} p_0(z|q)\, dz \qquad (2.362)$$

where now

$$\Pr\left\{ \max_{m \neq 2} e_m \leq \frac{z}{\gamma} \middle| x = 1 \right\}$$

$$= \Pr\left\{ e_1 \leq \frac{z}{\gamma} \middle| x = 1 \right\} \prod_{m=3}^M \Pr\left\{ e_m \leq \frac{z}{\gamma} \middle| x = 1 \right\}$$

$$= \left(1 - e^{-z/2\gamma\sigma_1^2}\right)\left[1 - e^{-z/2\gamma\sigma_0^2}\right]^{M-2}$$

$$= \left(1 - e^{-z/2\gamma\sigma_1^2}\right) \sum_{k=0}^{M-2} \binom{M-2}{k}(-1)^k e^{-kz/2\gamma\sigma_0^2} \qquad (2.363)$$

and thus

$$p((2,1)|1) = \sum_{k=0}^{M-2}\binom{M-2}{k}\frac{(-1)^k}{1+\dfrac{k}{\gamma}} - \sum_{k=0}^{M-2}\binom{M-2}{k}\frac{(-1)^k}{1+\dfrac{\sigma_0^2}{\gamma\sigma_1^2}+\dfrac{k}{\gamma}}$$

$$= P\left(1 + \frac{\sigma_0^2}{\gamma\sigma_1^2}, \frac{1}{\gamma}, M-1\right) - P\left(1, \frac{1}{\gamma}, M-1\right)$$

$$= \begin{cases} \dfrac{1}{1+\gamma\dfrac{\sigma_1^2}{\sigma_0^2}}, & M = 2 \\[4mm] \dfrac{1}{1+\gamma\dfrac{\sigma_1^2}{\sigma_0^2}} + P\left(1 + \dfrac{\sigma_0^2}{\gamma\sigma_1^2}, \dfrac{1}{\gamma}, M-1\right) - P\left(1, \dfrac{1}{\gamma}, M-1\right), & M > 2. \end{cases}$$

$$(2.364)$$

By similar steps we have

$$p((2,0)|1) = \Pr\left\{ \max_{m \neq 2} e_m \leq e_2 \leq \gamma \max_{m \neq 2} e_m \Big| x = 1 \right\}$$

$$= \Pr\left\{ e_2 \geq \max_{m \neq 2} e_m \Big| x = 1 \right\} - \Pr\left\{ e_2 \geq \gamma \max_{m \neq 2} e_m \Big| x = 1 \right\}$$

$$= \begin{cases} \dfrac{\sigma_0^2/\sigma_1^2}{1 + \sigma_0^2/\sigma_1^2} - \dfrac{\sigma_0^2/\sigma_1^2}{\gamma + \sigma_0^2/\sigma_1^2}, & M = 2 \\[3mm] \dfrac{\sigma_0^2/\sigma_1^2}{1 + \sigma_0^2/\sigma_1^2} - \dfrac{\sigma_0^2/\sigma_1^2}{\gamma + \sigma_0^2/\sigma_1^2} \\[3mm] + P\left(1 + \dfrac{\sigma_0^2}{\sigma_1^2}, 1, M - 1\right) - P(1, 1, M - 1) \\[3mm] + P\left(1, \dfrac{1}{\gamma}, M - 1\right) - P\left(1 + \dfrac{\sigma_0^2}{\gamma\sigma_1^2}, \dfrac{1}{\gamma}, M - 1\right), & M \geq 2 \end{cases} \qquad (2.365)$$

and by symmetry for $y = (m, b)$

$$p((m,b)|x) = \begin{cases} p((1,b)|1), & m = x \\ p((2,b)|1), & m \neq x \end{cases} \qquad (2.366)$$

$$b = 0, 1.$$

Assuming an ML metric where $m(y, x|J) = \ln p(y|x)$, then the coding parameter is

$$D(\boldsymbol{p}, \boldsymbol{J}) = \sum_{q=1}^{Q} P_q D_q(\boldsymbol{J}) \qquad (2.367)$$

where

$$D_q(\boldsymbol{J}) = \sum_{y} \sqrt{p(y|x)p(y|\hat{x})}$$

$$= 2\sqrt{p((1,1)|1)p((2,1)|1)} + 2\sqrt{p((1,0)|1)p((2,0)|1)}$$

$$+ (M - 2)p((2,1)|1) + (M - 2)p((2,0)|1). \qquad (2.368)$$

For a suboptimum metric of the general form

$$m(y, x) = \begin{cases} m_{c,1}, & y = (x, 1) \\ m_{c,0}, & y = (x, 0) \\ m_{e,1}, & y = (\hat{x}, 1), & \hat{x} \neq x \\ m_{e,0}, & y = (\hat{x}, 0), & \hat{x} \neq x \end{cases} \qquad (2.369)$$

the coding parameter is given by (2.367) with

$$D_q(J) = \min_{\lambda \geq 0} E\left\{ e^{\lambda[m(y,\hat{x})-m(y,x)]} \big| x \right\}$$

$$= \min_{\lambda \geq 0} \left\{ p\big((1,1)|1\big) e^{\lambda[m_{e,1}-m_{c,1}]} \right.$$

$$\left. + p\big((1,0)|1\big) e^{\lambda[m_{e,0}-m_{c,0}]} + p\big((2,1)|1\big) e^{\lambda[m_{c,1}-m_{e,1}]} \right.$$

$$\left. + p\big((2,0)|1\big) e^{\lambda[m_{c,0}-m_{e,0}]} \right.$$

$$\left. + (M-2)p\big((2,1)|1\big) + (M-2)p\big((2,0)|1\big) \right\}. \qquad (2.370)$$

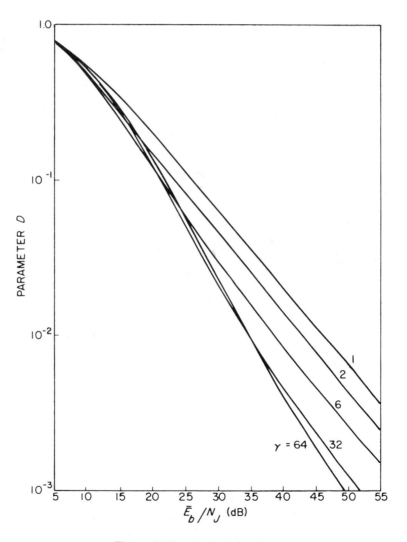

Figure 2.120. D for $M = 2$.

For the special case of the uniform channel with broadband noise jamming, the coding parameter D with various values of the parameter γ is shown in Figures 2.120 to 2.123. Shown in Figure 2.124 are the corresponding cutoff rates for $M = 2$. These figures also show the hard decision examples ($\gamma = 1$) so that the advantage of one additional quality bit can be seen.

2.7.3 List Metric

Another useful metric is based on a channel output that provides an ordered list of the M energy detector output samples. For the M energy detector

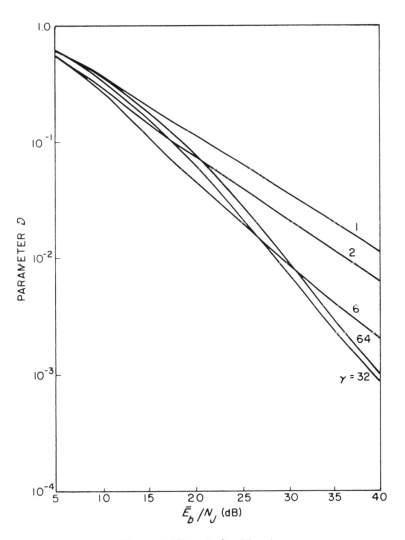

Figure 2.121. D for $M = 4$.

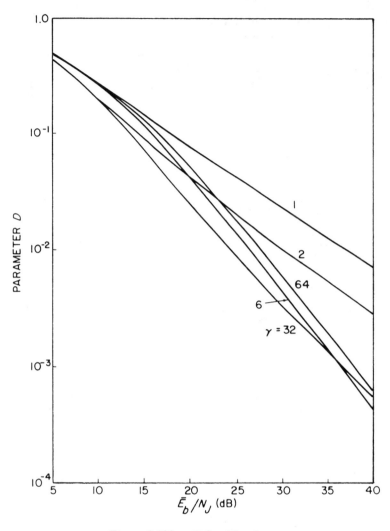

Figure 2.122. D for $M = 8$.

output samples $e = (e_1, e_2, \ldots, e_M)$, the demodulator-to-decoder interface function is the list

$$y = f(e) = (y_1, y_2, \ldots, y_M) \qquad (2.371)$$

where $y_l = k$ if e_l is the k-th largest energy term among the M energy detector output samples. For example, for $M = 4$ if $e_3 > e_2 > e_1 > e_4$ then $y = (3, 2, 1, 4)$. Here $y_1 = 3$ indicates e_1 was the third largest energy sample, $y_2 = 2$ indicates e_2 was the second largest, etc.

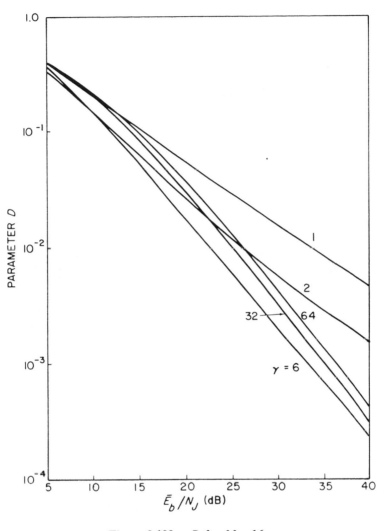

Figure 2.123. D for $M = 16$.

Note that list output can be obtained without using any AGC to maintain optimum thresholds such as in the energy quantizer. Previous work [28], [29] that considered list-of-L detection with optimum metrics has shown that this is inferior to energy quantization for AWGN channels. This, however, is not necessarily true in jamming and multiple access channels.

For the channel that outputs the ordered list of the M energy detector output samples, there are $M!$ distinct possible outputs or ordered lists. For this M input, $M!$ output coding channel we now derive expressions for the channel conditional probabilities $\{ p(y|x) \}$.

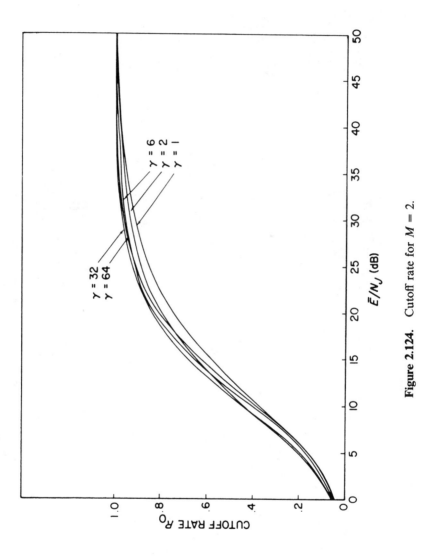

Figure 2.124. Cutoff rate for $M = 2$.

Suppose $x = 1$ and e_1 is the l-th largest energy detector output sample so that $y_1 = l$. Assume further the M energy detector output samples satisfy

$$e_2 > e_3 > ,\ldots, > e_l > e_1 > e_{l+1} > ,\ldots, > e_M \qquad (2.372)$$

so that the channel output becomes

$$y = (l, 1, 2, \ldots, l - 1, l + 1, \ldots, M). \qquad (2.373)$$

It is difficult to find an expression for $p(y|x)$ directly. By symmetry, since $\{e_m : m \neq 1\}$ are i.i.d. random variables and there are $(l - 1)!$ ways to order $e_2, e_3, \ldots, e_l$ and $(M - l)!$ ways to order $e_{l+1}, \ldots, e_M$, we have the relationship

$$p(y|1) = \frac{1}{(M - l)!(l - 1)!} \Pr\left\{ \min_{2 \leq k \leq l} e_k \geq e_1 \geq \max_{l \leq k \leq M} e_k \middle| x = 1\right\}$$

$$(2.374)$$

where y is given by (2.373). This merely states that different ordering of the largest $l - 1$ energy samples and the smallest $M - l$ energy samples all have the same conditional probability when ε_1 is the l-th largest and $x = 1$.

Next consider the probability

$$\Pr\left\{ \min_{2 \leq k \leq l} e_k \geq e_1 \geq \min_{l \leq k \leq M} e_k \middle| x = 1\right\}$$

$$= \int_0^\infty \Pr\{e_2 \geq z, \ldots, e_l \geq z, e_{l+1} \leq z, \ldots, e_M \leq z | x = 1\} p_1(z|q)\, dz$$

$$= \int_0^\infty (e^{-z/2\sigma_0^2})^{l-1} (1 - e^{-z/2\sigma_0^2})^{M-l} p_1(z|q)\, dz$$

$$= \sum_{k=0}^{M-l} \binom{M - l}{k} (-1)^k \int_0^\infty e^{-(k+l-1)z/2\sigma_0^2} p_1(z|q)\, dz$$

$$= \sum_{k=0}^{M-l} \binom{M - l}{k} \frac{(-1)^k}{1 + (k + l - 1)\dfrac{\sigma_1^2}{\sigma_0^2}}$$

$$= \frac{1}{1 + (l - 1)\dfrac{\sigma_1^2}{\sigma_0^2}} - P\left(1 + (l - 1)\frac{\sigma_1^2}{\sigma_0^2}, \frac{\sigma_1^2}{\sigma_0^2}, M - l + 1\right)[1 - \delta_{lM}].$$

$$(2.375)$$

Thus the conditional probability of the input x with output y where $y_x = l$

is given by

$$p(y|x) = \frac{1}{(M-l)!(l-1)!} \left\{ \frac{1}{1+(l-1)\frac{\sigma_1^2}{\sigma_0^2}} \right.$$

$$\left. -P\left(1+(l-1)\frac{\sigma_1^2}{\sigma_0^2}, \frac{\sigma_1^2}{\sigma_0^2}, M-l+1\right)[1-\delta_{lM}]\right\}. \quad (2.376)$$

Since there are $(M-1)!$ outputs y with $y_x = l$,

$$P_l = Pr\{y_x = l|x\}$$

$$= \binom{M-1}{l-1} \left\{ \frac{1}{1+(l-1)\frac{\sigma_1^2}{\sigma_0^2}} \right.$$

$$\left. -P\left(1+(l-1)\frac{\sigma_1^2}{\sigma_0^2}, \frac{\sigma_1^2}{\sigma_0^2}, M-l+1\right)[1-\delta_{lM}]\right\}. \quad (2.377)$$

This is the probability that the transmitted symbol has a corresponding energy detector output that is the l-th largest among the M energy detector outputs. Comparing (2.376) and (2.377)

$$p(y|x) = \frac{1}{(M-1)!}P_{y_x} \quad (2.378)$$

when y is any output with component y_x which is the place on the list that resulted for the energy detector output corresponding to input x.

We now have an expression for the conditional probabilities of the coding channel created by the demodulator-to-decoder interface function that provides an ordered list of the M energy detector output samples. Here the ML metric is clearly

$$m(y, x|J) = \log p_{y_x}. \quad (2.379)$$

This ML decision rule results in the coding parameter

$$D(p, J) = \sum_{q=1}^{Q} p_q D_q(J) \quad (2.380)$$

where

$$D_q(J) = \sum_y \sqrt{p(y|x)p(y|\hat{x})}$$

$$= \sum \frac{1}{(M-1)!}\sqrt{P_l P_{\hat{l}}}$$

$$= \sum_{k=1}^{M} \sum_{j \neq k} \frac{(M-2)!}{(M-1)!}\sqrt{P_k P_j}$$

$$= \sum_{k=1}^{M} \sum_{j=1}^{M} \frac{1}{M-1}\sqrt{P_k P_j} - \sum_{k=1}^{M} \frac{1}{M-1} P_k$$

$$= \frac{1}{M-1}\left[\sum_{k=1}^{M} \sum_{j=1}^{M} \sqrt{P_k P_j} - 1\right] \qquad (2.381)$$

with $\{P_k\}$ given by (2.377).

Next consider a suboptimum metric that assigns a value or number to each place on the list in the output. In particular let the metric be given by

$$m(y, x) = N_{y_x}. \qquad (2.382)$$

This assignment is defined by the vector

$$N = (N_1, N_2, \dots, N_M) \qquad (2.383)$$

where N_l is the assigned number placed on the energy detector output with the l-th largest value.

For the general metric defined by (2.382) the coding parameter is given by

$$D_q(J) = \min_{\lambda \geq 0} E\left\{ e^{\lambda[m(y,\hat{x}) - m(y,x)]} \big| x \right\}$$

$$= \min_{\lambda \geq 0} E\left\{ e^{\lambda[N_{y_{\hat{x}}} - N_{y_x}]} \big| x \right\}$$

$$= \min_{\lambda \geq 0} \sum_{k=1}^{M} \sum_{j \neq k} e^{\lambda[N_j - N_k]} Pr\{y_x = k, y_{\hat{x}} = j | x\} \qquad (2.384)$$

where

$$Pr\{y_x = k, y_{\hat{x}} = j | x\} = Pr\{y_{\hat{x}} = j | y_k = k, x\} Pr\{y_k = k | x\}$$

$$= \frac{1}{M-1} P_k. \qquad (2.385)$$

Thus

$$D_q(J) = \min_{\lambda \geq 0} \frac{1}{M-1} \sum_{k=1}^{M} P_k \left[\sum_{j=1}^{M} e^{\lambda [N_j - N_k]} - 1 \right]. \qquad (2.386)$$

Creighton [30] considered this list metric for the uniform channel with no background noise where there is Rayleigh fading and partial-band noise jamming. For the partial-band noise jammer that jams ρ fraction of the band, define the binary random variable Z where $Pr(Z = 1) = \rho$ and $Pr(Z = 0) = 1 - \rho$. Since the probabilities $\{P_k\}$ depend on the jammer noise level, we emphasize this by writing $P_k = P_k(Z)$ showing this dependence. Recall that each probability depended on the hopped sub-band that had a jammer noise level characterized by $\beta_q J_q$. Here there are only two levels of noise jamming characterized by the random variable Z. The general metric now has the form

$$m(y, x|Z) = N_{y_x}(Z) \qquad (2.387)$$

where for the ML metric with jammer state information (JSI) where Z is known at the receiver

$$N_{y_x}(Z) = \log P_{y_x}(Z). \qquad (2.388)$$

When the receiver does not have JSI, then we have average probability

$$\overline{P}_k = \rho P_k(1) + (1 - \rho) P_k(0) \qquad (2.389)$$

$$k = 1, 2, \ldots, M$$

and the ML metric without JSI given by

$$N_{y_x}(Z) = \log \overline{P}_{y_x}. \qquad (2.390)$$

In general an arbitrary assignment of numbers to the list values in y is possible. Here too we can consider the case with JSI where Z is known and different assignments are used for $Z = 1$ and $Z = 0$. In general, the ML metric requires knowledge of $P_k(Z)$ or $\overline{P}_k$ while having JSI requires knowledge of the random variable Z in the partial-band case and J for a more general noise power distribution.

For the ML metric with or without JSI the worst case partial-band noise jammer is again found to be the broadband noise jammer ($\rho = 1$). Figure 2.125 shows the cutoff rates for ML metrics (2.388) and (2.390) with worst case partial-band noise jamming and compares the results with no Rayleigh fading. Figures 2.126 to 2.130 compare cutoff rates for $M = 2, 4, 8, 16$, and 32 where the curves are labelled as follows:

1. Soft decision energy metric with JSI
2. Soft decision energy metric without JSI
3. ML list metric with JSI
4. ML list metric without JSI.

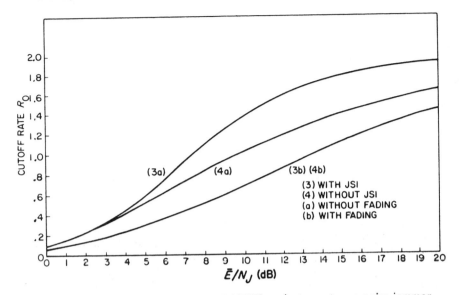

Figure 2.125. List metric R_0 for FH/4FSK against worst case noise jammer.

Although the unquantized soft decision energy metric with JSI gives better performance, its advantage over the list metrics decreases as the alphabet size increase. Furthermore, with other types of jamming or interference in the channel the list metric can outperform the soft decision energy metric. Multitone jamming and multiple access interference cases are where the list

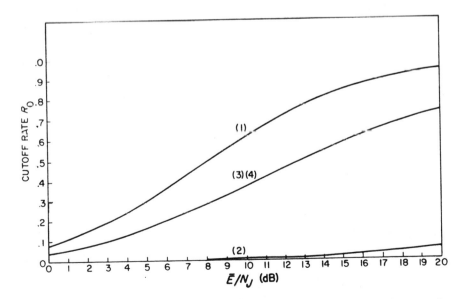

Figure 2.126. List and energy metrics R_0 for FH/8FSK against worst case noise jammer.

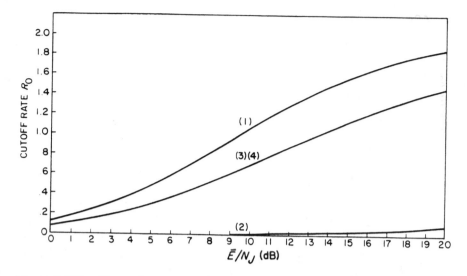

Figure 2.127. List and energy metrics R_0 for FH/4FSK against worst case noise jammer.

metric can do better than the energy metric [30], [31] (see Chapter 5, Volume III).

The results presented here generalize to arbitrary memoryless channels with FH/MFSK signals with list output as long as P_l is the probability that the transmitted symbol has a corresponding energy detector output sample

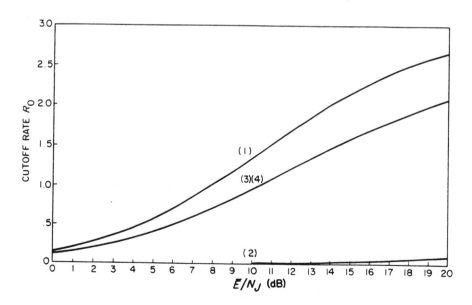

Figure 2.128. List and energy metrics R_0 for FH/8FSK against worst case noise jammer.

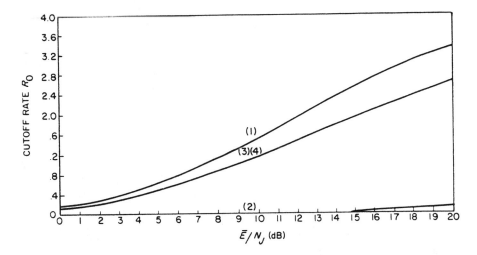

Figure 2.129. List and energy metrics R_0 for FH/16FSK against worst case noise jammer.

that is the l-th largest among all M energy detector samples. Indeed, with the usual symmetry assumptions, the same expressions apply for the coding parameter D.

In some circumstances it may be desirable to shorten the list to the L largest energy detector outputs where $L < M$. In this case the lowest $M - L$ energy detector outputs are treated indistinguishably as a single off-list group. Here the probability of the correct symbol energy detector sample

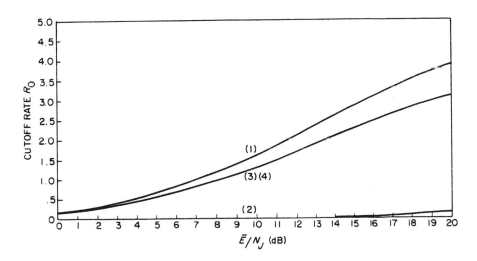

Figure 2.130. List and energy metrics R_0 for FH/32FSK against worst case noise jammer.

being *l*-th on the list is

$$P_l(L) = \begin{cases} P_l; & l = 1, 2, \ldots, L \\ \dfrac{\displaystyle\sum_{j=L+1}^{M} P_j}{M - L}; & l = L + 1, \ldots, M. \end{cases} \qquad (2.391)$$

With the ML metric the coding parameter is then

$$D_q(J) = \frac{1}{M-1}\left[\sum_{k=1}^{M}\sum_{j=1}^{M} \sqrt{P_k(L)P_j(L)} - 1 \right]$$

$$= \frac{1}{M-1}\left[\left(\sum_{k=1}^{M} \sqrt{P_k(L)} \right)^2 - 1 \right]$$

$$= \frac{1}{M-1}\left[\left(\sum_{k=1}^{L} \sqrt{P_k} + \sqrt{(M-L)\sum_{j=L+1}^{M} P_j} \right)^2 - 1 \right]. \quad (2.392)$$

This is the list-of-*L* detection scheme presented in [28], [29], and [32].

For the uniform channel with no receiver noise, Crepeau [32] has evaluated the performance of the ML metric with a list-of-*L* where there is worst case multitone jamming and broadband noise jamming with slowly varying Rayleigh fading. Figure 2.131 shows the $M = 16$ case with Rayleigh fading for R_0 in bits per symbol. There is little difference in the performance due to increasing the list size. For worst case multitone jamming, Figures 2.132 to 2.135 show a more dramatic difference in just going from $L = 1$ to $L = 2$. In general, the list metric is a useful metric when there are multiple tones in the channel due to a jammer and/or other users.

2.7.4 Metrics for Binary Codes

For FH/MFSK signals there are three well-known types of codes one can use that have *M*-ary code symbols. Reed-Solomon codes [21] are natural block codes to use where the decoders generally require a hard decision *M*-ary channel. *M*-ary alphabet convolutional codes include those found by Trumpis [18] and dual-*K* codes[9] [20] where $M = 2^K$. These codes can be decoded using the Viterbi algorithm with all metrics discussed in this chapter.

Today, however, most codes used in practice are binary codes, especially binary convolutional codes. These have been used primarily with coherent BPSK and QPSK modulations with one-bit (hard decision), two-bit, and three-bit quantized channel outputs. For binary convolutional codes LSI Viterbi decoders chips and sequential decoder chips have been developed. It

---

[9]These are convolutional codes using the Galois Field GF(2^K).

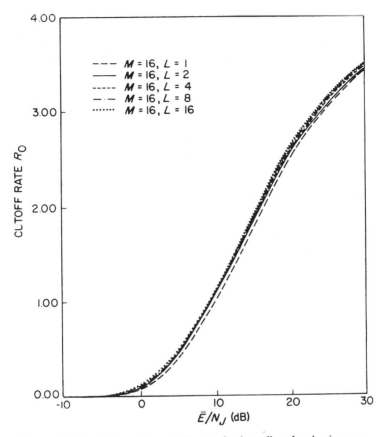

Figure 2.131. List metric cutoff rates for broadband noise jammer.

seems reasonable to exploit these developments and use binary convolutional codes with the FH/MFSK signals. Here we examine how the M-ary input channel with outputs of M energy detector samples can be converted into a binary input channel with one-bit, two-bit, and three-bit outputs that will be suitable for binary codes with corresponding decoders.

For $M = 2^K$ there are K bits associated with each M-ary channel input symbol. For example, with $K = 3$ ($M = 8$), we have a channel with eight input symbols and the output samples as seen in Table 2.23.

Between the M-ary input and the M energy detector outputs there exists MFSK modulation, frequency hopping, the real channel, dehopping and M energy detectors. To create a binary input channel with one-bit output, we merely make a hard decision as to which M-ary symbol was sent. For example, for $M = 8$ above if e_4 is the largest energy detector output then the channel output hard decision bits would be 011. Here the probability of a bit error is denoted δ_q if the signal hopped into the q-th sub-band. If the bits are interleaved before converting them into M-ary symbols and then deinterleaved, the hard decision bits obtained are the outputs of the usual

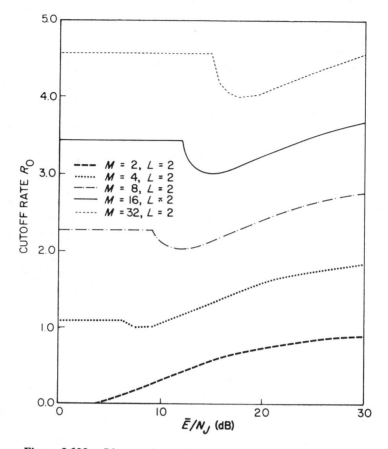

Figure 2.132.   List metric cutoff rates for worst case tone jammer.

binary symmetric channel (BSC) with crossover probability δ_q, which is the uncoded bit error probability for the FH/MFSK system.

For the above BSC a binary code can be used with a decoder that uses jammer state information (JSI) with metric given by (2.317) and the resulting coding parameter given by (2.318). Without JSI the decoder metric becomes (2.325) with the resulting coding parameter given by (2.327) and (2.328).

Suppose next we want a binary input channel with two-bit outputs. Here the hard decision with 1-bit quality measure suggested by Viterbi [27] and discussed in Section 2.7.2 can be used. For $M = 8$ there are 16 outputs here of the form $(m, 1)$ and $(m, 0)$ for $M = 1, 2, \ldots, 8$. We can now convert these into two-bit binary channel outputs by taking, for example, the output $(4, 1) \cdot$ and converting it into three two-bit binary outputs of the form

$$(4, 1) \rightarrow (0, 1)\,(1, 1),\,(1, 1) \qquad (2.393)$$

where recall $m = 4$ corresponds to the 011 bits. Similarly, if $(4, 0)$ is the

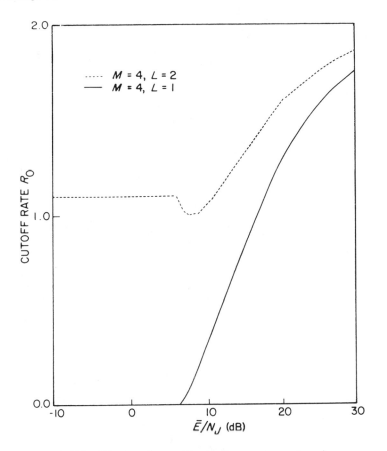

Figure 2.133. List metric cutoff rates for worst case tone jammer.

output of the channel in Section 2.7.2, then we have the conversion

$$(4,0) \rightarrow (0,0), (1,0), (1,0). \qquad (2.394)$$

With interleaving and deinterleaving the result is a binary input channel with two-bit outputs as shown in Figure 2.136 where the transition probabilities can be obtained using the type of analysis given in Section 2.7.2 (see Viterbi [27]).

This approach can be extended to three-bit outputs for each input bit by using three threshold levels $\gamma_1 > \gamma_2 > \gamma_3 \geq 1$ where (2.351) is generalized as

$$y = \begin{cases} (m,11), & e_m \geq \gamma_1 \max_{k \neq m} e_k \\ (m,10), & \gamma_1 \max_{k \neq m} e_k > e_m \geq \gamma_2 \max_{k \neq m} e_k \\ (m,01), & \gamma_2 \max_{k \neq m} e_k > e_m \geq \gamma_3 \max_{k \neq m} e_k \\ (m,00), & \gamma_3 \max_{k \neq m} e_k > e_m \geq \max_{k \neq m} e_k. \end{cases} \qquad (2.395)$$

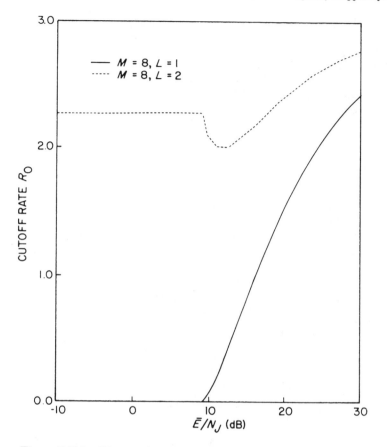

Figure 2.134. List metric cutoff rates for worst case tone jammer.

Again for the $M = 8$ example we have the conversion

$$(4, 01) \rightarrow (001), (101), (101) \qquad (2.396)$$

where the hard decision bits are followed by two "quality" bits. With interleaving and deinterleaving there are three-bit outputs for each binary input into the channel which can be used in the usual three-bit soft decision decoder for binary codes.

Another approach to obtaining a binary input channel with a quantized output is to consider the conditional probability of each transmitted bit given the M energy detector outputs. In general let $b_1, b_2, \ldots, b_K$ be the K coded bits that result in an M-ary symbol m with corresponding energy detector output e_m. Using Bayes rule (see Lee [33])

$$\Pr\{b_l | e\} = \frac{\Pr\{e | b_l\} \Pr\{b_l\}}{\Pr\{e\}}. \qquad (2.397)$$

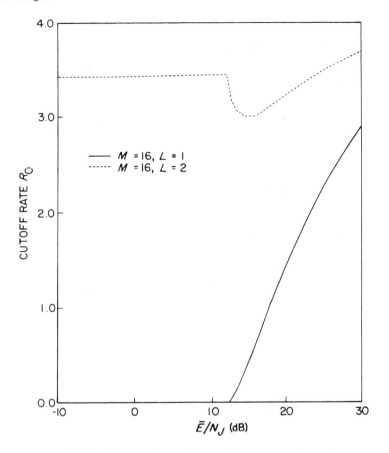

Figure 2.135. List metric cutoff rates for worst case tone jammer.

Table 2.23
Bits to symbol conversion, $K = 3$, $M = 8$.

Bits $b_1 b_2 b_3$	8-ary Input m	Energy Detector Output e
0 0 0	→ 1	e_1
0 0 1	→ 2	e_2
0 1 0	→ 3	e_3
0 1 1	→ 4	e_4
1 0 0	→ 5	e_5
1 0 1	→ 6	e_6
1 1 0	→ 7	e_7
1 1 1	→ 8	e_8

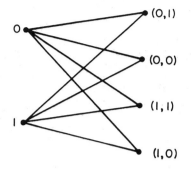

Figure 2.136. Binary input 2-bit output channel.

Next observing that there are several possible M-ary symbols having the l-th bit, b_l,

$$\Pr\{e|b_l\} = \sum_{m=1}^{M} \Pr\{e|m, b_l\}\Pr\{m|b_l\}. \tag{2.398}$$

Defining

$$\mathscr{M}(b_l) = \{m: l\text{-th bit is } b_l\}, \tag{2.399}$$

by symmetry

$$\Pr\{m|b_l\} = \begin{cases} \dfrac{2}{M}, & m \in \mathscr{M}(b_l) \\ 0, & m \notin \mathscr{M}(b_l). \end{cases} \tag{2.400}$$

Recall

$$\Pr\{e|m, b_l\} = \Pr\{e|m\}$$

$$= \prod_{k=1}^{M} p(e_k|m, q) \tag{2.401}$$

where $p(e_k|m, q)$ is given by (2.330) for the general slowly fading non-uniform channel with jammer power distribution J. Defining

$$r(e_k) = \frac{N_0 + \beta_q J_q}{\alpha_q \bar{E} + N_0 + \beta_q J_q} \exp\left\{\frac{\alpha_q \bar{E} e_k}{(N_0 + \beta_q J_q)(\alpha_q \bar{E} + N_0 + \beta_q J_q)}\right\} \tag{2.402}$$

the final form is

$$Pr\{b_l|e\} = G(e) \sum_{m \in \mathscr{M}(b_l)} r(e_m) \tag{2.403}$$

where

$$G(e) = \frac{\dfrac{1}{M} \displaystyle\prod_{k=1}^{M} \left(\dfrac{1}{N_0 + \beta_q J_q} e^{-1/(N_0 + \beta_q J_q)} \right)}{\Pr\{e\}}. \tag{2.404}$$

But $\Pr\{b_l | e\}$ is a sum of exponential functions of e_m, $m \in \mathcal{M}(b_l)$ and for signal-to-noise ratios of interest such a sum is dominated by the largest term. Thus, using the approximation

$$\Pr\{b_l | e\} \cong G(e) \max_{m \in \mathcal{M}(b_l)} r(e_m) \tag{2.405}$$

gives an approximate ML metric (taking logarithm) given by

$$m(e, b_l) = \max_{m \in \mathcal{M}(b_l)} e_m. \tag{2.406}$$

For the $K = 3$ ($M = 8$) example of Table 2.23

$$\mathcal{M}(b_1 = 0) = \{1, 2, 3, 4\}$$
$$\mathcal{M}(b_1 = 1) = \{5, 6, 7, 8\}$$
$$\mathcal{M}(b_2 = 0) = \{1, 2, 5, 6\}$$
$$\mathcal{M}(b_2 = 1) = \{3, 4, 7, 8\}$$
$$\mathcal{M}(b_3 = 0) = \{1, 3, 5, 7\}$$
$$\mathcal{M}(b_3 = 1) = \{2, 4, 6, 8\} \tag{2.407}$$

and the metrics for each of the three coded binary inputs,

$$m(e, b_1 = 0) = \max\{e_1, e_2, e_3, e_4\}$$
$$m(e, b_1 = 1) = \max\{e_5, e_6, e_7, e_8\}$$
$$m(e, b_2 = 0) = \max\{e_1, e_2, e_5, e_6\}$$
$$m(e, b_2 = 1) = \max\{e_3, e_4, e_7, e_8\}$$
$$m(e, b_3 = 0) - \max\{e_1, e_3, e_5, e_7\}$$
$$m(e, b_3 = 1) = \max\{e_2, e_4, e_6, e_8\}. \tag{2.408}$$

The above metrics are approximate ML metrics with no quantizations. They result in a binary input channel with real valued outputs. Various quantized outputs can be obtained for each coded input bit in the form

$$y_l = f\big(m(e, b_l = 0) - m(e, b_l = 1)\big)$$
$$= f\left(\max_{m \in \mathcal{M}(b_l = 0)} e_m - \max_{m \in \mathcal{M}(b_l = 1)} e_m \right) \tag{2.409}$$

where $f(\cdot)$ is typically a uniform quantizer function. Here the difference

$$\max_{m \in \mathcal{M}(b_l = 0)} e_m - \max_{m \in \mathcal{M}(b_l = 1)} e_m \tag{2.410}$$

is treated like the correlator output of the coherent BPSK demodulator. With interleaving and deinterleaving we have a discrete memoryless binary input quantized output coding channel suitable for the usual binary codes with corresponding decoders.

Although the above conversion of an FH/MFSK channel into a discrete memoryless binary input channel was based on a Rayleigh fading channel with receiver and jammer Gaussian noise statistics, it can certainly be used in other cases as well. It is a straightforward conversion and should be robust in the sense of being good for all cases of interest. Generally, for Gaussian noise and optical signals with M-ary pulse position modulation good binary convolutional codes with these converted binary input channels perform as well or better than M-ary input channels for roughly the same complexity (see Lee [33]). For very high speeds and large M, Reed-Solomon codes look attractive for the hard decision M-ary channels. Reed-Solomon codes at data rates over 100 Mbps exist for $M = 2^8 = 256$ [21].

2.8 REFERENCES

[1] R. C. Dixon, *Spread Spectrum Systems*, New York: John Wiley, 1976.

[2] A. J. Viterbi and I. M. Jacobs, "Advances in Coding and Modulation for Noncoherent Channels Affected by Fading, Partial Band, and Multiple Access Interference," in *Advances in Communication Systems*, Vol. 4. New York: Academic Press, 1975, pp. 279–308.

[3] A. J. Viterbi, *Principles of Coherent Communication*, New York: McGraw-Hill, 1966.

[4] J. I. Marcum, "Table of Q-Functions," Rand Corporation Report RM-339, Jan. 1, 1950.

[5] S. W. Houston, "Modulation Techniques for Communication, Part I: Tone and Noise Jamming Performance of Spread Spectrum M-ary FSK and 2, 4-ary DPSK Waveforms," *Proceedings of the IEEE National Aerospace and Electronics Conference (NAECON '75)*, Dayton, Ohio, June 10–12, 1975, pp. 51–58.

[6] A. J. Viterbi, "Spread spectrum communications—Myths and realities," *IEEE Commun. Mag.*, vol. 17, pp. 11–18, May 1979.

[7] B. D. Trumpis, "On the optimum detection of fast frequency hopped MFSK signals in worst case jamming," TRW internal memorandum, June 1981.

[8] J. K. Omura and B. K. Levitt, "Coded error probability evaluation for antijam communication systems," *IEEE Trans. Commun.*, COM-30, pp. 896–903, May 1982.

[9] J. M. Wozencraft and I. M. Jacobs, *Principles of Communication Engineering*, New York: John Wiley, 1976.

[10] I. M. Jacobs, "Probability of error bounds for binary transmission on the slowly fading Rician channel," *IEEE Trans. Inform. Theory*, IT-12, pp. 431–441, October 1966.

[11] B. K. Levitt, "Strategies for FH/MFSK signaling with diversity in worst case partial band noise," in *Record of Joint USC/ARO Workshop on "Research Trends in Military Communications,"* Wickenburg, AZ, May 1–4, 1983.

[12] D. J. Torrieri, *Principles of Military Communication Systems*, Dedham, MA: Artech House, 1981.

[13] J. K. Omura and T. Kailath, "Some useful probability distributions," Technical Report No. 7050-6, Stanford Electronics Laboratories, Stanford University, p. 83, September 1965.

[14] B. K. Levitt and J. K. Omura, "Coding tradeoffs for improved performance of FH/MFSK systems in partial band noise," *Record of the National Telecommunications Conference (NTC '81)*, pp. D 9.1.1–D 9.1.5, November 1981.

[15] W. E. Stark, "Coding for Frequency-Hopped Spread Spectrum Channels with Partial-Band Interference," Ph.D. Dissertation, University of Illinois, 1982.

[16] H. H. Ma and M. A. Poole, "Error-correcting codes against the worst-case partial-band jammer," *IEEE Trans. Commun.*, COM-32, pp. 124–133, February 1984.

[17] J. P. Odenwalder, "Optimal Decoding of Convolutional Codes," Ph.D. Dissertation, University of California, Los Angeles, 1970, pp. 62–68.

[18] B. D. Trumpis, "Convolutional Coding for *M*-ary Channels," Ph.D. Dissertation, University of California, Los Angeles, 1975.

[19] A. J. Viterbi and J. K. Omura, *Principles of Digital Communication and Coding*, New York: McGraw-Hill, 1979, p. 253.

[20] J. P. Odenwalder, "Dual-*K* convolutional codes for noncoherently demodulated channels," in *Proceedings of the IEEE International Telecommunications Conference (ITC '76)*, pp. 165–174, October 1976.

[21] E. R. Berlekamp, "The technology of error-correcting codes," *Proceedings of the IEEE*, vol. 68, pp. 570–572, May 1980.

[22] R. S. Orr, "Quasi-independent frequency hopping—A new spread spectrum multiple access technique," *Record of the IEEE International Conference on Communications (ICC '81)*, pp. 76.2.1–76.2.6, June 1981.

[23] R. J. McEliece and W. E. Stark, "The optimal code rate vs. a partial band noise jammer," *Record of the IEEE Military Communications Conference (MILCOM '82)*, pp. 8.5-1 to 8.5-4, October 1982.

[24] J. G. Proakis, *Digital Communications*, McGraw-Hill, New York: 1983.

[25] D. Avidor, "Anti-Jam Analysis of FH/MFSK Communication Systems in HF Rayleigh Fading Channels," Ph.D. Dissertation, University of California at Los Angeles, 1981.

[26] D. Avidor and J. K. Omura, "Analysis of FH/MFSK systems in non uniform Rayleigh fading channels," *MILCOM '82*, October 17–20, 1982.

[27] A. J. Viterbi, "A robust ratio-threshold technique to mitigate tone and partial-band jamming in coded MFSK systems," *MILCOM '82*, October 18–20, 1982.

[28] K. L. Jordan, Jr., "The performance of sequential decoding in conjunction with efficient modulation," *IEEE Trans. Commun. Tech.*, COM-14 (3), pp. 283–297, June 1966.

[29] J. M. Wozencraft and R. S. Kennedy, "Modulation and demodulation for probabilistic decoding," *IEEE Trans. Inform. Theory*, IT-12, pp. 291–297, July 1966.

[30] M. A. Creighton, "Analysis of List Decoding Metrics for Jamming and Multiple Access Channels," Ph.D. Dissertation, University of California, Los Angeles, 1985.

[31] P. J. Crepeau, M. A. Creighton, and J. K. Omura, "Performance of FH/MFSK with list metric detection against partial band noise and random tone jamming," *MILCOM '83*, October 31–November 2, 1983.

[32] P. J. Crepeau, "Generalized list detection for coded MFSK/FH signaling on fading and jamming channels," NRL Report 8708, Naval Research Laboratory, Washington, DC, June 1983.

[33] P. J. Lee, "Decomposition of $2^L \sim$ ary Input Channel into L Parallel Binary Input Component Channels and Independent Binary Coding," Ph.D. Dissertation, University of California, Los Angeles, 1985.

APPENDIX 2A: JUSTIFICATION OF FACTOR OF ONE-HALF FOR FH / MFSK SIGNALS WITH DIVERSITY IN PARTIAL-BAND NOISE

Chapter 2 employed a factor of $1/2$ in the Chernoff performance bounds for FH/MFSK signals with diversity in partial-band noise. This factor was originally introduced by Viterbi and Jacobs [2] based on an earlier existence proof by Jacobs [10]. However, some analysts were left with the erroneous belief that the result was valid only for maximum-likelihood metrics, which would disqualify the suboptimum linear combination metric of (2.57). Furthermore, Jacobs did not stress that his result does not usually apply to discrete-valued metrics such as those we encountered in the multitone jamming scenarios neglecting thermal noise.

Appendix 4B in Volume I derived many general conditions under which the factor of $1/2$ is valid. In particular, using (4B.31), (4B.38), (4B.39b), and (4B.43), we conclude that

$$\Pr\{x \geq 0\} \leq \min_{\lambda} \left(\tfrac{1}{2} \overline{e^{\lambda x}} \right) \qquad (2A.1)$$

provided that

$$p(x) \leq p(-x); \qquad \forall\, x \geq 0 \qquad (2A.2)$$

where $p(x)$ is the probability density function of the random variable x. (2A.1) is distinguished from the conventional Chernoff bound by the factor of $1/2$ and the absence of the restriction that the Chernoff parameter λ be non-negative. (In fact, the condition of (2A.2) guarantees that the minimizing $\lambda \geq 0$.) Since (2A.2) must be satisfied for every non-negative x, it cannot apply to discrete-valued random variables which are asymmetric about $x = 0$. We will now prove that (2A.2) is valid for the particular cases of interest.

Consider the reception of FH/MFSK signals with m diversity chips per M-ary symbol over the partial-band noise channel. Assume non-coherent energy detection of the diversity chips, linear summation of the detected chip energies for each symbol, and perfect jamming state side information. Let the random variable x above represent the detection metric: (2.61) shows that x is the difference of a non-central and a central chi-square random variable, each with $2m$ degrees of freedom. In particular, since the received signal energy over all m chips is KE_b, where $M = 2^K$, and the side information assumption requires that all m chips be jammed with noise of power spectral density N_J/ρ for an error to occur (designated by the event H in (2.61)),

$$x = \sum_{i=1}^{2m} \left(y_i^2 - z_i^2 \right) \tag{2A.3}$$

where the y_i's and z_i's are independent Gaussian random variables, statistically defined by

$$y_i = N\left(0, \frac{N_J}{2\rho}\right)$$

$$z_i = N\left(a_i, \frac{N_J}{2\rho}\right) \tag{2A.4}$$

$$\sum_{i=1}^{2m} a_i^2 = KE_b.$$

From [13], we can express the probability density function of x as

$$p(x) = b^m e^{-c} |x|^{m-1} \sum_{k=0}^{\infty} \frac{\left(bc^2 |x|\right)^{k/2}}{k!}$$

$$\times \begin{cases} \dfrac{W_{-k/2, \, m+(k-1)/2}(4bx)}{m!}; & x \geq 0 \\[2ex] \dfrac{W_{k/2, \, m+(k-1)/2}(-4bx)}{(k+m-1)!}; & x < 0 \end{cases} \tag{2A.5}$$

where

$$b \equiv \frac{\rho}{2N_J}, \qquad c \equiv \frac{\rho KE_b}{N_J}$$

and $W(\cdot)$ is the Whittaker function defined by

$$W_{\alpha, \beta}(x) \equiv \frac{x^{\beta+1/2} e^{-x/2}}{(\beta - \alpha - \frac{1}{2})!} \int_0^{\infty} dy \, e^{-xy} y^{\beta-\alpha-1/2} (1+y)^{\beta+\alpha-1/2};$$

$$\beta - \alpha + \tfrac{1}{2} > 0. \tag{2A.6}$$

Since $m \geq 1$ and $k \geq 0$ in (2A.5), the constraint on the indices of the Whittaker function in (2A.6) is satisfied.

Now we need to demonstrate that (2A.5) is compatible with (2A.2); that is, we must prove that

$$\sum_{k=0}^{\infty} \frac{\left(bc^2 x\right)^{k/2}}{k!(k+m-1)!} W_{k/2, \, m+(k-1)/2}(4bx)$$

$$\geq \sum_{k=0}^{\infty} \frac{\left(bc^2 x\right)^{k/2}}{k!m!} W_{-k/2, \, m+(k-1)/2}(4bx); \qquad \forall \, x \geq 0.$$

$$\tag{2A.7}$$

A sufficient condition is simply

$$m! \ W_{k/2, \, m+(k-1)/2}(4bx)$$
$$\geq (k + m - 1)! \ W_{-k/2, \, m+(k-1)/2}(4bx); \qquad \forall \, k, x \geq 0.$$

$$(2A.8)$$

Substituting (2A.6) into (2A.8), we need to show that

$$\int_0^\infty dy \, e^{-xy} [y(1 + y)]^{m-1} [m(1 + y)^k - y^k] \geq 0; \qquad \forall \, k, x \geq 0.$$

$$(2A.9)$$

However, since $m \geq 1$, the integrand is non-negative whenever $y \geq 0$. Consequently, (2A.9) is valid, completing the desired proof.

APPENDIX 2B: COMBINATORIAL COMPUTATION FOR $n = 1$ BAND MULTITONE JAMMING

This is a combinatorial problem required to compute the exact performance for FH/MFSK signals with diversity in $n = 1$ band multitone jamming, assuming non-coherent linear combining and perfect side information. The problem is deceptively simple to state, and difficult to resolve under certain conditions.

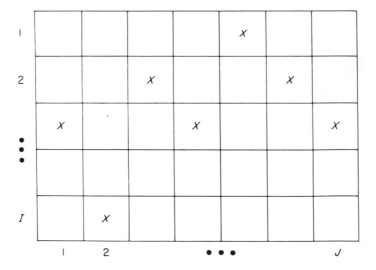

Figure 2B.1. Frequency (row)-time (column) grid for computing probabilities associated with $n = 1$ band multitone jamming.

Consider a grid containing I rows and J columns, as shown in Figure 2B.1. One X is placed in each column: the row location is random, equally likely, and independent from column to column. The number of X's in the i-th row is denoted by $l_i \in [0, J]$, and $l_{max} \equiv \max_{1 \le i \le I}(l_i)$. There are I^J distinct patterns of X's that can arise; we want to compute subsets of these defined by

$$N_L \equiv \text{number of patterns} \ni l_{max} = L. \tag{2B.1}$$

The difficulty here is that while the columns are statistically independent, the rows (specifically, the l_i's) are not.

Clearly, $1 \le l_{max} \le J$, so that $N_L = 0$ for $L \notin [1, J]$. For $L = 1$, each X must lie in a different row, which can only occur if $I \ge J$:

$$N_1 = \begin{cases} \dfrac{I!}{(I-J)!}; & I \ge J \\ 0; & I < J \end{cases}. \tag{2B.2}$$

For $J/2 < L \le J$, only *one* row can have L X's: there are I ways to choose that row, $\binom{J}{L}$ ways to choose the columns that contain the X's in that row, and $(I-1)^{J-L}$ ways to place the remaining X's (all of these $I - 1$ other l_i's will be less than L since $J - L < J/2$), so that

$$N_L = \binom{J}{L} I (I-1)^{J-L}; \qquad J/2 < L \le J. \tag{2B.3}$$

For smaller values of L, it is more difficult to derive a general, explicit expression for N_L, but not impossible. For example, consider the next level of complexity, $J/3 < L \le J/2$. There are now $\binom{I}{2}\binom{J}{L}\binom{J-L}{L}(I-2)^{J-2L}$ patterns with two rows of L X's and $I - 2$ other rows with less than L X's; these patterns are counted twice in (2B.3). Furthermore, if $2L + 1 \le J$ (which always occurs in the region $J/3 < L \le J/2$ except when J is even and $L = J/2$ exactly), we must eliminate those patterns which contain L X's in one row and $L + 1$ or more X's in another row. The resulting modification to (2B.3) is[1]

$$N_L = \binom{J}{L} I (I-1)^{J-L} - \binom{I}{2}\binom{J}{L}\binom{J-L}{L}(I-2)^{J-2L}$$

$$- I(I-1)\binom{J}{L} \sum_{j=L+1}^{J-L} \binom{J-L}{j}(I-2)^{J-L-j};$$

$$\frac{J}{3} < L \le \frac{J}{2} \tag{2B.4}$$

where the summation is understood to be zero for the singular case $L > (J-1)/2$ discussed above.

---

[1] Derived by Dr. Laif Swanson of the Jet Propulsion Laboratory, Pasadena, California.

For the band multitone jamming application, we do not actually need the individual N_L's. A sufficient parameter is

$$S_{I,J}(L) \equiv \text{number of patterns} \ni \bigcup_{i=1}^{I} (l_i \geq L)$$

$$= \text{number of patterns} \ni l_{\max} \geq L$$

$$= \sum_{i=L}^{J} N_i \tag{2B.5}$$

where we have made use of the disjointness of the N_L's (i.e., none of the patterns in N_i coincide with any pattern in N_j for $i \neq j$). As noted earlier, before $S_{I,J}(L)$ was defined,

$$S_{I,J}(1) = I^J. \tag{2B.6}$$

Also,

$$S_{I,J}(2) = S_{I,J}(1) - N_1$$

$$= \begin{cases} I^J - \dfrac{I!}{(I-J)!}; & I \geq J \\[2mm] I^J; & I < J. \end{cases} \tag{2B.7}$$

Finally, adopting the conventions that $\lfloor x$ denotes the integer portion of x and $\sum_{i=a}^{b} c_i = 0$ if $a > b$, we can write

$$S_{I,J}(L) = I \sum_{i=L}^{J} \binom{J}{i} (I-1)^{J-i}$$

$$- \binom{I}{2} \sum_{i=L}^{\lfloor J/2 \rfloor} \binom{J}{i} \binom{J-i}{i} (I-2)^{J-2i} \tag{2B.8}$$

$$- I(I-1) \sum_{i=L}^{\lfloor J/2 \rfloor} \binom{J}{i} \sum_{j=i+1}^{J-i} \binom{J-i}{j} (I-2)^{J-i-j};$$

$$\frac{J}{3} < L \leq J.$$

So for $J \leq 8$, we can compute $S_{I,J}(L)$ for the entire range of $L \in [1, J]$; for $J > 8$, we are missing the region $3 \leq L \leq \lfloor J/3 \rfloor$. In principle, we could extend (2B.4) to the next region, $J/4 < L \leq \lfloor J/3 \rfloor$, and so on, but this is increasingly more complex and not really a worthwhile exercise.

Part 3

OTHER FREQUENCY-HOPPED SYSTEMS

Chapter 3

COHERENT MODULATION TECHNIQUES

Thus far in our discussions, we have considered only those frequency-hopped (FH) systems in which the carrier phases of the individual transmitted hop frequency pulses bear no relation to one another. Such a spread-spectrum (SS) technique was referred to as *non-coherent FH* and the method for recovering the data modulation appropriately employed some form of energy detection. Clearly, another possibility exists for implementing the FH modulator, wherein phase continuity is maintained from one hop pulse to another thereby resulting in so-called *coherent FH*. Assuming that the coherent frequency synthesizer in the receiver is capable of estimating and correcting for the phase errors associated with the electrical path length over the transmission channel and the Doppler shift, then following the dehopping operation one could employ coherent detection techniques for recovery of the data modulation.

Of all the modulation techniques which lend themselves to coherent detection, those that are most commonly found in present-day applications are the so-called quadrature modulations which include quadriphase-shift-keying (QPSK), quadrature amplitude-shift-keying (QASK), and quadrature partial response (QPR).[1] We shall consider binary phase-shift-keying (BPSK) as a degenerate case of QPSK. The performance of these modulation techniques over the additive white Gaussian noise (AWGN) channel is well documented in the literature [1], [2], [3]. This chapter presents the comparable results when an FH/SS modulation is superimposed on these conventional techniques in order to combat the intentional interference introduced by a jammer [4], [5]. The receiver structures that will be analyzed will be those normally employed for coherent detection of these modulations in an AWGN background. As such, our intent here is not to consider optimum

---

[1] The QPR Class I modulation, which is the only case we shall consider, is also referred to in the literature as duobinary-encoded QPSK [1].

267

receiver structures based upon jamming state information, but rather to demonstrate the effect of the despread jammer on the conventional structures.

The scenario under which the jammer is assumed to operate here is the same as that proposed in the previous chapters for the design of non-coherent FH/SS systems. In particular, we assume an intelligent jammer that has knowledge of the form of data and SS modulations, including such items as data rate, spreading bandwidth and hop rate, but no knowledge of the code selected for determining the spectrum-spreading hop frequencies. Once again, the strategy employed is to design for the worst case jammer in the sense that, given the modulation form of the communicator, the jammer is assumed to employ the type of jamming which is most deleterious to the communication receiver. The two most common jammer types are still the partial-band noise and partial-band multitone jammers. Hence, for each form of coherent data modulation considered, we shall direct all of our attention to these two jammer types, developing in each case the performance corresponding to the worst case jamming strategy.

Although the primary emphasis of this chapter is on pure FH quadrature modulation systems, later on we shall consider the additional improvement in jam resistance offered by superimposing a balanced pseudonoise (PN) modulation on the individual hop frequencies. This hybrid SS technique is referred to as FH/PN modulation and is normally employed in situations where the system requires anti-jam protection beyond that which FH or PN modulation is capable of producing on its own.

3.1 PERFORMANCE OF FH/QPSK IN THE PRESENCE OF PARTIAL-BAND MULTITONE JAMMING

An FH/QPSK signal is characterized by transmitting (Figure 3.1)

$$s^{(i)}(t) = \sqrt{2S} \sin\left(\omega_h^{(i)}t + \theta^{(i)}\right) \tag{3.1}$$

in the i-th signalling interval $(i - 1)T_s \le t \le iT_s$, where $\omega_h^{(i)}$ is the particular carrier radian frequency selected by the frequency hopper for this interval.[2] According to the designated SS code, $\theta^{(i)}$ is the information symbol which ranges over the set of possible values

$$\theta_m = \frac{m\pi}{4}; \qquad m = 1, 3, 5, 7 \tag{3.2}$$

and S is the transmitted average power.

---

[2] We assume here the case of slow frequency hopping (SFH), i.e., the hop rate is equal to or a submultiple of the information symbol rate and that the frequency hopper and symbol clock are synchronous. Thus, in a given symbol interval, the signal frequency is constant and the jammer, if transmitting a tone at that frequency, affects the entire symbol interval.

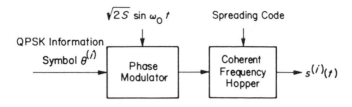

Figure 3.1. Block diagram of a coherent FH/QPSK modulator.

At the receiver (Figure 3.2), the sum of additive white Gaussian noise $n(t)$, the jammer $J(t)$, and a random phase-shifted version of the transmitted signal $s^{(i)}(t; \theta)$ are first frequency dehopped, then coherently demodulated by a conventional QPSK demodulator.

The band-pass noise $n(t)$ has the usual narrowband representation

$$n(t) = \sqrt{2} \left[N_c(t) \cos\left(\omega_h^{(i)} t + \theta \right) - N_s(t) \sin\left(\omega_h^{(i)} t + \theta \right) \right] \quad (3.3)$$

where $N_c(t)$ and $N_s(t)$ are statistically independent low-pass white Gaussian noise processes with single-sided noise spectral density N_0 w/Hz. The partial-band multitone jammer $J(t)$ is assumed to have a total power J which is evenly divided among Q jammer tones. Thus, each tone has power

$$J_0 = \frac{J}{Q}. \quad (3.4)$$

Furthermore, since the jammer is assumed to have knowledge of the exact location of the spreading bandwidth W_{ss} and the number N of hops in this bandwidth, then, as was done in our previous discussions, we shall assume that he will randomly locate each of his Q tones coincident with Q of the N hop frequencies. Thus,

$$\rho \triangleq \frac{Q}{N} \quad (3.5)$$

represents the fraction of the total band which is continuously jammed with tones, each having power J_0. Once again, the jammer's strategy is to distribute his total power J (i.e., choose ρ and J_0) in such a way as to cause the communicator to have maximum probability of error.

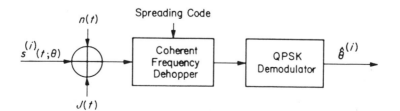

Figure 3.2. Block diagram of a coherent FH/QPSK demodulator.

In view of the foregoing, the total received signal in a signalling interval which contains a jamming tone at the hop frequency is given by

$$y^{(i)}(t) = s^{(i)}(t; \theta) + n(t) + J(t) \qquad (3.6)$$

where

$$s^{(i)}(t; \theta) = \sqrt{2S} \, \sin\left(\omega_h^{(i)}t + \theta^{(i)} + \theta\right), \qquad (3.7)$$

$n(t)$ is given by (3.3) and

$$J(t) = \sqrt{2J_0} \, \cos\left(\omega_h^{(i)}t + \theta_J + \theta\right) \qquad (3.8)$$

with θ_J uniformly distributed on $(0, 2\pi)$ and independent of the information symbol phase $\theta^{(i)}$. Over an integral number of hop bands, the fraction ρ of the total number of signalling intervals will have a received signal characterized by (3.6). In the remaining fraction $(1 - \rho)$ of the signalling intervals, the received signal is simply characterized by

$$y^{(i)}(t) = s^{(i)}(t; \theta) + n(t). \qquad (3.9)$$

After ideal coherent demodulation by the frequency hopper, the in-phase and quadrature components of the received signal become[3]

$$\varepsilon_I(t) \triangleq y^{(i)}(t)\left[\sqrt{2} \, \sin\left(\omega_h^{(i)}t + \theta\right)\right] = \sqrt{S} \, \cos\theta^{(i)} - \sqrt{J_0} \, \sin\theta_J - N_s(t)$$

$$\varepsilon_Q(t) \triangleq y^{(i)}(t)\left[\sqrt{2} \, \cos\left(\omega_h^{(i)}t + \theta\right)\right] = \sqrt{S} \, \sin\theta^{(i)} + \sqrt{J_0} \, \cos\theta_J + N_c(t).$$

$$(3.10)$$

These signals are then passed through integrate-and-dump filters of duration equal to the information symbol interval T_s to produce the in-phase and quadrature decision variables

$$z_I \triangleq \int_{(i-1)T_s}^{iT_s} \varepsilon_I(t) \, dt = \sqrt{S} \, T_s \cos\theta^{(i)} - \sqrt{J_0} \, T_s \sin\theta_J + N_I$$

$$= a_i \sqrt{\frac{S}{2}} \, T_s - \sqrt{J_0} \, T_s \sin\theta_J + N_I$$

$$z_Q \triangleq \int_{(i-1)T_s}^{iT_s} \varepsilon_Q(t) \, dt = \sqrt{S} \, T_s \sin\theta^{(i)} + \sqrt{J_0} \, T_s \cos\theta_J + N_Q$$

$$= b_i \sqrt{\frac{S}{2}} \, T_s + \sqrt{J_0} \, T_s \cos\theta_J + N_Q \qquad (3.11)$$

where

$$N_I \triangleq - \int_{(i-1)T_s}^{iT_s} N_s(t) \, dt$$

$$N_Q \triangleq \int_{(i-1)T_s}^{iT_s} N_c(t) \, dt \qquad (3.12)$$

---

[3] We ignore double-harmonic terms.

are zero mean Gaussian random variables with variance $N_0 T_s/2$ and, in view of the possible values for $\theta^{(i)}$ given in (3.2), $\{a_i\}$ and $\{b_i\}$ are the equivalent independent in-phase and quadrature binary information sequences which take on values ± 1.

The receiver estimates of a_i and b_i are obtained by passing z_I and z_Q through hard limiters, giving

$$\hat{a}_i = \operatorname{sgn} z_I; \qquad \hat{b}_i = \operatorname{sgn} z_Q. \tag{3.13}$$

Hence, given a_i, b_i and θ_J, the probability that the i-th symbol is in error is the probability that either $\hat{a}_i$ or $\hat{b}_i$ is in error, i.e.,

$$
\begin{aligned}
P_{s_i}(\theta_J) &= \Pr\{\hat{a}_i \neq a_i \text{ or } \hat{b}_i \neq b_i\} \\
&= \Pr\{\hat{a}_i \neq a_i\} + \Pr\{\hat{b}_i \neq b_i\} \\
&\quad - \Pr\{\hat{a}_i \neq a_i\}\Pr\{\hat{b}_i \neq b_i\}.
\end{aligned} \tag{3.14}
$$

Since the signal set is symmetric, we can compute (3.14) for any of the four possible signal points and obtain the average probability of symbol error conditioned on the jammer phase $P_s(\theta_J)$. Thus, assuming for simplicity that $a_i = 1$, $b_i = 1$, we compute $P_s(\theta_J)$ from (3.14), combined with (3.11) and (3.13), as

$$
\begin{aligned}
P_s(\theta_J) &= \Pr\{z_I < 0 | a_i = 1\} + \Pr\{z_Q < 0 | b_i = 1\} \\
&\quad - \Pr\{z_I < 0 | a_i = 1\}\Pr\{z_Q < 0 | b_i = 1\} \\
&= P_I(\theta_J) + P_Q(\theta_J) - P_I(\theta_J)P_Q(\theta_J)
\end{aligned} \tag{3.15}
$$

where

$$
\begin{aligned}
P_I(\theta_J) &= \Pr\left\{ N_I < -\sqrt{\frac{S}{2}}\, T_s + \sqrt{J_0}\, T_s \sin\theta_J \right\} \\
&= Q\left[\sqrt{\frac{ST_s}{N_0}}\left(1 - \sqrt{\frac{2J_0}{S}}\, \sin\theta_J\right) \right] \\
P_Q(\theta_J) &= \Pr\left\{ N_Q < -\sqrt{\frac{S}{2}}\, T_s - \sqrt{J_0}\, T_s \cos\theta_J \right\} \\
&= Q\left[\sqrt{\frac{ST_s}{N_0}}\left(1 + \sqrt{\frac{2J_0}{S}}\, \cos\theta_J\right) \right]
\end{aligned} \tag{3.16}
$$

with $Q(x)$ the Gaussian probability integral as used in previous chapters.

Finally, the unconditional average probability of symbol error P_{s_J} for symbol intervals which are jammed is obtained by averaging $P_s(\theta_J)$ of (3.15) over the uniform distribution of θ_J. Thus,

$$P_{s_J} = \frac{1}{2\pi} \int_0^{2\pi} \left[P_I(\theta_J) + P_Q(\theta_J) - P_I(\theta_J)P_Q(\theta_J) \right] d\theta_J. \tag{3.17}$$

Recognizing that, for a QPSK signal, the symbol time T_s is twice the bit time T_b, letting $E_b = ST_b$ denote the bit energy, we then have

$$\frac{ST_s}{N_0} = \frac{2ST_b}{N_0} = \frac{2E_b}{N_0}. \tag{3.18}$$

Furthermore, from (3.4) and (3.5),

$$\frac{2J_0}{S} = \frac{2J}{\rho NS}. \tag{3.19}$$

Now, if the hop frequency slots are $1/T_s$ wide, in terms of the total hop frequency band W_{ss} and the number of hop slots N in that band, we then have

$$N = \frac{W_{ss}}{1/T_s} = W_{ss}T_s = 2W_{ss}T_b. \tag{3.20}$$

Substituting (3.20) into (3.19) gives

$$\frac{2J_0}{S} = \frac{J/W_{ss}}{\rho ST_b} = \frac{J/W_{ss}}{\rho E_b} \triangleq \frac{N_J}{\rho E_b}. \tag{3.21}$$

As in previous chapters, the quantity J/W_{ss} represents the *effective jammer power spectral density* in the hop band; thus, we have again introduced the notation N_J to represent this quantity.

Finally, rewriting (3.16) using (3.18) and (3.21) gives

$$P_I(\theta_J) = Q\left[\sqrt{\frac{2E_b}{N_0}}\left(1 - \sqrt{\frac{N_J}{\rho E_b}}\sin\theta_J\right)\right]$$

$$P_Q(\theta_J) = Q\left[\sqrt{\frac{2E_b}{N_0}}\left(1 + \sqrt{\frac{N_J}{\rho E_b}}\cos\theta_J\right)\right]. \tag{3.22}$$

For the fraction $(1 - \rho)$ of symbol (hop) intervals where the jammer is absent, the average symbol error probability is given by the well-known result [1]

$$P_{s_0} = 2Q\left(\sqrt{\frac{2E_b}{N_0}}\right) - Q^2\left(\sqrt{\frac{2E_b}{N_0}}\right). \tag{3.23}$$

Thus, the average error probability over all symbols (jammed and un-jammed) is simply

$$P_s = \rho P_{s_J} + (1 - \rho)P_{s_0}, \tag{3.24}$$

where P_{s_J} is given by (3.17), together with (3.22), and P_{s_0} is given in (3.23).

Before presenting numerical results illustrating the evaluation of (3.24), it is of interest to examine its limiting behavior as $N_0 \to 0$. Clearly, from (3.23) we have

$$\lim_{E_b/N_0 \to \infty} P_{s_0} = 0 \tag{3.25}$$

Also,

$$\lim_{E_b/N_0 \to \infty} P_{s_J} = \begin{cases} 0; & \dfrac{\rho E_b}{N_J} > 1 \\[2ex] \dfrac{2}{\pi} \cos^{-1} \sqrt{\dfrac{\rho E_b}{N_J}}; & \dfrac{1}{2} < \dfrac{\rho E_b}{N_j} \leq 1 \\[2ex] \dfrac{1}{\pi} \cos^{-1} \sqrt{\dfrac{\rho E_b}{N_J}} + \dfrac{1}{4}; & 0 < \dfrac{\rho E_b}{N_J} \leq \dfrac{1}{2}. \end{cases} \quad (3.26)$$

This result can be obtained directly from the graphical interpretation given in Figure 3.3. Finally, substituting (3.25) and (3.26) in (3.24) gives the

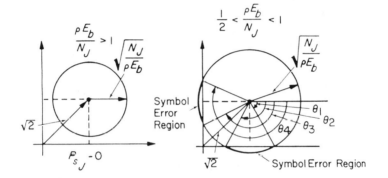

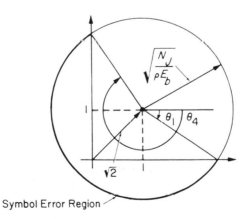

Figure 3.3. Graphical interpretation of (3.26).

desired limiting behavior for the average symbol error probability, namely,

$$
\lim_{E_b/N_0 \to \infty} P_s = \begin{cases} 0; & \dfrac{\rho E_b}{N_J} > 1 \\[2ex] \dfrac{2\rho}{\pi} \cos^{-1} \sqrt{\dfrac{\rho E_b}{N_J}}; & \dfrac{1}{2} < \dfrac{\rho E_b}{N_J} \le 1 \\[2ex] \dfrac{\rho}{\pi} \cos^{-1} \sqrt{\dfrac{\rho E_b}{N_J}} + \dfrac{\rho}{4}; & 0 < \dfrac{\rho E_b}{N_J} \le \dfrac{1}{2}. \end{cases} \qquad (3.27)
$$

The partial-band fraction ρ corresponding to the worst case jammer (maximum P_s) can be obtained by differentiating (3.27) with respect to ρ and equating to zero. Assuming that, for a fixed E_b/N_J, this worst case ρ occurs when $1/2 < \rho E_b/N_J \le 1$, then

$$
\frac{d}{d\rho} \left[\frac{2\rho}{\pi} \cos^{-1} \sqrt{\frac{\rho E_b}{N_J}} \right] = 0 \qquad (3.28)
$$

implies

$$
\tan^{-1} Z = \frac{1}{2Z} \qquad (3.29)
$$

where

$$
Z \triangleq \sqrt{\frac{1 - \rho E_b/N_J}{\rho E_b/N_J}}. \qquad (3.30)
$$

The solution to (3.29) may be numerically found to be

$$
Z = 0.7654 \qquad (3.31)
$$

or

$$
\rho_{wc} = \begin{cases} \dfrac{0.6306}{E_b/N_J}; & E_b/N_J > 0.6306 \\[2ex] 1; & E_b/N_J \le 0.6306. \end{cases} \qquad (3.32)
$$

Note that, since Q and N are integers, then ρ as defined in (3.5) is not a continuous variable. Thus, for a given N, the true worst case ρ would be the rational number nearest to (3.32) which yields an integer value of Q. Also, the second part of (3.32) comes about from the fact that Q is constrained to be less than or equal to N. Thus, when E_b/N_J is such that the solution of (3.30) and (3.31) gives a value of $\rho > 1$, we take $\rho = 1$ (full-band jamming) as the worst case jammer. Substituting (3.32) into (3.27) gives the limiting average symbol error probability performance corresponding to the worst case jammer, namely,

$$\lim_{E_b/N_0 \to \infty} P_{s_{\max}} = \begin{cases} \dfrac{0.2623}{E_b/N_J}; & E_b/N_J > 0.6306 \\[2ex] \dfrac{2}{\pi}\cos^{-1}\sqrt{\dfrac{E_b}{N_J}}; & 0.5 < E_b/N_J \leq 0.6306 \\[2ex] \dfrac{1}{\pi}\cos^{-1}\sqrt{\dfrac{E_b}{N_J}} + \dfrac{1}{4}; & 0 < E_b/N_J \leq 0.5. \end{cases}$$

$$(3.33)$$

The final step in the characterization of the performance of FH/QPSK in the presence of multitone jamming is the conversion of average symbol error probability to average bit error probability. If one encodes the information symbols using a Gray code, the average bit error probability, P_b, for a multiple phase-shift-keyed (MPSK) signal is then approximated for large

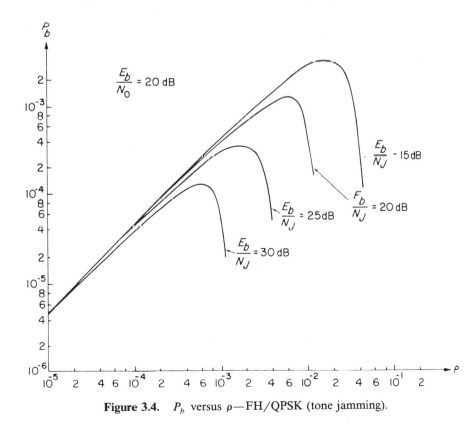

Figure 3.4. P_b versus ρ—FH/QPSK (tone jamming).

E_b/N_0 by

$$P_b \cong \frac{P_s}{\log_2 M} \tag{3.34}$$

where $\log_2 M$ is the number of bits/symbol. The approximation in (3.34) refers to the fact that only errors in symbols whose corresponding signal phases are adjacent to that of the transmitted signal are accounted for. Since a Gray code has the property that adjacent symbols differ in only a single bit, then an error in an adjacent symbol is accompanied by one, and only one, bit error.

Since QPSK is the particular case of MPSK corresponding to $M = 4$, then from (3.34),

$$P_b \cong \tfrac{1}{2} P_s \tag{3.35}$$

where P_s is given by (3.24) or its limiting form in (3.27).

Fortunately, for the case of QPSK, it is straightforward to account for the diagonal symbol errors which result in two bit errors and, thus, arrive at an *exact* expression for P_b. In fact, P_b for QPSK is identical to P_b for binary PSK (BPSK) and is given by

$$P_b = \rho \left[\frac{1}{2\pi} \int_0^{2\pi} P_I(\theta_J)\, d\theta_J \right] + (1 - \rho) Q\left(\sqrt{\frac{2E_b}{N_0}} \right)$$

$$= \rho \left[\frac{1}{2\pi} \int_0^{2\pi} P_Q(\theta_J)\, d\theta_J \right] + (1 - \rho) Q\left(\sqrt{\frac{2E_b}{N_0}} \right) \tag{3.36}$$

where $P_I(\theta_J)$ and $P_Q(\theta_J)$ are given in (3.22). Thus, comparing the approximate result of (3.35) (using (3.17), (3.22), and (3.24)) with the exact result of (3.36), we observe that the difference between the two resides in the terms resulting from the *product* of error probabilities, namely,

$$\frac{1}{2\pi} \int_0^{2\pi} P_I(\theta_J) P_Q(\theta_J)\, d\theta_J$$

and $Q^2(\sqrt{2E_b/N_0})$. Also, by analogy with (3.27), the exact limiting form of P_b becomes

$$\lim_{E_b/N_0 \to \infty} P_b = \begin{cases} 0; & \dfrac{\rho E_b}{N_J} > 1 \\[2ex] \dfrac{\rho}{\pi} \cos^{-1} \sqrt{\dfrac{\rho E_b}{N_J}} ; & 0 < \dfrac{\rho E_b}{N_J} \leq 1 \end{cases} \tag{3.37}$$

with a worst case ρ as in (3.32) and corresponding maximum error probability

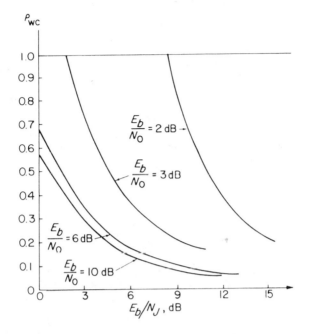

Figure 3.5. Worst case ρ versus E_b/N_J—FH/QPSK (tone jamming).

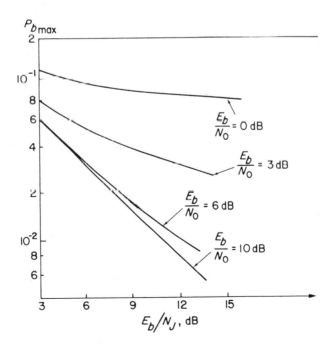

Figure 3.6. Worst case P_b versus E_b/N_J—FH/QPSK (tone jamming).

$$\lim_{E_b/N_0 \to \infty} P_{b_{max}} = \begin{cases} \dfrac{0.1311}{E_b/N_J}; & E_b/N_J > 0.6306 \\[3mm] \dfrac{1}{\pi} \cos^{-1} \sqrt{\dfrac{E_b}{N_J}}; & E_b/N_J \le 0.6306. \end{cases} \qquad (3.38)$$

Figure 3.4 is a typical plot of P_b versus ρ, with E_b/N_J as a parameter for the case $E_b/N_0 = 20$ dB. It is seen that, for fixed E_b/N_0 and E_b/N_J, there exists a value of ρ which maximizes P_b and, thus, represents the worst case multitone jammer situation. In the limit, as E_b/N_0 approaches infinity, this value of ρ becomes equal to that given by (3.32). Figure 3.5 is a plot of worst case ρ versus E_b/N_J, with E_b/N_0 as a parameter. Figure 3.6 illustrates the corresponding plot of $P_{b_{max}}$ versus E_b/N_J, with E_b/N_0 fixed.

3.2 PERFORMANCE OF FH/QASK IN THE PRESENCE OF PARTIAL-BAND MULTITONE JAMMING

An FH/QASK-M signal is characterized by transmitting

$$s^{(i)}(t) = \sqrt{2}\,\delta\left[b_n \cos\omega_h^{(i)}t + a_m \sin\omega_h^{(i)}t\right] \qquad (3.39)$$

in the i-th signalling interval. The total number of signals possible M is typically the square of an even number K, and the quadrature amplitudes a_m and b_n take on equally likely values m and n, respectively, with $m, n = \pm 1, \pm 3, \ldots \pm(K-1)$. Also, δ is a parameter which is related to the average power S of the signal set by

$$S = \tfrac{2}{3}(K^2 - 1)\delta^2. \qquad (3.40)$$

Analogous to the step leading to (3.11), we can arrive at expressions for the in-phase and quadrature decision variables, namely,

$$z_I = a_i \delta T_s - \sqrt{J_0}\,T_s \sin\theta_J + N_I$$

$$z_Q = b_i \delta T_s + \sqrt{J_0}\,T_s \cos\theta_J + N_Q. \qquad (3.41)$$

The QASK receiver estimates of a_i and b_i are obtained by passing z_I and z_Q through K-level quantizers

$$\hat{a}_i = Q_K(z_I); \qquad \hat{b}_i = Q_K(z_Q) \qquad (3.42)$$

where

$$Q_K(x) = \begin{cases} 1; & 0 \le x \le 2\delta T_s \\ 3; & 2\delta T_s \le x \le 4\delta T_s \\ \vdots & \\ (K-3); & (K-4)\delta T_s \le x \le (K-2)\delta T_s \\ (K-1); & (K-2)\delta T_s \le x \le \infty \end{cases} \qquad (3.43)$$

and $Q_K(x) = -Q_K(-x)$. Hence, given a_i, b_i, and θ_J, the probability that the i-th symbol is in error is the probability that $\hat{a}_i$ or $\hat{b}_i$ is in error. Thus, once again, (3.14) is valid. Here, however, we must compute (3.14) for the $K^2/4$ points in any quadrant in order to obtain the average probability of symbol error conditioned on the jammer phase. Thus, using QASK-16 ($K = 4$) as an example, we have

$$\Pr\{\hat{a}_i \neq a_i\} \triangleq P_I(\theta_J)$$
$$= \tfrac{1}{2}\Pr\{0 > z_I > 2\delta T_s | a_i = 1\} + \tfrac{1}{2}\Pr\{z_I < 2\delta T_s | a_i = 3\}$$
$$\Pr\{\hat{b}_i \neq b_i\} \triangleq P_Q(\theta_J)$$
$$= \tfrac{1}{2}\Pr\{0 > z_Q > 2\delta T_s | b_i = 1\} + \tfrac{1}{2}\Pr\{z_Q < 2\delta T_s | b_i = 3\}$$

$$(3.44)$$

or

$$P_I(\theta_J) = \tfrac{1}{2}\Pr\left\{\delta T_s + \sqrt{J_0}\,T_s \sin\theta_J < N_I < -\delta T_s + \sqrt{J_0}\,T_s \sin\theta_J\right\}$$
$$+ \tfrac{1}{2}\Pr\left\{N_I < -\delta T_s + \sqrt{J_0}\,T_s \sin\theta_J\right\}$$
$$= Q\left[\sqrt{\frac{2\delta^2 T_s}{N_0}}\left(1 - \sqrt{\frac{J_0}{\delta^2}}\sin\theta_J\right)\right] + \frac{1}{2}Q\left[\sqrt{\frac{2\delta^2 T_s}{N_0}}\left(1 + \sqrt{\frac{J_0}{\delta^2}}\sin\theta_J\right)\right]$$

$$(3.45)$$

and

$$P_Q(\theta_J) = \tfrac{1}{2}\Pr\left\{\delta T_s - \sqrt{J_0}\,T_s \cos\theta_J < N_Q < -\delta T_s - \sqrt{J_0}\,T_s \cos\theta_J\right\}$$
$$+ \tfrac{1}{2}\Pr\left\{N_Q < -\delta T_s - \sqrt{J_0}\,T_s \cos\theta_J\right\}$$
$$= Q\left[\sqrt{\frac{2\delta^2 T_s}{N_0}}\left(1 + \sqrt{\frac{J_0}{\delta^2}}\cos\theta_J\right)\right] + \frac{1}{2}Q\left[\sqrt{\frac{2\delta^2 T_s}{N_0}}\left(1 - \sqrt{\frac{J_0}{\delta^2}}\cos\theta_J\right)\right].$$

$$(3.46)$$

Letting $K = 4$ in (3.40) and recognizing that $T_s = 4T_b$, we now have

$$\frac{2\delta^2 T_s}{N_0} = \frac{ST_s}{5N_0} = 2\left(\frac{2}{5}\frac{ST_b}{N_0}\right) \triangleq \frac{2E_b'}{N_0} \qquad (3.47)$$

where

$$E_b' \triangleq \tfrac{2}{5}E_b. \qquad (3.48)$$

Also,

$$\frac{J_0}{\delta^2} = \frac{10J}{\rho NS} = \frac{10J}{\rho(4W_{ss}T_b)S} = \frac{J/W_{ss}}{\rho E_b'} = \frac{N_J}{\rho E_b'}. \qquad (3.49)$$

Finally, then, the unconditional average probability of symbol error P_{s_J} for symbol intervals which are jammed is given by (3.17) with (for QASK-16)

$$P_I(\theta_J) = Q\left[\sqrt{\frac{2E_b'}{N_0}}\left(1 - \sqrt{\frac{N_J}{\rho E_b'}}\sin\theta_J\right)\right]$$

$$+ \frac{1}{2}Q\left[\sqrt{\frac{2E_b'}{N_0}}\left(1 + \sqrt{\frac{N_J}{\rho E_b'}}\sin\theta_J\right)\right]$$

$$P_Q(\theta_J) = Q\left[\sqrt{\frac{2E_b'}{N_0}}\left(1 + \sqrt{\frac{N_J}{\rho E_b'}}\cos\theta_J\right)\right]$$

$$+ \frac{1}{2}Q\left[\sqrt{\frac{2E_b'}{N_0}}\left(1 - \sqrt{\frac{N_J}{\rho E_b'}}\cos\theta_J\right)\right]. \tag{3.50}$$

For symbol intervals which are not jammed, the average symbol error probability is given by the well-known result [2]

$$P_{s_0} = 3Q\left(\sqrt{\frac{2E_b'}{N_0}}\right) - \frac{9}{4}Q^2\left(\sqrt{\frac{2E_b'}{N_0}}\right). \tag{3.51}$$

Thus, the average error probability over all symbols is once again given by (3.24) with, however, P_{s_J} of (3.17) together with (3.50) and P_{s_0} of (3.51).

As was done for FH/QPSK, one can compute the limiting performance of FH/QASK as E_b/N_0 approaches infinity. In particular, using a graphical interpretation analogous to Figure 3.3, we obtain the following result:

$$\lim_{E_b/N_0 \to \infty} P_{s_J} = \begin{cases} 0; & \dfrac{\rho E_b'}{N_J} > 1 \\[2mm] \dfrac{3}{\pi}\cos^{-1}\sqrt{\dfrac{\rho E_b'}{N_J}}; & \dfrac{1}{2} < \dfrac{\rho E_b'}{N_J} \le 1 \\[2mm] \dfrac{3}{4\pi}\cos^{-1}\sqrt{\dfrac{\rho E_b'}{N_J}} + \dfrac{9}{16}; & 0 < \dfrac{\rho E_b'}{N_J} \le \dfrac{1}{2}. \end{cases} \tag{3.52}$$

Finally, realizing that (3.25) also applies to P_{s_0} of (3.51), substituting (3.25) and (3.52) into (3.24) then gives the desired limiting behavior of the average symbol error probability of QASK-16, namely,

$$\lim_{E_b/N_0 \to \infty} P_s = \begin{cases} 0; & \dfrac{\rho E_b'}{N_J} > 1 \\[2mm] \dfrac{3\rho}{\pi}\cos^{-1}\sqrt{\dfrac{\rho E_b'}{N_J}}; & \dfrac{1}{2} < \dfrac{\rho E_b'}{N_J} \le 1 \\[2mm] \dfrac{3\rho}{4\pi}\cos^{-1}\sqrt{\dfrac{\rho E_b'}{N_J}} + \dfrac{9\rho}{16}; & 0 < \dfrac{\rho E_b'}{N_J} \le \dfrac{1}{2}. \end{cases} \tag{3.53}$$

To determine the worst case jamming situation, we again differentiate P_s, now given by (3.53), with respect to ρ and equate to zero. Recognizing that the expression for P_s of QASK-16 in the interval $1/2 < \rho E_b'/N_J \le 1$ is $3/2$ times that for P_s of QPSK in the interval $1/2 < \rho E_b/N_J \le 1$, we can immediately observe that the worst case ρ is now

$$
\rho_{wc} = \begin{cases} \dfrac{0.6306}{E_b'/N_J} = \dfrac{1.5765}{E_b/N_J}; & E_b'/N_J > 0.6306 \\[2mm] 1; & E_b'/N_J \le 0.6306 \end{cases}
\tag{3.54}
$$

and the corresponding worst case average symbol error probability performance is

$$
\lim_{E_b/N_0 \to \infty} P_{s_{max}} = \begin{cases} \dfrac{3}{2}\left(\dfrac{0.2623}{E_b'/N_J}\right) = \dfrac{0.9835}{E_b/N_J}; & E_b'/N_J > 0.6306 \\[3mm] \dfrac{3}{\pi}\cos^{-1}\sqrt{\dfrac{E_b'}{N_J}} & 0.5 < E_b'/N_J \le 0.6306 \\[3mm] \dfrac{3}{4\pi}\cos^{-1}\sqrt{\dfrac{E_b'}{N_J}} + \dfrac{9}{16}; & 0 < E_b'/N_J \le 0.5 \end{cases}
$$

$$\tag{3.55}$$

where we have also made use of (3.48).

If one encodes the QASK symbols with a perfect Gray code, then accounting only for adjacent symbol errors (which is equivalent to one bit error per symbol error), the average bit error probability for large E_b/N_0 and $\rho E_b/N_J$ is related to the average symbol error probability by (3.34), where now $M = K^2$ is the total number of symbols or $\log_2 M = \log_2 K^2$ is the number of bits/symbol. Clearly, for QASK-16,

$$
P_b \cong \tfrac{1}{4}P_s \qquad (\text{for } E_b/N_0 \gg 1, \rho E_b/N_J \gg 1).
\tag{3.56}
$$

(3.56) provides an optimistic estimate of P_b. The exact expression can be calculated via the fact that QASK-16 is obtained from independent amplitude-shift-keying on two quadrature components of a carrier. Assuming a perfectly coherent receiver, no interchannel effects exist in the demodulation process. Hence, the bit error probability P_b for QASK-16 is identical to P_b for each individual channel and is given by

$$
P_b = \rho P_{b_J} + (1 - \rho)P_{b_0}
\tag{3.57}
$$

where

$$P_{b_J} = \frac{1}{4\pi} \int_0^{2\pi} \left[P_Q(\theta_J) + P_Q^*(\theta_J) \right] d\theta_J \tag{3.58}$$

with $P_Q(\theta_J)$ as in (3.50) and $P_Q^*(\theta_J)$ equal to

$$P_Q^*(\theta_J) = Q\left[\sqrt{\frac{2E_b'}{N_0}} \left(3 + \sqrt{\frac{N_J}{\rho E_b'}} \cos\theta_J \right) \right]$$

$$- \frac{1}{2} Q\left[\sqrt{\frac{2E_b'}{N_0}} \left(5 + \sqrt{\frac{N_J}{\rho E_b'}} \cos\theta_J \right) \right]. \tag{3.59}$$

Furthermore, P_{b_0} of (3.57) represents the average bit error probability in the presence of noise only and is given by

$$P_{b_0} = \frac{3}{4} Q\left[\sqrt{\frac{2E_b'}{N_0}} \right] + \frac{1}{2} Q\left[3\sqrt{\frac{E_b'}{N_0}} \right] - \frac{1}{4} Q\left[5\sqrt{\frac{2E_b'}{N_0}} \right]. \tag{3.60}$$

Once again before presenting numerical results illustrating the evaluation of (3.57), it is of interest to examine its limiting behavior as $N_0 \to 0$. Following the approach taken for FH/QPSK, we can arrive at the following result:

$$\lim_{E_b/N_0 \to \infty} P_b = \begin{cases} 0; & \dfrac{\rho E_b'}{N_J} > 1 \\[2ex] \dfrac{3\rho}{4\pi} \cos^{-1}\sqrt{\dfrac{\rho E_b'}{N_J}}; & \dfrac{1}{9} < \dfrac{\rho E_b'}{N_J} \le 1 \\[2ex] \dfrac{3\rho}{4\pi} \cos^{-1}\sqrt{\dfrac{\rho E_b'}{N_J}} + \dfrac{2\rho}{4\pi} \cos^{-1}\sqrt{\dfrac{9\rho E_b'}{N_J}}; & \dfrac{1}{25} < \dfrac{\rho E_b'}{N_J} \le \dfrac{1}{9} \\[2ex] \dfrac{3\rho}{4\pi} \cos^{-1}\sqrt{\dfrac{\rho E_b'}{N_J}} + \dfrac{2\rho}{4\pi} \cos^{-1}\sqrt{\dfrac{9\rho E_b'}{N_J}} - \dfrac{\rho}{\pi} \cos^{-1}\sqrt{\dfrac{25\rho E_b'}{N_J}}; \\[2ex] \hspace{5cm} 0 < \dfrac{\rho E_b'}{N_J} \le \dfrac{1}{25}. \end{cases} \tag{3.61}$$

The partial-band fraction ρ corresponding to the worst case jammer (maximum P_b) is obtained by differentiating (3.61) with respect to ρ and equating to zero. Assuming that, for a fixed E_b'/N_J this worst case ρ occurs when $1/9 < \rho E_b'/N_J < 1$, then the solution to the transcendental equation which results from the differentiation is identical to (3.54). Substituting (3.54) into (3.61) gives the limiting average bit error probability performance corre-

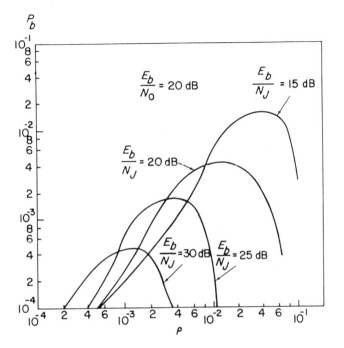

Figure 3.7. P_b versus ρ for FH/QASK-16 in tone jamming with $E_b/N_0 = 20$ dB.

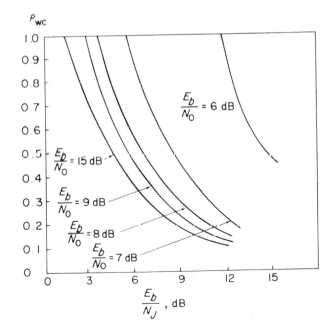

Figure 3.8. Worst case ρ versus E_b/N_J—FH/QASK-16 (tone jamming).

sponding to the worst case jammer, namely,

$$\lim_{E_b/N_0 \to \infty} P_{b_{max}} = \begin{cases} \dfrac{0.0984}{E_b'/N_J} = \dfrac{0.2459}{E_b/N_J}; & E_b'/N_J > 0.6306 \\[2em] \dfrac{3}{4\pi} \cos^{-1}\sqrt{\dfrac{E_b'}{N_J}}; & 1/9 < E_b'/N_J \le 0.6306 \\[2em] \dfrac{3}{4\pi} \cos^{-1}\sqrt{\dfrac{E_b'}{N_J}} + \dfrac{2}{4\pi} \cos^{-1}\sqrt{\dfrac{9E_b'}{N_J}}; \\[1em] \qquad\qquad\qquad\qquad 1/25 < E_b'/N_J \le 1/9 \\[2em] \dfrac{3}{4\pi} \cos^{-1}\sqrt{\dfrac{E_b'}{N_J}} + \dfrac{2}{4\pi} \cos^{-1}\sqrt{\dfrac{9E_b'}{N_J}} - \dfrac{1}{\pi} \cos^{-1}\sqrt{\dfrac{25E_b'}{N_J}}; \\[1em] \qquad\qquad\qquad\qquad 0 < E_b'/N_J \le 1/25. \end{cases}$$

$$(3.62)$$

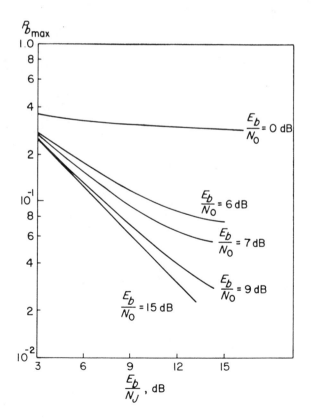

Figure 3.9. Worst case P_b versus E_b/N_J—FH/QASK-16 (tone jamming).

Figures 3.7–3.9 are the numerical evaluations of FH/QASK-16 perfor-mance which are analogous to those in Figures 3.4–3.6 characterizing FH/QPSK.

3.3 PERFORMANCE OF FH/QPSK IN THE PRESENCE OF PARTIAL-BAND NOISE JAMMING

Now assume that the jammer $J(t)$ spreads his total power J uniformly across a fraction ρ of the total hop frequency band W_{ss}. Then, insofar as the data demodulation process is concerned, the jammer appears as an additional additive noise source of power spectral density

$$N_J' = \frac{J}{\rho W_{ss}} = \frac{N_J}{\rho}. \tag{3.63}$$

Note that the power spectral density of the noise jammer defined in (3.63) is identical to the effective power spectral density defined for the multitone jammer in (3.21).

Since the jammer noise can be assumed to be independent of the background AWGN, one can add their power spectral densities and use this sum to represent the total noise perturbing the receiver. Thus, the error probability performance of FH/QPSK in the presence of partial-band noise jamming is characterized by taking the well-known results for just an AWGN background and replacing N_0 by $N_0 + N_J'$.

Without going into great detail, we then see that the average symbol error probability is once again given by (3.24), with P_{s_0} as in (3.23); however,

$$P_{s_J} = 2Q\left(\sqrt{\frac{2E_b}{N_0 + N_J'}}\right) - Q^2\left(\sqrt{\frac{2E_b}{N_0 + N_J'}}\right). \tag{3.64}$$

Substituting (3.63) into (3.64), we can rewrite P_{s_J} in the alternate form:

$$P_{s_J} = 2Q\left(\left[\left(\frac{2E_b}{N_0}\right)^{-1} + \left(\frac{2\rho E_b}{N_J}\right)^{-1}\right]^{-1/2}\right)$$
$$- Q^2\left(\left[\left(\frac{2E_b}{N_0}\right)^{-1} + \left(\frac{2\rho E_b}{N_J}\right)^{-1}\right]^{-1/2}\right). \tag{3.65}$$

Also, the average bit error probability is now

$$P_b = \rho Q\left(\left[\left(\frac{2E_b}{N_0}\right)^{-1} + \left(\frac{2\rho E_b}{N_J}\right)^{-1}\right]^{-1/2}\right) + (1 - \rho)Q\left(\sqrt{\frac{2E_b}{N_0}}\right) \tag{3.66}$$

which is identical to the result for noise jamming of FH/BPSK.

The limiting behavior of (3.66) as E_b/N_0 approaches infinity is easily seen to be

$$\lim_{E_b/N_0 \to \infty} P_b = \rho Q\left(\sqrt{\frac{2\rho E_b}{N_J}}\right). \tag{3.67}$$

Differentiating (3.67) with respect to ρ and equating the result to zero gives the transcendental equation

$$Q\left(\sqrt{\frac{2\rho E_b}{N_J}}\right) = \sqrt{\frac{\rho E_b}{N_J}}\left(\frac{e^{-\rho E_b/N_J}}{2\sqrt{\pi}}\right), \tag{3.68}$$

whose solution for ρ corresponds to the worst case jammer situation. In particular, letting

$$x = \sqrt{\frac{\rho E_b}{N_J}}, \tag{3.69}$$

(3.68) then becomes

$$Q(\sqrt{2}\,x) = \frac{xe^{-x^2}}{2\sqrt{\pi}} \tag{3.70}$$

which, when solved numerically, yields

$$x = 0.842. \tag{3.71}$$

Equating (3.69) and (3.71) gives the partial-band fraction for the worst case

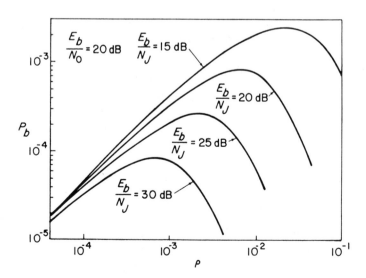

Figure 3.10. P_b versus ρ for FH/QPSK in noise jamming with $E_b/N_0 = 20$ dB.

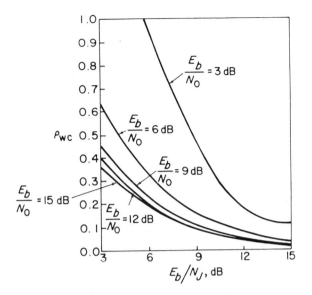

Figure 3.11. Worst case ρ versus E_b/N_J—FH/QPSK (noise jamming).

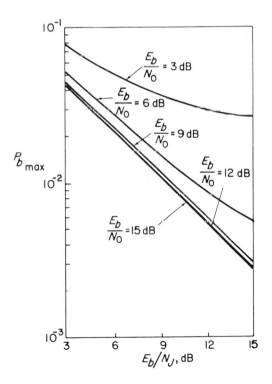

Figure 3.12. Worst case P_b versus E_b/N_J—FH/QPSK (noise jamming).

jammer, namely,

$$\rho_{\text{wc}} = \begin{cases} \dfrac{0.7090}{E_b/N_J}; & E_b/N_J > 0.7090 \\ 1; & E_b/N_J \leq 0.7090 \end{cases} \tag{3.72}$$

and a corresponding maximum average bit error probability

$$\lim_{E_b/N_0 \to \infty} P_{b_{\max}} = \begin{cases} \dfrac{0.0829}{E_b/N_J}; & E_b/N_J > 0.7090 \\ Q\left(\sqrt{2E_b/N_J}\right); & E_b/N_J \leq 0.7090. \end{cases} \tag{3.73}$$

Figures 3.10–3.12 characterize the performance of FH/QPSK in the presence of noise jamming, as computed from (3.66).

3.4 PERFORMANCE OF FH/QASK IN THE PRESENCE OF PARTIAL-BAND NOISE JAMMING

Assuming the same noise jammer model as that discussed in the previous section for FH/QPSK, the average symbol error probability of FH/QASK-16 is then given by (3.24), with P_{s_0} as in (3.51), but now

$$P_{s_J} = 3Q\left(\left[\left(\frac{2E_b'}{N_0}\right)^{-1} + \left(\frac{2\rho E_b'}{N_J}\right)^{-1}\right]^{-1/2}\right)$$
$$- \frac{9}{4}Q^2\left(\left[\left(\frac{2E_b'}{N_0}\right)^{-1} + \left(\frac{2\rho E_b'}{N_J}\right)^{-1}\right]^{-1/2}\right). \tag{3.74}$$

The limiting behavior of (3.24), together with (3.51) and (3.74) as E_b/N_0 approaches infinity is simply

$$\lim_{E_b/N_0 \to \infty} P_s = 3\rho Q\left(\sqrt{\frac{2\rho E_b'}{N_J}}\right) - \frac{9\rho}{4}Q^2\left(\sqrt{\frac{2\rho E_b'}{N_J}}\right) \tag{3.75}$$

which yields a partial-band fraction for the worst case jammer given by the solution to

$$3Q(\sqrt{2}\,x) - 3x\frac{e^{-x^2}}{\sqrt{\pi}} - \frac{9}{4}Q^2(\sqrt{2}\,x) + \frac{9}{2}Q(\sqrt{2}\,x)\left[\frac{xe^{-x^2}}{2\sqrt{\pi}}\right] = 0,$$

$$\tag{3.76}$$

with

$$x = \sqrt{\frac{\rho E_b'}{N_J}} \,. \tag{3.77}$$

Numerical solution of (3.76) yields

$$\rho_{wc} = \begin{cases} \dfrac{0.7921}{(2/5) E_b/N_J} = \dfrac{1.9802}{E_b/N_J}; & E_b/N_J > 1.9802 \\[2mm] 1; & E_b/N_J \le 1.9802 \end{cases} \tag{3.78}$$

and a corresponding worst case average symbol error probability

$$\lim_{E_b/N_0 \to \infty} P_{s_{max}} = \begin{cases} \dfrac{0.5700}{E_b/N_J}; & E_b/N_J > 1.9802 \\[3mm] 3Q\left(\sqrt{\dfrac{2E_b'}{N_J}}\right) - \dfrac{9}{4}Q^2\left(\sqrt{\dfrac{2E_b'}{N_J}}\right); & E_b/N_J \le 1.9802. \end{cases} \tag{3.79}$$

The average bit error probability P_b is obtained from (3.57), where P_{b_0} is given by (3.60) and P_{b_J} is given by

$$P_{b_J} = \tfrac{3}{4}Q(x) + \tfrac{1}{2}Q(3x) - \tfrac{1}{4}Q(5x) \tag{3.80}$$

with

$$x = \left[\left(\frac{2E_b'}{N_0}\right)^{-1} + \left(\frac{2\rho E_b'}{N_J}\right)^{-1} \right]^{-1/2}. \tag{3.81}$$

The limiting behavior of P_b as E_b/N_0 approaches infinity is then

$$\lim_{E_b/N_0 \to \infty} P_b = \rho\left[\tfrac{3}{4}Q(x_\infty) + \tfrac{1}{2}Q(3x_\infty) - \tfrac{1}{4}Q(5x_\infty)\right] \tag{3.82}$$

where x_∞ is the corresponding limit of x in (3.81), namely,

$$x_\infty = \sqrt{\frac{2\rho E_b'}{N_J}} \,. \tag{3.83}$$

The worst case jammer is found through a procedure analogous to the one in Section 3.3:

$$\rho_{wc} = \begin{cases} \dfrac{1.758}{E_b/N_J}; & E_b/N_J > 1.758 \\[2mm] 1; & E_b/N_J \le 1.758 \end{cases} \tag{3.84}$$

and a corresponding maximum average bit error probability of

$$\lim_{E_b/N_0 \to \infty} P_{b_{max}} = \begin{cases} \dfrac{0.1555}{E_b/N_J}; & E_b/N_J > 1.758 \\[2mm] \dfrac{3}{4}Q\left[\sqrt{\dfrac{2E_b'}{N_J}}\right] + \dfrac{1}{2}Q\left[3\sqrt{\dfrac{2E_b'}{N_J}}\right] - \dfrac{1}{4}Q\left[5\sqrt{\dfrac{2E_b'}{N_J}}\right]; \\[2mm] & E_b/N_J \le 1.758. \end{cases}$$

$$(3.85)$$

It is interesting to note that the ratio $(\lim P_{b_{max}}/\lim P_{s_{max}})$, as computed from (3.79) and (3.85) is 0.2728, which is slightly higher than the approximate value of $1/4$ as in (3.56). Furthermore, those limits are achieved for values of ρ (see (3.78) and (3.84)) which are different but relatively close.

Figures 3.13–3.15 characterize the performance of FH/QASK-16 in the presence of noise jamming, as computed from the results given in this section. In Figure 3.13 the dashed line represents the locus of the maxima of P_b versus ρ, with E_b/N_J as a parameter. It can be seen that the value of 20 dB for E_b/N_0 is sufficiently high to warrant an almost linear relationship between $P_{b_{max}}$ and ρ, which closely conforms with the theoretically predicted line (see (3.84) and (3.85))

$$\lim_{E_b/N_0 \to \infty} P_{b_{max}} = 0.0878\rho. \qquad (3.86)$$

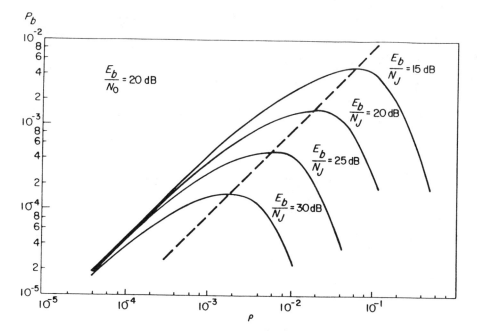

Figure 3.13. P_b versus ρ for FH/QASK-16 in noise jamming with $E_b/N_0 = 20$ dB.

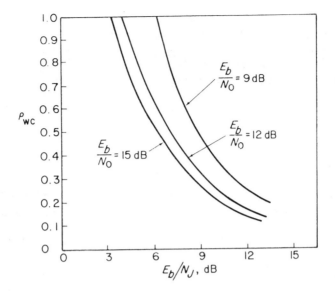

Figure 3.14.   Worst case ρ versus E_b/N_J—FH/QASK-16 (noise jamming).

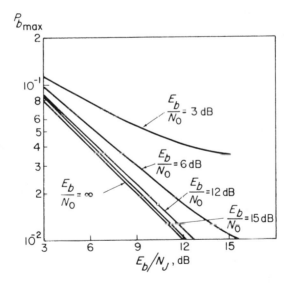

Figure 3.15.   Worst case P_b versus E_b/N_J—FH/QASK-16 (noise jamming).

3.5   PERFORMANCE OF FH / PN / QPSK IN THE PRESENCE OF PARTIAL-BAND MULTITONE JAMMING

When a pseudonoise (PN) balanced modulation is superimposed on an FH/QPSK signal, each jammer tone of power J_0 is then spread over a bandwidth equal to the PN chip rate R_c. Let $R_s = 1/T_s$ be the information symbol rate and R_c/R_s represent the *processing gain associated with the PN spreading*, herein taken to be a larger integer I. It follows from the assumption of a large processing gain that the spread tone jammer possesses a fairly flat power spectral density within the data modulation bandwidth and is now caused to behave like a white noise jammer of spectral density

$$N_J' = J_0/R_c = J_0 T_c. \qquad (3.87)$$

Using (3.4) and (3.5), we can rewrite (3.87) as

$$N_J' = J/\rho N R_c. \qquad (3.88)$$

Since the hop frequency slots (assumed to be *contiguous*) must now be R_c wide to accommodate the PN modulation, in terms of the total hop frequency band W_{ss} and the number of hop slots N in that band, we then have

$$N = W_{ss}/R_c. \qquad (3.89)$$

Combining (3.88) and (3.89) gives

$$N_J' = \frac{J/W_{ss}}{\rho} = \frac{N_J}{\rho}, \qquad (3.90)$$

which is identical to (3.63) (the case of FH/QPSK in the presence of partial-band noise jamming) *independent of the chip rate R_c.*

The above discussion concerns the spectral characteristics of the spread tone jammer which, as concluded, resembles a white noise jammer with identical spectral density. We shall now discuss the assumptions and theoretical adjustments under which the spread tone jammer can also be treated as a *Gaussian* noise interference.

Consider an FH/PN/QPSK demodulator, similar to the one in Figure 3.2, where the despreading process now also includes a PN code correlator following the frequency dehopper. Let p represent the PN code length (in number of chips). Accounting for the effects of the spread jammer only (i.e., neglecting thermal noise), it follows that the decision variable z_I for the in-phase channel (see (3.11)) becomes

$$z_I = a_i \sqrt{\frac{S}{2}} T_s - \sqrt{J_0} \sin\theta_J C_{PN} \qquad (3.91)$$

where

$$C_{PN} = \int_{(i-1)T_s}^{iT_s} c(t)\, dt \qquad (3.92)$$

and $c(t)$ is the ± 1-valued PN waveform. In deriving (3.91), ideal PN code synchronization at the receiver has been assumed. Recalling that the processing gain is the ratio of the PN chip rate to the data symbol rate, or equivalently $T_s/T_c = I$, then it follows that the integral in (3.92) amounts to a partial correlation of the PN code, starting from some random phase, provided that the number I of integrated chips does not equal the code length p. Phrased differently, the conclusions to follow hold when $1 \ll I \ll p$ (or in a weaker sense when $1 \ll I$ modulo $p \ll p$). This is because, as is well known, the full-period integration of a PN code equals the constant $1/p$, in which case, no randomness about C_{PN} exists. On the other hand, when p is very large, successive code chips can be considered almost independent, identically distributed ± 1-valued random variables, in which case the condition $I \ll p$ would provide an approximate binomial distribution for the random variable C_{PN}. The additional constraint $I \gg 1$ then causes this binomial distribution to behave like a Gaussian distribution.

Arguments similar to the above were made in Chapter 1 in connection with the evaluation of the performance of pulse-jammed direct-sequence spread-spectrum systems. A more rigorous treatment of the validity of the Gaussian assumption for C_{PN} has been examined in [6] for a variety of PN and Gold codes, with sufficient evidence that it holds, at least approximately, for a wide class of such codes.

Let us now return to (3.91). Conditioned on the $(0, 2\pi)$-uniformly distributed phase θ_J, z_I is, according to the above, a Gaussian random variable whose conditional variance is

$$\text{var}\{z_I|\theta_J\} = T_s T_c J_0 \sin^2\theta_J. \tag{3.93}$$

Hence, when the jammer is present, the average bit error probability for the I channel (identically for the Q channel) is given by

$$P_{b_I}^t = \frac{1}{2\pi} \int_0^{2\pi} Q\left(\sqrt{\frac{ST_s}{2T_c J_0}} \frac{1}{|\sin\theta_J|}\right) d\theta_J = \frac{2}{\pi} \int_0^{\pi/2} Q\left(\sqrt{\frac{\rho E_b}{N_J}} \frac{1}{\sin\theta_J}\right) d\theta_J \tag{3.94}$$

where (3.18), (3.87), (3.90), and (3.93) have been used in arriving at this result. Finally, multiplying (3.94) by ρ gives the average bit error probability of FH/PN/QPSK in the presence of partial-band multitone jamming, namely,

$$P_b^t = \frac{2\rho}{\pi} \int_0^{\pi/2} Q\left(\sqrt{\frac{\rho E_b}{N_J}} \frac{1}{\sin\theta_J}\right) d\theta_J. \tag{3.95}$$

Although (3.95) can be used in assessing the effect of a spread tone jammer on system performance, we shall now indicate an even simpler way of evaluating the tone interference effect by means of converting the spread tone jammer to an *equivalent* AWGN interference, where the equivalence is understood *in terms of its effect on bit error probability*. Let N_{0_J} represent the one-sided power spectral density of AWGN interference whose statistical characteristics remain the same after the PN despreader. Clearly, the bit error probability in this case is given by

$$P_{b_J}'' = Q\left(\sqrt{\frac{2E_b}{N_{0_J}}}\right). \tag{3.96}$$

Equations (3.94) and (3.96) have been plotted in Figure 3.16. The abscissa is the signal-to-jammer power ratio SJR in decibels where, for the spread tone case $(SJR)^t = E_b/N_J$ while, for the noise case, $(SJR)^n = E_b/N_{0_J}$. A careful examination of Figure 3.16 reveals that, for values of $(SJR)^t$ up to approximately 12 dB, the difference $(SJR)^t_{(dB)} - (SJR)^n_{(dB)}$ between the signal-to-jammer ratios which achieve the same performance is, to a high degree of accuracy, linearly increasing with $(SJR)^t_{(dB)}$, the slope of the line being 0.2. It is therefore concluded that, given a spread tone jammer of $(SJR)^t_{(dB)}$, an

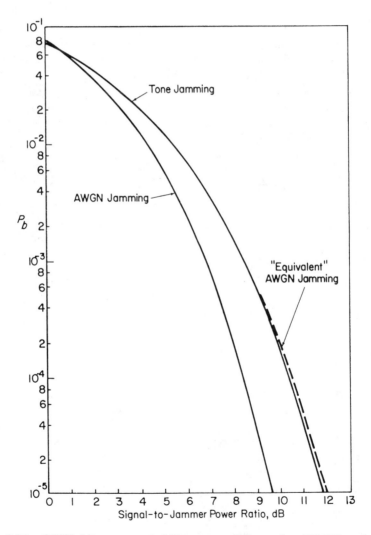

Figure 3.16. QPSK bit error probability versus $SJR_{(dB)}$ for AWGN and tone jamming and spread tone "equivalent" AWGN jamming.

"equivalent" AWGN jammer can be devised for which

$$(\text{SJR})^{n}_{\text{equiv(dB)}} = (0.8) \times (\text{SJR})^{t}_{\text{(dB)}}. \tag{3.97}$$

If the SJR's are measured in ordinary numbers rather than decibels, then (3.97) implies that

$$(\text{SJR})^{n}_{\text{equiv}} = \left((\text{SJR})^{t}\right)^{0.8}. \tag{3.98}$$

From (3.96) and (3.98), it then follows that the spread tone jammer can be conveniently thought of as an *"equivalent"* AWGN jammer with corresponding bit error probability

$$P^{t}_{b_J} = Q\left(\sqrt{2\left(\frac{\rho E_b}{N_J}\right)^{0.8}}\right). \tag{3.99}$$

Equation (3.99) has been plotted in Figure 3.16 (dashed lines), from which the high degree of agreement with the exact expression for $P^{t}_{b_J}$ can be witnessed. The range of applicability of (3.99) is, for the current purposes, more than adequate since, in the frequency bands where the jammer is present, the system is forced to operate at high bit error probabilities (this is especially true when the jammer strategy, i.e., choice of ρ, has been optimized). From (3.99), it follows that the overall average bit error probability (accounting for partial-band jamming) is given by

$$P^{t}_{b} = \rho Q\left(\sqrt{2\left(\frac{\rho F_{h}}{N_J}\right)^{0.8}}\right). \tag{3.100}$$

The worst case jammer can be found from (3.100), with the result

$$\rho_{\text{wc}} = \begin{cases} \dfrac{0.9220}{E_b/N_J}; & E_b/N_J \geq 0.9220 \\[2mm] 1; & E_b/N_J < 0.9220 \end{cases} \tag{3.101}$$

with corresponding

$$P^{t}_{b_{\text{max}}} = \begin{cases} \dfrac{0.0789}{E_b/N_J}; & E_b/N_J \geq 0.9220 \\[2mm] Q\left(\sqrt{2\left(\dfrac{E_b}{N_J}\right)^{0.8}}\right); & E_b/N_J < 0.9220. \end{cases} \tag{3.102}$$

A comparison of (3.102) with (3.73) indicates that the worst tone jammer for FH/PN/QPSK is slightly less effective than the worst noise jammer for FH/QPSK.

3.6 PERFORMANCE OF FH / PN / QASK IN THE PRESENCE OF PARTIAL-BAND MULTITONE JAMMING

A line of thought similar to that of Section 3.6 also applies here. The tone/noise jamming equivalence discussed in Section 3.6 is also applicable because the first Q function in (3.82) dominates over the other terms in the range of interest. Without belaboring the point, we summarize here the pertinent results (also in the absence of thermal noise)

$$P_b^t = \rho P_{b_J} \tag{3.103}$$

where P_{b_J} is given by (3.80) with

$$x = \sqrt{2\left(\frac{2\rho E_b}{5N_J}\right)^{0.8}}. \tag{3.104}$$

The corresponding worst case results are

$$\rho_{\text{wc}} = \begin{cases} \dfrac{2.3050}{E_b/N_J}; & E_b/N_J \geq 2.3050 \\ 1; & E_b/N_J < 2.3050 \end{cases} \tag{3.105}$$

and

$$P_{b_{\max}}^t = \begin{cases} \dfrac{0.1483}{E_b/N_J}; & E_b/N_J \geq 2.3050 \\ \tfrac{3}{4}Q(x_1) + \tfrac{1}{2}Q(3x_1) - \tfrac{1}{4}Q(5x_1); & E_b/N_J < 2.3050 \end{cases} \tag{3.106}$$

where

$$x_1 = \sqrt{2\left(\frac{2E_b}{5N_J}\right)^{0.8}}. \tag{3.107}$$

Thus far in this chapter, we have focussed our entire attention on how augmenting "ideal" QPSK and QASK with a FH/SS technique allows them to combat a partial-band multitone or noise jammer. The word "ideal" is used here to indicate that although the underlying modulations are so-called "bandwidth efficient" because of the structure of their signal constellations, we have not considered any additional bandwidth conservation produced by transmitter and receiver filtering. Stated another way, we have assumed throughout that the basic transmission pulse is rectangular in shape. In the remainder of this chapter, we examine the detection efficiency of quadrature partial response Class I signals in the presence of the identical jamming scenario postulated thus far. Consideration of this form of modulation represents a departure from the basic assumption mentioned above as discussed in the next section.

3.7 PERFORMANCE OF FH / QPR IN THE PRESENCE OF PARTIAL-BAND MULTITONE JAMMING

In the design of most digital data transmission systems, the occurrence of intersymbol interference is ordinarily treated as an undesirable phenomenon. Certain signalling systems, however, utilize a controlled amount of intersymbol interference to achieve certain beneficial effects. These systems have been called, variously, *duobinary*, *polybinary*, and *partial response* [7]–[10]. When such a modulation type is transmitted on quadrature carriers over a common channel, the acronym *quadrature partial response* (QPR) has been applied. A typical modulator, which generates an FH three-level QPR signal,[4] is illustrated in Figure 3.17. The transmitted signal in the i-th hop interval is

$$s^{(i)}(t) = \sqrt{2}\,A\left[\sum_{n=-\infty}^{\infty} c_n h_T(t - 2nT_b)\right]\cos \omega_h^{(i)}t$$
$$+ \sqrt{2}\,A\left[\sum_{n=-\infty}^{\infty} d_n h_T(t - (2n+1)T_b)\right]\sin \omega_h^{(i)}t, \quad (3.108)$$

where again $\omega_h^{(i)}$ is the particular carrier frequency selected by the frequency hopper for this interval according to the designated SS code, $h_T(t)$ is the impulse response of the transmit filter $H_T(\omega)$, and T_b is the bit time interval. The amplitude A will soon be related to the average transmitted power S.

The conventional demodulator for an FH three-level QPR signal corrupted by additive Gaussian noise $n(t)$ is illustrated in Figure 3.18. From a noise power standpoint, it is advantageous to split the overall partial response (duobinary) signal shaping equally between the transmit and receive filters. Since, for a three-level signal, the overall shaping characteristic is given by

$$H(\omega) = \begin{cases} 4T_b\cos \omega T_b; & |\omega| < \dfrac{\pi}{2T_b} \\ 0; & \text{otherwise,} \end{cases} \quad (3.109)$$

then, based on the above statement, we have

$$H_T(\omega) = H_R(\omega) = (H(\omega))^{1/2} = \begin{cases} (4T_b\cos \omega T_b)^{1/2}; & |\omega| < \dfrac{\pi}{2T_b} \\ 0; & \text{otherwise.} \end{cases}$$
$$(3.110)$$

---

[4] We shall concentrate our efforts on only FH three-level QPR. Extension to the case of FH $(2L - 1)$-level QPR is straightforward.

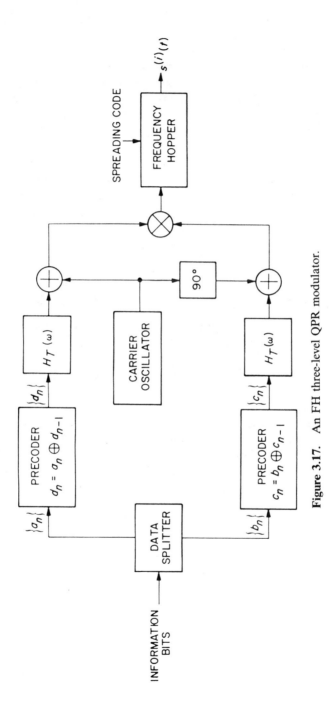

Figure 3.17. An FH three-level QPR modulator.

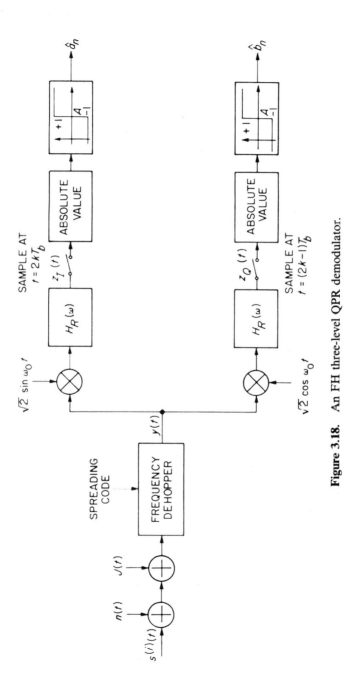

Figure 3.18.   An FH three-level QPR demodulator.

Also, the impulse response $h(t)$ corresponding to $H(\omega)$ of (3.109) is given by

$$h(t) = \frac{4}{\pi} \left[\frac{\cos \dfrac{\pi t}{2T_b}}{1 - \dfrac{t^2}{T_b^2}} \right]. \tag{3.111}$$

Notice that, if the response sampling time t_0 is chosen to be $-T_b$, then

$$h_n \triangleq h(t)|_{t=2nT_b+t_0} = \begin{cases} 1; & n = 0, 1 \\ 0; & \text{otherwise,} \end{cases} \tag{3.112}$$

and the controlled intersymbol interference comes from the preceding symbol only.

In a hop interval which contains the partial-band multitone jammer $J(t)$, the total received signal $y(t)$ after dehopping can be expressed in the form

$$y(t) = \sqrt{2} \left[A \sum_{n=-\infty}^{\infty} c_n h_T(t - 2nT_b) + N_c(t) \right] \cos \omega_0 t$$

$$+ \sqrt{2} \left[A \sum_{n=-\infty}^{\infty} d_n h_T(t - (2n+1)T_b) - N_s(t) \right] \sin \omega_0 t$$

$$+ \sqrt{2J_0} \cos(\omega_0 t + \theta_J). \tag{3.113}$$

After demodulation by the in-phase and quadrature reference signals,

$$r_I(t) = \sqrt{2} \sin \omega_0 t$$

$$r_Q(t) = \sqrt{2} \cos \omega_0 t, \tag{3.114}$$

and receive filtering by $H_R(\omega)$, the following signals result:

$$z_Q(t) \triangleq \left[y(t)(\sqrt{2} \cos \omega_0 t) \right] \circledast h_R(t)$$

$$= A \sum_{n=-\infty}^{\infty} c_n h(t - 2nT_b) + \sqrt{J_0} H_R(0) \cos \theta_J + \hat{N}_c(t)$$

$$z_I(t) \triangleq \left[y(t)(\sqrt{2} \sin \omega_0 t) \right] \circledast h_R(t)$$

$$= A \sum_{n=-\infty}^{\infty} d_n h(t - (2n+1)T_b) - \sqrt{J_0} H_R(0) \sin \theta_J - \hat{N}_s(t),$$

$$\tag{3.115}$$

where

$$\hat{N}_c(t) \triangleq H_R(p)N_c(t)$$

$$\hat{N}_s(t) \triangleq H_R(p)N_s(t) \tag{3.116}$$

with p the Heaviside operator and

$$\hat{\sigma}^2 \triangleq E\{\hat{N}_c^2(t)\} = E\{\hat{N}_s^2(t)\}$$

$$= \frac{N_0}{2}\left[\frac{1}{2\pi}\int_{-\infty}^{\infty}|H_R(\omega)|^2\,d\omega\right] = \frac{2N_0}{\pi}. \tag{3.117}$$

The sampled values of $z_c(t)$ and $z_s(t)$ are given by

$$z_{Qk} \triangleq z_Q\big[(2k-1)T_b\big] - A(c_k \mid c_{k-1}) + \sqrt{4J_0T_b}\,\cos\theta_J + \hat{N}_{ck}$$

$$z_{Ik} \triangleq z_I(2kT_b) = A(d_k + d_{k-1}) - \sqrt{4J_0T_b}\,\sin\theta_J - \hat{N}_{sk} \tag{3.118}$$

where

$$\hat{N}_{ck} \triangleq \hat{N}_c\big[(2k-1)T_b\big]$$

$$\hat{N}_{sk} \triangleq \hat{N}_s(2kT_b) \tag{3.119}$$

and we have made use of the fact (see (3.110)) that $H_R(0) = \sqrt{4T_b}$.

Table 3.1 lists the four possible combinations of precoded symbols c_k and c_{k-1} (or d_k and d_{k-1}) and the corresponding duobinary values $c_k + c_{k-1}$ (or $d_k + d_{k-1}$). Here we use $+1$ and -1 symbols rather than zeros and ones; hence, the modulo-2 operation in the precoder is replaced by arithmetic multiplication.

From Table 3.1, it is clear that the detection criteria should be

$$\hat{b}_k = \begin{cases} 1 & \text{if } |z_{ck}| > A \\ -1 & \text{if } |z_{ck}| < A \end{cases}; \qquad \hat{a}_k = \begin{cases} 1 & \text{if } |z_{sk}| > A \\ -1 & \text{if } |z_{sk}| < A \end{cases}. \tag{3.120}$$

Thus, since the a_k's and b_k's are equally likely, the in-phase and quadrature

Table 3.1
Transformation of the input symbols into their duobinary equivalents.

Transmitted Symbol b_k	Received Symbols		Duobinary Value $c_k + c_{k-1}$
	c_k	c_{k-1}	
-1	$+1$	-1	0
-1	-1	$+1$	0
$+1$	$+1$	$+1$	$+2$
$+1$	-1	-1	-2

bit error probabilities are given by

$$P_Q(\theta_J) \triangleq \Pr\{\hat{b}_k \neq b_k\} = \tfrac{1}{2}\Pr\{|z_{Qk}| > A| \ c_k + c_{k-1} = 0\}$$
$$+ \tfrac{1}{4}\Pr\{|z_{Qk}| < A| \ c_k + c_{k-1} = 2\}$$
$$+ \tfrac{1}{4}\Pr\{|z_{Qk}| < A| \ c_k + c_{k-1} = -2\}$$

$$P_I(\theta_J) \triangleq \Pr\{\hat{a}_k \neq a_k\} = \tfrac{1}{2}\Pr\{|z_{Ik}| > A| \ d_k + d_{k-1} = 0\}$$
$$+ \tfrac{1}{4}\Pr\{|z_{Ik}| < A| \ d_k + d_{k-1} = 2\}$$
$$+ \tfrac{1}{4}\Pr\{|z_{Ik}| < A| \ d_k + d_{k-1} = -2\} \tag{3.121}$$

or, using (3.118),

$$P_Q(\theta_J) = \frac{3}{4}Q\left(\frac{A + \sqrt{4J_0 T_b}\,\cos\theta_J}{\hat{\sigma}}\right) - \frac{1}{4}Q\left(\frac{3A + \sqrt{4J_0 T_b}\,\cos\theta_J}{\hat{\sigma}}\right)$$
$$+ \frac{3}{4}Q\left(\frac{A - \sqrt{4J_0 T_b}\,\cos\theta_J}{\hat{\sigma}}\right) - \frac{1}{4}Q\left(\frac{3A - \sqrt{4J_0 T_b}\,\cos\theta_J}{\hat{\sigma}}\right)$$

$$P_I(\theta_J) = \frac{3}{4}Q\left(\frac{A - \sqrt{4J_0 T_b}\,\sin\theta_J}{\hat{\sigma}}\right) - \frac{1}{4}Q\left(\frac{3A - \sqrt{4J_0 T_b}\,\sin\theta_J}{\hat{\sigma}}\right)$$
$$+ \frac{3}{4}Q\left(\frac{A + \sqrt{4J_0 T_b}\,\sin\theta_J}{\hat{\sigma}}\right) - \frac{1}{4}Q\left(\frac{3A + \sqrt{4J_0 T_b}\,\sin\theta_J}{\hat{\sigma}}\right)$$

$$\tag{3.122}$$

To express (3.122) in terms of more meaningful parameters such as signal-to-Gaussian noise ratio and signal-to-jamming noise ratio, we must relate the signal amplitude A to the average transmitted power S. Since,

$$\overline{c_n c_m} = \overline{d_n d_m} = \begin{cases} 1; & n = m \\ 0; & n \neq m \end{cases} \tag{3.123}$$

then from (3.108), we have that

$$S = (\sqrt{2}\,A)^2 \left[\frac{1}{2T_b}\int_0^{2T_b}\sum_{n=-\infty}^{\infty} h_T^2(t - 2nT_b)\,dt\right]$$
$$= 2A^2\left[\frac{1}{2T_b}\int_{-\infty}^{\infty} h_T^2(t)\,dt\right] = 2A^2\left[\frac{1}{2T_b}\left(\frac{1}{2\pi}\int_{-\infty}^{\infty}|H_T(\omega)|^2\,d\omega\right)\right].$$

$$\tag{3.124}$$

Substituting (3.110) into (3.124) and evaluating the integral gives

$$S = \frac{4A^2}{\pi T_b}. \tag{3.125}$$

Combining (3.117) and (3.125) gives an effective rms signal-to-Gaussian noise ratio of

$$\frac{A}{\hat{\sigma}} = \sqrt{\frac{\pi^2}{16}\left(\frac{2E_b}{N_0}\right)}. \tag{3.126}$$

Also, from (3.21) and (3.125), we have

$$\frac{\sqrt{4J_0 T_b}}{A} = \sqrt{\frac{8N_J}{\rho\pi E_b}} = \sqrt{\frac{\pi}{2}\left(\frac{16}{\pi^2}\right)\left(\frac{N_J}{\rho E_b}\right)}. \tag{3.127}$$

Finally, substituting (3.126) and (3.127) into (3.122) results in

$$P_Q(\theta_J) = \frac{3}{4}\left\{Q\left[\sqrt{\frac{2E_b'}{N_0}}\left(1 + \sqrt{\frac{\pi}{2}\left(\frac{N_J}{\rho E_b'}\right)}\cos\theta_J\right)\right]\right.$$

$$+ Q\left[\sqrt{\frac{2E_b'}{N_0}}\left(1 - \sqrt{\frac{\pi}{2}\left(\frac{N_J}{\rho E_b'}\right)}\cos\theta_J\right)\right]\right\}$$

$$- \frac{1}{4}\left\{Q\left[\sqrt{\frac{2E_b'}{N_0}}\left(3 - \sqrt{\frac{\pi}{2}\left(\frac{N_J}{\rho E_b'}\right)}\cos\theta_J\right)\right]\right.$$

$$+ Q\left[\sqrt{\frac{2E_b'}{N_0}}\left(3 + \sqrt{\frac{\pi}{2}\left(\frac{N_J}{\rho E_b'}\right)}\cos\theta_J\right)\right]\right\}$$

$$P_I(\theta_J) = \frac{3}{4}\left\{Q\left[\sqrt{\frac{2E_b'}{N_0}}\left(1 - \sqrt{\frac{\pi}{2}\left(\frac{N_J}{\rho E_b'}\right)}\sin\theta_J\right)\right]\right.$$

$$+ Q\left[\sqrt{\frac{2E_b'}{N_0}}\left(1 + \sqrt{\frac{\pi}{2}\left(\frac{N_J}{\mu E_b'}\right)}\sin\theta_J\right)\right]\right\}$$

$$- \frac{1}{4}\left\{Q\left[\sqrt{\frac{2E_b'}{N_0}}\left(3 + \sqrt{\frac{\pi}{2}\left(\frac{N_J}{\rho E_b'}\right)}\sin\theta_J\right)\right]\right.$$

$$+ Q\left[\sqrt{\frac{2E_b'}{N_0}}\left(3 - \sqrt{\frac{\pi}{2}\left(\frac{N_J}{\rho E_b'}\right)}\sin\theta_J\right)\right]\right\} \tag{3.128}$$

where we have further introduced the notation

$$E_b' = \frac{\pi^2}{16}E_b. \tag{3.129}$$

The average probability of bit error P_{b_J} in a detection interval which is jammed is obtained by averaging $P_Q(\theta_J)$ or $P_I(\theta_J)$ of (3.128) over the uniform distribution of θ_J; namely,

$$P_{b_J} = \frac{1}{2\pi} \int_0^{2\pi} P_Q(\theta_J) \, d\theta_J = \frac{1}{2\pi} \int_0^{2\pi} P_I(\theta_J) \, d\theta_J. \qquad (3.130)$$

Since, on the average, the fraction ρ of the total number of detection intervals is jammed, the average bit error probability over all detection intervals (jammed and unjammed) is given by (3.57) where

$$P_{b_0} = \frac{3}{2} Q\left(\sqrt{\frac{2E_b'}{N_0}} \right) - \frac{1}{2} Q\left(3\sqrt{\frac{2E_b'}{N_0}} \right) \qquad (3.131)$$

represents the average bit error probability of duobinary QPSK in the absence of jamming. Note that the average bit error probability performance of duobinary, as given by [7, Eq. (4.114)], agrees only with the leading term of (3.131) and is thus only asymptotically correct as E_b/N_0 approaches infinity.

The symbol error probability of duobinary QPSK is, as for any QPSK system, the probability that either the in-phase or the quadrature bit is in error. Thus,

$$P_s = \Pr\{ \hat{a}_i \neq a_i \text{ or } \hat{b}_i \neq b_i \} = \Pr\{ \hat{a}_i \neq a_i \} + \Pr\{ \hat{b}_i \neq b_i \}$$

$$- \Pr\{ \hat{a}_i \neq a_i \} \Pr\{ \hat{b}_i \neq b_i \}, \qquad (3.132)$$

which when averaged over all symbols (jammed and unjammed) gives the result in (3.24) where P_{s_J} is given by (3.17) together with now (3.122) and

$$P_{s_0} = 3Q\left(\sqrt{\frac{2E_b'}{N_0}} \right) - \frac{9}{4} Q^2\left(\sqrt{\frac{2E_b'}{N_0}} \right) - Q\left(3\sqrt{\frac{2E_b'}{N_0}} \right)$$

$$+ \frac{3}{2} Q\left(\sqrt{\frac{2E_b'}{N_0}} \right) Q\left(3\sqrt{\frac{2E_b'}{N_0}} \right) - \frac{1}{4} Q^2\left(3\sqrt{\frac{2E_b'}{N_0}} \right). \quad (3.133)$$

Note that the first two terms of (3.133) resemble the functional form of the average symbol error probability expression for QASK-16 as in (3.51). There, however, E_b' is related to the true bit energy E_b by $E_b' = (2/5)E_b$.

Figure 3.19 is a typical plot of P_b (as evaluated from (3.57)) versus ρ, with E_b/N_J as a parameter and $E_b/N_0 = 20$ dB. One again observes that, by fixing E_b/N_0 and E_b/N_J, there exists a value of ρ which maximizes P_b and thus represents the worst case multitone jammer situation. Figure 3.20 is a

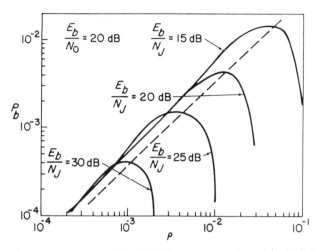

Figure 3.19. P_b versus ρ for FH/QPR in tone jamming with $E_b/N_0 = 20$ dB.

plot of worst case ρ versus E_b/N_J, with E_b/N_0 as a parameter. Figure 3.21 illustrates the corresponding plot of $P_{b_{max}}$ versus E_b/N_J, with E_b/N_0 fixed.

As was true for the FH/QPSK and FH/QASK modulations considered previously, it is of interest to study the limiting behavior of FH/QPR in the presence of multitone jamming as the Gaussian noise (e.g., N_0) goes to zero. Let

$$N_J' \triangleq \frac{\pi}{2} N_J. \qquad (3.134)$$

Then, for FH/QPR, we obtain the following results:

$$\lim_{E_b/N_0 \to \infty} P_{b_J}$$

$$= \begin{cases} 0; & \dfrac{\rho E_b'}{N_J'} > 1 \\[2ex] \dfrac{3}{2\pi} \cos^{-1} \sqrt{\dfrac{\rho E_b'}{N_J'}}; & \dfrac{1}{9} < \dfrac{\rho E_b'}{N_J'} < 1 \\[2ex] \dfrac{1}{2\pi} \left[3 \cos^{-1} \sqrt{\dfrac{\rho E_b'}{N_J'}} - \cos^{-1}\left(3\sqrt{\dfrac{\rho E_b'}{N_J'}} \right) \right]; & 0 < \dfrac{\rho E_b'}{N_J'} < \dfrac{1}{9}. \end{cases}$$

$$(3.135)$$

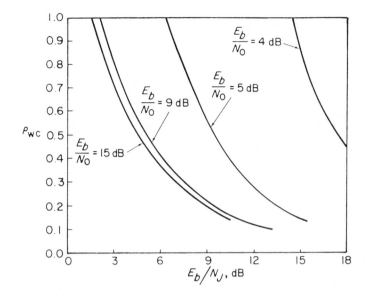

Figure 3.20. Worst case ρ versus E_b/N_J—FH/QPR (tone jamming).

Also, recognizing that $P_{b_0} \to 0$ in the limit as $E_b/N_0 \to \infty$, we obtain from (3.135) and (3.57) the desired limiting behavior to the average bit error probability, namely,

$$
\lim_{E_b/N_0 \to \infty} P_b
$$

$$
= \begin{cases} 0; & \dfrac{\rho E_b'}{N_J'} > 1 \\[3ex] \dfrac{3\rho}{2\pi} \cos^{-1} \sqrt{\dfrac{\rho E_b'}{N_J'}} \; ; & \dfrac{1}{9} < \dfrac{\rho E_b'}{N_J'} < 1 \\[3ex] \dfrac{\rho}{2\pi}\left[3\cos^{-1}\sqrt{\dfrac{\rho E_b'}{N_J'}} - \cos^{-1}\left(3\sqrt{\dfrac{\rho E_b'}{N_J'}}\right)\right]; & 0 < \dfrac{\rho E_b'}{N_J'} < \dfrac{1}{9}. \end{cases}
$$

$$(3.136)$$

The partial-band fraction ρ corresponding to the worst case jammer (maximum P_b) can be obtained by differentiating (3.136) with respect to ρ and equating to zero. Assuming that, for a fixed E_b'/N_J', this worst case ρ occurs when $1/9 < \rho E_b'/N_J' < 1$, then analogous to (3.32), we immediately

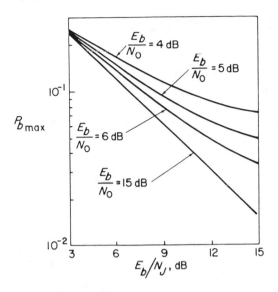

Figure 3.21.   Worst case P_b versus E_b/N_J—FH/QPR (tone jamming).

obtain

$$\rho_{wc} = \begin{cases} \dfrac{0.6306}{E_b'/N_J'} = \dfrac{1.6058}{E_b/N_J}; & E_b/N_J \geq 1.6058 \\[2mm] 1; & E_b/N_J < 1.6058. \end{cases} \tag{3.137}$$

Substituting (3.137) into (3.136) gives the limiting average bit error probability performance corresponding to the worst case jammer, namely,

$$\lim_{E_b/N_0 \to \infty} P_{b_{max}} = \begin{cases} \dfrac{3}{4}\left(\dfrac{0.2623}{E_b'/N_J'}\right) = \dfrac{0.5009}{E_b/N_J}; & E_b/N_J \geq 1.6058 \\[3mm] \dfrac{3}{2\pi}\cos^{-1}\sqrt{\dfrac{\pi E_b}{8N_J}}; & 8/9\pi \leq E_b/N_J < 1.6058 \\[3mm] \dfrac{1}{2\pi}\left[3\cos^{-1}\sqrt{\dfrac{\pi E_b}{8N_J}} - \cos^{-1}\left(3\sqrt{\dfrac{\pi E_b}{8N_J}}\right)\right]; \\[3mm] \qquad\qquad\qquad\qquad 0 \leq E_b/N_J < 8/9\pi. \end{cases}$$

$$\tag{3.138}$$

3.8 PERFORMANCE OF FH/QPR IN THE PRESENCE OF PARTIAL-BAND NOISE JAMMING

Once again, modeling the partial-band noise jammer as an additional additive noise source with power spectral density given by (3.63), the average bit error probability performance in the presence of partial-band noise jamming is obtained from (3.131) by replacing N_0 by $N_0 + N_J/\rho$, namely,

$$
P_{b_J} = \frac{3}{2} Q \left(\left[\left(\frac{2E_b'}{N_0} \right)^{-1} + \left(\frac{2\rho E_b'}{N_J} \right)^{-1} \right]^{-1/2} \right)
$$

$$
- \frac{1}{2} Q \left(3 \left[\left(\frac{2E_b'}{N_0} \right)^{-1} + \left(\frac{2\rho E_b'}{N_J} \right)^{-1} \right]^{-1/2} \right). \qquad (3.139)
$$

A similar result can be obtained from (3.133) for average symbol error probability in the presence of partial-band noise jamming. Finally, substituting (3.139) and (3.131) in (3.57) then gives the average bit error probability over all detection intervals, namely,

$$
P_b = \frac{3\rho}{2} Q \left(\left[\left(\frac{2E_b'}{N_0} \right)^{-1} + \left(\frac{2\rho E_b'}{N_J} \right)^{-1} \right]^{-1/2} \right)
$$

$$
- \frac{\rho}{2} Q \left(3 \left[\left(\frac{2E_b'}{N_0} \right)^{-1} + \left(\frac{2\rho E_b'}{N_J} \right)^{-1} \right]^{-1/2} \right)
$$

$$
+ \frac{3(1-\rho)}{2} Q \left(\sqrt{\frac{2E_b'}{N_0}} \right) - \frac{(1-\rho)}{2} Q \left(3\sqrt{\frac{2E_b'}{N_0}} \right). \qquad (3.140)
$$

The limiting behavior of (3.140) as E_b/N_0 approaches infinity is easily seen to be

$$
\lim_{E_b/N_0 \to \infty} P_b = \frac{3\rho}{2} Q \left(\sqrt{\frac{2\rho E_b'}{N_J}} \right) - \frac{\rho}{2} Q \left(3\sqrt{\frac{2\rho E_b'}{N_J}} \right) \qquad (3.141)
$$

which yields a partial-band fraction for the worst case jammer given by

$$
\rho_{wc} = \begin{cases} \dfrac{0.7114}{E_b'/N_J} = \dfrac{1.1532}{E_b/N_J}; & E_b/N_J \geq 1.1532 \\[2mm] 1; & E_b/N_J < 1.1532 \end{cases} \qquad (3.142)
$$

and a corresponding worst case average bit error probability

$$
\lim_{E_b/N_0 \to \infty} P_{b_{\max}}
$$

$$
= \begin{cases} \dfrac{0.2014}{E_b/N_J}; & E_b/N_J \geq 1.1532 \\[3mm] \dfrac{3}{2} Q \left(\sqrt{\dfrac{\pi^2 E_b}{8N_J}} \right) - \dfrac{1}{2} Q \left(3\sqrt{\dfrac{\pi^2 E_b}{8N_J}} \right); & E_b/N_J < 1.1532. \end{cases} \qquad (3.143)
$$

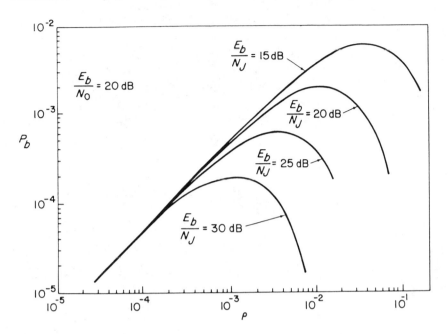

Figure 3.22. P_b versus ρ for FH/QPR in noise jamming with $E_b/N_0 = 20$ dB.

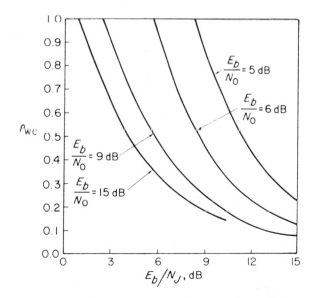

Figure 3.23. Worst case ρ versus E_b/N_J—FH/QPR (noise jamming).

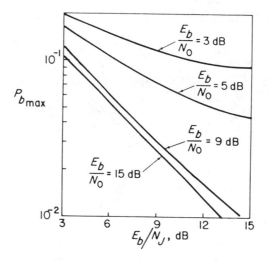

Figure 3.24. Worst case P_b versus E_b/N_J—FH/QPR (noise jamming).

Table 3.2

Worst case jammer performance for various modulations and jammer types.

$$\text{Worst case partial-band fraction } \rho_{wc} = \frac{K_\rho}{E_b/N_J}$$

$$\text{Maximum average bit error probability } P_{b_{max}} = \frac{K_P}{E_b/N_J}$$

Spread-Spectrum Modulation	Jammer Type	K_ρ	K_P
FH/QPSK	Multitone	0.6306	0.1311
	Noise	0.7090	0.0829
FH/QASK-16	Multitone	1.5765	0.2459
	Noise	1.758	0.1555
FH/PN/QPSK	Multitone	0.9220	0.0789
	Noise	0.7090	0.0829
FH/PN/QASK-16	Multitone	2.3050	0.1483
	Noise	1.758	0.1555
FH/QPR-3	Multitone	1.6058	0.5009
	Noise	1.1532	0.2014

Figures 3.22–3.24 characterize the performance of FH/QPR in the presence of partial-band noise jamming as computed from the results given in this section.

3.9 SUMMARY AND CONCLUSIONS

We conclude this chapter by summarizing the asymptotic worst case jammer performance results for the various modulations and jammer types in the form of Table 3.2. In all cases, we consider the limiting behavior as E_b/N_0 approaches infinity so that both the worst case partial-band fraction and corresponding maximum average bit error probability are inversely related to E_b/N_J.

Based on the table entries, the following conclusions can be reached:

- FH/QASK-16 is more susceptible to jamming (noise or multitone) than FH/QPSK by about 2.73 dB.
- Both FH/QPSK and FH/QASK are more susceptible to multitone jamming by about 2 dB.
- FH three-level QPR is more susceptible to multitone jamming than noise jamming by about 4 dB.
- FH three-level QPR is more susceptible to tone and noise jammers than FH/QASK-16 by 3.09 dB and 1.12 dB, respectively.

3.10 REFERENCES

[1] D. P. Taylor and D. Cheung, "The effect of carrier phase error on the performance of a duobinary shaped QPSK signal," *IEEE Trans. Commun.*, COM-25, no. 7, pp. 738–744, July 1977.

[2] W. C. Lindsey and M. K. Simon, *Telecommunication Systems Engineering*, Englewood Cliffs, NJ: Prentice-Hall, 1973, chapter 5.

[3] M. K. Simon and J. G. Smith, "Carrier synchronization and detection of QASK signal sets," *IEEE Trans. Commun.*, COM-22, no. 2, February 1974.

[4] M. K. Simon and A. Polydoros, "Coherent detection of frequency-hopped quadrature modulations in the presence of jamming; Part I: QPSK and QASK modulations," *IEEE Trans. Commun.*, COM-29, no. 11, pp. 1644–1660, November 1981.

[5] M. K. Simon, "Coherent detection of frequency-hopped quadrature modulations in the presence of jamming; Part II: QPR Class I modulation," *IEEE Trans. Commun.*, COM-29, no. 11, pp. 1660–1668, November 1981.

[6] N. E. Bekir, "Bounds on the Distribution of Partial Correlation for PN and Gold Sequences," Ph.D. Dissertation, Electrical Engineering Department, University of Southern California, January 1978.

[7] R. W. Lucky, J. Salz and E. J. Weldon, Jr., *Principles of Data Communications*, New York: McGraw-Hill, pp. 85–86, 1968.

[8] A. Lender, "The duobinary technique for high-speed data transmission," *AIEE Trans. Commun. Electr.*, vol. 82, pp. 214–218, May 1963.

[9] E. R. Kretzmer, "Generalization of a technique for binary data communication," *IEEE Trans. Commun. Syst.*, COM-14, pp. 67–68, February 1966.

[10] P. Kabal and S. Pasupathy, "Partial Response Signaling," *IEEE Trans. Commun.*, COM-23, pp. 921–935, September 1975.

Chapter 4

DIFFERENTIALLY COHERENT MODULATION TECHNIQUES

Traditionally coherent modulations such as multiple phase-shift-keying (MPSK) and quadrature amplitude-shift-keying (QASK) can also be detected using differentially coherent techniques. These techniques are useful in applications where the receiver is unable to provide an exact carrier reference phase for demodulating each data symbol but is capable of establishing a phase reference to within an arbitrary number of radians, say ϕ_a, of the exact phase. By differentially encoding the transmitted phase information and using a form of phase difference detection, the ϕ_a phase ambiguity can be resolved and the system is then capable of credible data detection. Obviously, because of the lack of perfect phase information per symbol, some degradation in system performance over coherent detection of the same modulation will exist.

In the spread-spectrum (SS) application, the need for differentially coherent detection comes about as follows [1]–[4]. Conventional frequency hopping (FH) as used to protect communication systems from radio frequency interference (RFI) or jamming has non-continuous carrier phase from hop to hop. For the fast hop rates needed in most RFI and jamming environments, the receiver dehopper does not have a chance to acquire the phase of each new carrier in the hop sequence. Differentially coherent detection provides a possible solution to the effect of phase discontinuities introduced by frequency hopping since the *absolute* phase of each carrier in the hop sequence is irrelevant insofar as the data detection process is concerned. More explicitly, the case of interest in FH differentially coherent modulation systems is where there are many (at least two) data symbols per hop, i.e., the so-called slow frequency hopping (SFH) case. Thus, for all the data symbols on a given hop, the carrier phase is constant and, regardless of its value, the combination of differential phase encoding and phase difference detection is sufficient to produce reliable communication. Of course, as in any differentially coherent system, the first data symbol in the sequence

313

(here the first data symbol in each hop) is lost since a phase reference has not yet been established for its detection. This again emphasizes the necessity of having many symbols per hop so as to reduce this effective loss in information rate to a minimum.

As in the previous chapter, we shall again examine the performance of the various modulations considered in the presence of partial-band noise and partial-band multitone jamming. In each case, the worst case jamming strategy will be determined, which, as before, consists of specifying the worst case partial-band fraction and the corresponding maximum average error probability. To simplify the ensuing analysis, we shall ignore the presence of the additive white Gaussian noise (AWGN) background in favor of the degradation produced by the jamming interference. Modification of the results to include the quiescent noise contribution would follow along the same lines as the approach taken in the previous chapter for coherent detection of these same modulations.

4.1 PERFORMANCE OF FH / MDPSK IN THE PRESENCE OF PARTIAL-BAND MULTITONE JAMMING

In a differentially encoded frequency-hopped M-ary differential phase-shift-keyed (FH/MDPSK) system, the information to be transmitted in the i-th signalling interval $(i - 1)T_s \leq t \leq iT_s$ is conveyed by appropriately selecting one of M phases

$$\theta_m = \frac{(2m - 1)\pi}{M}; \qquad m = 1, 2, \ldots, M \qquad (4.1)$$

and adding it to the total accumulated phase in the $(i - 1)$-st signalling interval of a constant amplitude (A), fixed frequency (assumed known at the receiver) sinusoid. Typically $M = 2^K$ with K integer, and these are the only cases we shall consider in detail. Furthermore, since the derivation of the performance of FH/MDPSK in the presence of a partial-band multitone jammer will rely largely on certain geometric relations, it is expedient to deal with both the signal and the jammer as phasors. Thus, the transmitted signal $s^{(i)}(t)$ in the i-th signalling interval is conveniently represented in complex form by

$$S^{(i)} = Ae^{j(\theta^{(i)} + \theta_T^{(i-1)})} \qquad (4.2)$$

where $\theta_T^{(i-1)}$ is the total accumulated phase in the $(i - 1)$-st signalling interval and $\theta^{(i)}$ ranges over the set $\{\theta_m\}$ of (4.1).

In the presence of multitone jamming interference as characterized in the previous chapter, a jamming tone $J(t)$, constant in both phase and magnitude (amplitude), is added to the transmitted signal. Since, when the jammer "hits," he is assumed[1] to be of the same frequency as the signal, then we may also represent the jammer in complex form, namely,

$$J = Ie^{j\theta_J} \tag{4.3}$$

where θ_J is a random phase uniformly distributed in the interval $(0, 2\pi)$. Thus, in any hop interval which is hit by the jammer (the probability of this occurring is the partial-band fraction ρ), the signals on which a decision for the i-th signalling interval is to be based are given (in complex form) by

$$Y^{(i-1)} = Ae^{j\theta_T^{(i-1)}} + Ie^{j\theta_J}$$

$$Y^{(i)} = Ae^{j(\theta^{(i)} + \theta_T^{(i-1)})} + Ie^{j\theta_J}. \tag{4.4}$$

Assuming a receiver structure that is optimum in the absence of the jammer, i.e., it employs the optimum decision rule for MDPSK against wideband noise, then in the presence of the on-tune jammer this rule would result in the estimate

$$\hat{\theta}^{(i)} = \theta_k \tag{4.5}$$

where k is such that

$$|\arg(Y^{(i)} - Y^{(i-1)}) - \theta_k| \le \frac{\pi}{M}. \tag{4.6}$$

Then if θ_n is indeed the true value of $\theta^{(i)}$, a symbol (phase) error is made, i.e., $\hat{\theta}^{(i)} \ne \theta^{(i)}$ whenever

$$|\arg(Y^{(i)} - Y^{(i-1)}) - \theta_n| > \frac{\pi}{M}. \tag{4.7}$$

Without loss in generality, we shall, for convenience, rotate the actual transmitted signal vectors by π/M radians so that the possible transmitted signal phases of (4.1) become

$$\theta_m = \frac{2\pi m}{M}; \quad m = 0, \pm 1, \pm 2, \ldots, \pm\left(\frac{M-2}{2}\right), \frac{M}{2}. \tag{4.8}$$

Finally, letting $Q_{2\pi n/M}$; $n = 0, \pm 1, \pm 2, \ldots \pm (M-2)/2, M/2$ denote the probability of the error event in (4.7), namely,

$$Q_{2\pi n/M} = \Pr\left\{ |\arg(Y^{(i)} - Y^{(i-1)}) - \theta_n| > \frac{\pi}{M} \right\} \tag{4.9}$$

and noting that since we have assumed the absence of an AWGN background, the probability of error in hop intervals which are not hit by the jammer is zero, then the average symbol error probability for MDPSK in

---

[1] The assumption of on-tune jamming is made solely to simplify the analysis, as has been done in previous chapters. Both the analytical technique and the sensitivity of the results that follow from its application depend heavily on this assumption. Some evidence of this statement will be discussed at the end of this section.

the presence of multitone jamming is given by

$$P_s(M) = \frac{\rho}{M} \sum_n Q_{2\pi n/M} \tag{4.10}$$

where the summation on n ranges over the set $n = 0, \pm 1, \pm 2, \ldots,$ $\pm(M-2)/2, M/2$. Since θ_J is uniformly distributed, we can recognize the symmetry

$$Q_{2\pi n/M} = Q_{-2\pi n/M}; \quad n = 1, 2, \ldots, \frac{M-2}{2}. \tag{4.11}$$

Also if $\theta_n = 0$ is transmitted, then, from (4.4), $Y^{(i-1)}$ and $Y^{(i)}$ are identical vectors. Equivalently,

$$|\arg(Y^{(i)} - Y^{(i-1)}) - \theta_0| = 0 \tag{4.12}$$

and, from (4.9),

$$Q_0 = 0. \tag{4.13}$$

Thus, using (4.11) and (4.13), $P_s(M)$ of (4.10) simplifies to

$$P_s(M) = \frac{\rho}{M} \left[Q_\pi + 2 \sum_{n=1}^{\frac{M-2}{2}} Q_{2\pi n/M} \right]. \tag{4.14}$$

Finally, using the relation between average symbol and bit error probabilities, namely,

$$P_b(M) = \left[\frac{M}{2(M-1)} \right] P_s(M), \tag{4.15}$$

the average bit error probability for MDPSK in the presence of multitone jamming is given by

$$P_b(M) = \frac{\rho}{2(M-1)} \left[Q_\pi + 2 \sum_{n=1}^{\frac{M-2}{2}} Q_{2\pi n/M} \right]. \tag{4.16}$$

Actually, the relation in (4.15) holds as an equality only for orthogonal signal sets [5]. However, for low signal-to-jammer noise ratios, the right-hand side of (4.15) becomes a tight upper bound for the average bit error probability performance of FH/MDPSK. For binary DPSK ($M = 2$), the equality in (4.15) is exact.

We shall see shortly that, for the evaluation of $Q_{2\pi n/M}$, it is convenient to renormalize the problem in terms of the ratio of jamming power per tone $J_0 = J/Q$ to signal power S. Let β^2 denote this ratio, i.e.,

$$\beta^2 = \frac{J/Q}{S}. \tag{4.17}$$

Then, recalling from the previous chapter that the number of hop slots N in the total hop frequency band W_{ss} is

$$N = \frac{W_{ss}}{1/T_s} = W_{ss}T_b\log_2 M \tag{4.18}$$

then, the partial-band fraction ρ can be expressed in terms of β^2 and the bit energy-to-jammer noise spectral density ratio E_b/N_J by

$$\rho \triangleq \frac{Q}{N} = \frac{J}{\beta^2 S W_{ss}T_b\log_2 M} = \frac{1}{(\log_2 M)\beta^2 E_b/N_J}. \tag{4.19}$$

Using (4.19), we can rewrite (4.16) in the form

$$P_b(M) = \frac{1}{2(M-1)(\log_2 M)\beta^2 E_b/N_J}\left[Q_\pi + 2\sum_{n=1}^{\frac{M-2}{2}}Q_{2\pi n/M}\right]. \tag{4.20}$$

Before proceeding to the evaluation of $Q_{2\pi n/M}$; $n = 1, 2, \ldots, M/2$ we make the final observation that the per tone jamming-to-signal power ratio β^2 can also be expressed in terms of the vector definitions of the signal and tone jamming interference. Since from (4.2) and (4.3) the signal power and jammer power per tone are given by

$$S = \frac{A^2}{2}; \qquad \frac{J}{Q} = \frac{I^2}{2}, \tag{4.21}$$

then equivalently from (4.17) we have that

$$\beta^2 = \frac{I^2}{A^2}. \tag{4.22}$$

4.1.1 Evaluation of $Q_{2\pi n/m}$

In view of (4.22), $Q_{2\pi n/M}$ of (4.9) may be restated in the normalized form

$$Q_{2\pi n/M} = \Pr\{|\arg(Z^{(i)} - Z^{(i-1)}) - 2n\theta| > \theta\} \tag{4.23}$$

where

$$\theta \triangleq \frac{\pi}{M} \tag{4.24}$$

and

$$Z^{(i-1)} = e^{-jn\theta} + \beta e^{j\theta_J} \triangleq R_1 e^{j(-n\theta+\psi_1)}$$
$$Z^{(i)} = e^{jn\theta} + \beta e^{j\theta_J} \triangleq R_2 e^{j(n\theta+\psi_2)}. \tag{4.25}$$

Note that, in obtaining (4.25) from (4.4), we have substituted for $\theta^{(i)}$ its assumed true value, namely, $\theta_n = 2\pi n/M - 2n\theta$, and, since θ_J is uniformly distributed, we have arbitrarily established the symmetry $\theta_T^{(i-1)} = -\pi n/M = -n\theta$. Figure 4.1 is a graphical representation of (4.25) where we have further introduced the notation

$$\psi \triangleq \arg(Z^{(i)} - Z^{(i-1)}). \tag{4.26}$$

Thus, using (4.25) and (4.26),

$$\arg(Z^{(i)} - Z^{(i-1)}) - 2n\theta = \psi - 2n\theta$$

$$= n\theta + \psi_2 - (-n\theta + \psi_1) - 2n\theta$$

$$= \psi_2 - \psi_1 \tag{4.27}$$

and, hence,

$$Q_{2\pi n/M} = \Pr\{|\psi_2 - \psi_1| > \theta\} = 1 - \Pr\{|\psi_2 - \psi_1| \le \theta\}. \tag{4.28}$$

Consider the product (asterisk denotes complex conjugate)

$$(Z^{(i-1)})^* Z^{(i)} e^{-j2n\theta} = R_1 R_2 e^{j(n\theta - \psi_1)} e^{j(n\theta + \psi_2)} e^{-j2n\theta}$$

$$= R_1 R_2 e^{j(\psi_2 - \psi_1)}. \tag{4.29}$$

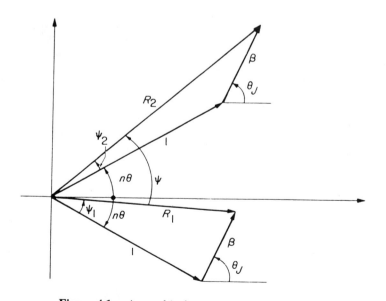

Figure 4.1. A graphical representation of (4.25).

The above product can also be written in the form

$$(Z^{(i-1)})^* Z^{(i)} e^{-j2n\theta} = (e^{jn\theta} + \beta e^{-j\theta_J})(e^{jn\theta} + \beta e^{j\theta_J}) e^{-j2n\theta}$$
$$= [e^{j2n\theta} + \beta^2 + \beta e^{jn\theta}(e^{j\theta_J} + e^{-j\theta_J})] e^{-j2n\theta}$$
$$= 1 + \beta^2 e^{-j2n\theta} + 2\beta \cos\theta_J e^{-jn\theta}. \tag{4.30}$$

Thus, using (4.29) and (4.30) in (4.28) results in the equivalent relation

$$Q_{2\pi n/M} = 1 - \Pr\{-\theta \le \arg[(Z^{(i-1)})^* Z^{(i)} e^{-j2n\theta}] \le \theta\}$$
$$= 1 - \Pr\{-\theta \le \arg[1 + \beta^2 e^{-j2n\theta} + 2\beta \cos\theta_J e^{-jn\theta}] \le \theta\}. \tag{4.31}$$

Equation (4.31) can be given a geometric interpretation as in Figure 4.2. Here the vector **OQ** (a line drawn from point *O* to either point *Q*)

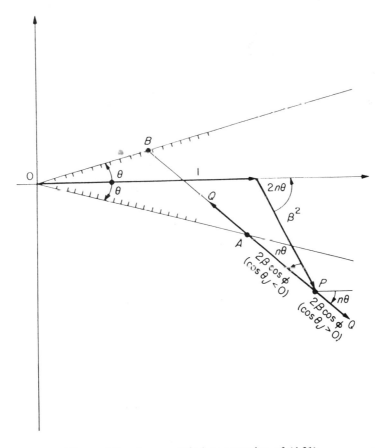

Figure 4.2.   A geometric interpretation of (4.31).

represents the complex number whose argument is required in (4.31), i.e.,

$$OQ = 1 + \beta^2 e^{-j2n\theta} + 2\beta \cos\theta_j e^{jn\theta}. \tag{4.32}$$

Thus, in terms of the geometry in Figure 4.2, (4.31) may be written in the alternate form

$$Q_{2\pi n/M} = 1 - \Pr\{\text{point } Q \text{ is within the } 2\theta \text{ wedge}\}$$

$$= 1 - \Pr\{\text{point } Q \text{ lies along the line } AB\}. \tag{4.33}$$

Considering separately the cases where point P falls outside and inside the 2θ wedge, which expressed mathematically corresponds to the inequalities (see Figure 4.3)

$$\beta^2 \gtrless \frac{\sin\theta}{\sin[(2n - 1)\theta]}, \tag{4.34}$$

then after much routine trigonometry, it can be shown that

$$Q_{2\pi n/M} = Q_{2n\theta}$$

$$= \begin{cases} \dfrac{1}{\pi} \cos^{-1}\left[\dfrac{\beta^2\sin[(2n + 1)\theta] + \sin\theta}{2\beta\sin[(n + 1)\theta]}\right] u(\beta - \beta_n) \\[2ex] + \dfrac{1}{\pi} \cos^{-1}\left[\dfrac{\sin\theta - \beta^2\sin[(2n - 1)\theta]}{2\beta\sin[(n - 1)\theta]}\right] u(\beta - \beta_{n-1}); \quad 0 < \beta < 1 \\[3ex] 1; \quad \beta \geq 1 \quad n = 2, 3, \ldots, \dfrac{M}{2} - 1, \end{cases}$$

$$\tag{4.35}$$

where $u(\beta)$ is the unit step function and

$$\beta_{n-1} \triangleq \frac{-\sin[(n - 1)\theta] + \sin n\theta}{\sin[(2n - 1)\theta]}. \tag{4.36}$$

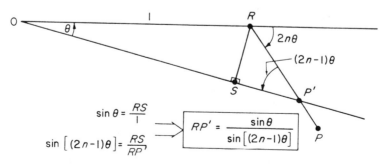

Figure 4.3. The geometry needed to establish (4.34).

Note that

$$\lim_{\beta \to 1} Q_{2n\theta} = \frac{1}{\pi} \left\{ \cos^{-1} \left[\frac{\sin[(2n+1)\theta] + \sin\theta}{2\sin[(n+1)\theta]} \right] \right.$$

$$\left. + \cos^{-1} \left[\frac{\sin\theta - \sin[(2n-1)\theta]}{2\sin[(n-1)\theta]} \right] \right\}$$

$$= 1 \tag{4.37}$$

For $n = 1$, the appropriate result analogous to (4.35) is

$$Q_{2\theta} = \begin{cases} 0; & 0 < \beta < \beta_1 \\ \dfrac{1}{\pi} \cos^{-1} \left[\dfrac{\beta^2 \sin 3\theta + \sin\theta}{2\beta \sin 2\theta} \right]; & \beta_1 \le \beta < 1 \\ 1; & \beta \ge 1 \end{cases}$$

$$\beta_1 = \frac{\sin 2\theta - \sin\theta}{\sin 3\theta} = \frac{\sin\dfrac{2\pi}{M} - \sin\dfrac{\pi}{M}}{\sin\dfrac{3\pi}{M}}. \tag{4.38}$$

Here

$$\lim_{\beta \to 1} Q_{2\theta} = \frac{1}{\pi} \cos^{-1} \left[\frac{\sin 3\theta + \sin\theta}{2\sin 2\theta} \right] < 1. \tag{4.39}$$

Finally, for $n = M/2$, we have the result

$$Q_{M\theta} = Q_\pi = \begin{cases} 0; & 0 < \beta < \beta_{M/2-1} \\ \dfrac{2}{\pi} \cos^{-1} \left[\dfrac{\left(\sin\dfrac{\pi}{M}\right)(1 - \beta^2)}{2\beta \cos\dfrac{\pi}{M}} \right]; & \beta_{M/2-1} \le \beta < 1 \\ 1; & \beta \ge 1 \end{cases}$$

$$\beta_{M/2-1} = \frac{1 - \cos\dfrac{\pi}{M}}{\sin\dfrac{\pi}{M}} \tag{4.40}$$

Also,

$$\lim_{\beta \to 1} Q_\pi = 1. \tag{4.41}$$

As an example, Figure 4.4 is a plot of $Q_{2n\theta}$; $n = 1, 2, 3, 4, 8$ versus β for $M = 16$. These probabilities are computed from (4.35), (4.38), and (4.40). Using these results in (4.20), Figure 4.5 illustrates the product $(E_b/N_J) \times P_b$ (16) versus β. This curve has a maximum value of 1.457 at $\beta = 0.1614$, which, from (4.19), corresponds to the optimal (worst case) jamming strategy.

$$
\rho_{\text{wc}} = \begin{cases} \dfrac{9.597}{E_b/N_J}; & E_b/N_J \geq 9.597 \\ 1; & E_b/N_J < 9.597 \end{cases} \tag{4.42}
$$

Thus, the average bit error probability performance of FH/MDPSK ($M = 16$) in the presence of the worst case tone jammer is given by

$$
P_{b_{\text{max}}} = \begin{cases} \left. \dfrac{1}{30} \left[Q_\pi + 2 \sum_{n=1}^{7} Q_{n\pi/8} \right] \right|_{\beta = 1/(2\sqrt{E_b/N_J})}; & 0.25 < E_b/N_J < 9.597 \\ \dfrac{1.457}{E_b/N_J}; & E_b/N_J \geq 9.597 \\ 0.5; & E_b/N_J < 0.25. \end{cases} \tag{4.43}
$$

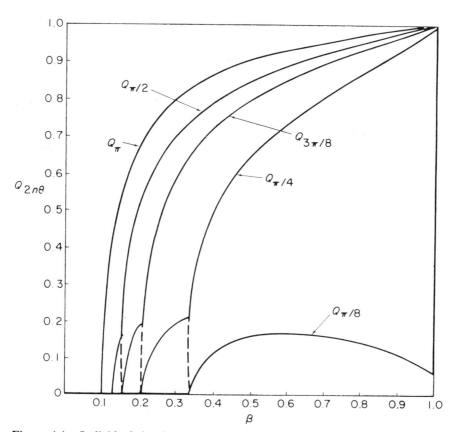

Figure 4.4. Individual signal point error probability components as a function of square root of jamming (per tone)-to-signal power ratio.

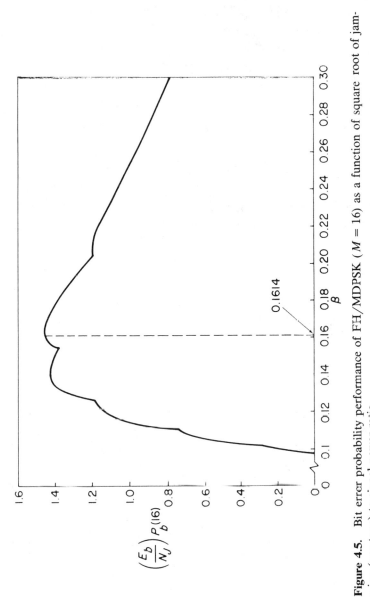

Figure 4.5. Bit error probability performance of FH/MDPSK ($M = 16$) as a function of square root of jamming (per tone)-to-signal power ratio.

Table 4.1

Asymptotic performance of FH/MDPSK for
worst case partial-band multitone jamming.

$$\text{Worst case partial-band fraction } \rho_{\text{wc}} = \frac{K_\rho}{E_b/N_J}$$

$$\text{Maximum average bit error probability } P_{b_{\max}} = \frac{K_P}{E_b/N_J}$$

M	β	K_ρ	K_P
2	1	1	0.50
4	0.5220	1.835	0.2593
8	0.2760	4.376	0.5280
16	0.1614	9.597	1.457

Similar results can be obtained[2] for FH/MDPSK with $M = 2, 4$, and 8. The asymptotic behavior of these results (i.e., ρ and P_b inversely related to E_b/N_J) is given in Table 4.1. Using the results in Table 4.1 and the fact that, for any M,

$$P_{b_{\max}} = P_b(M)\big|_{\beta=1/\sqrt{(\log_2 M) E_b/N_J}}; \quad E_b/N_J \le K_\rho \qquad (4.44)$$

with $P_b(M)$ given by (4.20), Figure 4.6 is an illustration of the average bit error probability performance of FH/MDPSK for worst case partial-band multitone jamming.

Before concluding this section, we wish to alert the reader to a point of pathological behavior that is directly attributable to the assumption of an on-tune tone jammer and is perhaps not obvious from the analytical or graphical results given. In particular, we observe from Figure 4.4 that $Q_{\pi/8}$ has a jump discontinuity at $\beta = 1$ and thus $P_{b_{\max}}$ of (4.43) will have a similar jump discontinuity at $E_b/N_J = 0.25$ (-6 dB). In fact, since from (4.39), $Q_{\pi/8} = .0625$ as β approaches one from below, then at $E_b/N_J = -6$ dB, $P_{b_{\max}}$ jumps from .4375 to .5. For other values of $M \ge 4$, a similar jump

---

[2] It should be noted here that the results in [1] for the performance of FH/MDPSK ($M = 4$) in the presence of the worst case tone jammer are partially incorrect. In particular, Houston finds $\beta = 0.52$ as the maximizing value. However, since the fraction of the band jammed, which is given by $\rho = 1/(2\beta^2 E_b/N_J)$, cannot exceed one, the value $\beta = 0.52$ can only be achieved if $E_b/N_J > 1.85$. For smaller values of E_b/N_J, the relation $\beta = 1/\sqrt{2E_b/N_J}$ must be used. Thus, we arrive at the following corrected results for $M = 4$:

$$P_{b_{\max}} = \begin{cases} \dfrac{0.2592}{E_b/N_J}; & E_b/N_J > 1.85 \\[2ex] \dfrac{1}{3\pi}\left[\cos^{-1}\left(\dfrac{2E_b/N_J - 1}{2\sqrt{2E_b/N_J}}\right) + \cos^{-1}\left(\dfrac{2E_b/N_J + 1}{4\sqrt{E_b/N_J}}\right)\right]; & 0.5 < E_b/N_J < 1.85 \\[2ex] 0.5; & E_b/N_J < 0.5 \end{cases}$$

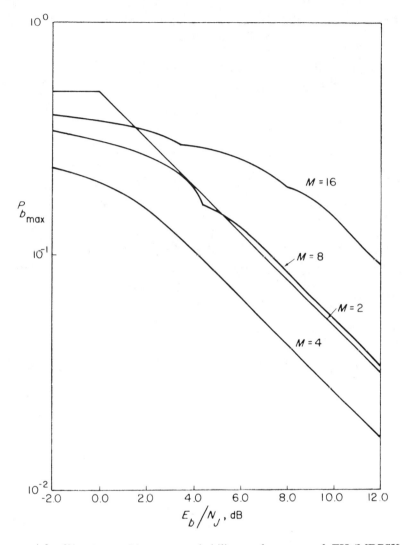

Figure 4.6. Worst case bit error probability performance of FH/MDPSK for partial-band multitone jamming.

discontinuity in the worst case tone jammer bit error probability will occur at $\beta = 1$ and $\rho = 1$, or equivalently, from (4.19), $E_b/N_J = 1/\log_2 M$. Since the range of E_b/N_J in Figure 4.6 extends only down to -2 dB, these jump discontinuities are not visible on this plot, i.e., the largest value of E_b/N_J at which a discontinuity occurs would correspond to $M = 4$ ($E_b/N_J = 1/2 = -3$ dB).

4.2 PERFORMANCE OF FH/MDPSK IN THE PRESENCE OF PARTIAL-BAND NOISE JAMMING

The average symbol error probability performance of MDPSK on an AWGN channel is given by [5], [6]

$$P_s(M) = \int_{\pi/M}^{\pi} \int_0^{\pi/2} \frac{\sin\alpha}{\pi} \left[1 + (\log_2 M) \frac{E_b}{N_0} (1 + \cos\psi \sin\alpha) \right]$$

$$\times \exp\left[-(\log_2 M) \frac{E_b}{N_0} (1 - \cos\psi \sin\alpha) \right] d\alpha \, d\psi \qquad (4.45)$$

where E_b/N_0 is the bit energy-to-noise ratio. More recently, the double integral in (4.45) has been shown [7] to be expressible as a single integral, namely,

$$P_s(M) = \left(\frac{\sin\dfrac{\pi}{M}}{\pi} \right) \int_0^{\pi/2} \frac{\exp\left[-(\log_2 M)\dfrac{E_b}{N_0}\left(1 - \cos\dfrac{\pi}{M}\cos\alpha \right) \right]}{1 - \cos\dfrac{\pi}{M}\cos\alpha} d\alpha$$

$$(4.46)$$

which lends itself to more convenient numerical evaluation.

To obtain the performance of FH/MDPSK in partial-band noise jamming, one has merely to replace E_b/N_0 by $\rho E_b/N_J$ in (4.46) and multiply the result by ρ where again ρ is the partial-band fraction. Thus, also using (4.15), the average bit error probability performance of FH/MDPSK in partial-band noise jamming is given by

$$P_b(M) = \frac{\rho M}{2(M-1)} \left(\frac{\sin\dfrac{\pi}{M}}{\pi} \right)$$

$$\times \int_0^{\pi/2} \frac{\exp\left[-(\log_2 M)\dfrac{\rho E_b}{N_J}\left(1 - \cos\dfrac{\pi}{M}\cos\alpha \right) \right]}{1 - \cos\dfrac{\pi}{M}\cos\alpha} d\alpha$$

$$= \left(\frac{E_b}{N_J} \right)^{-1} \left(\frac{M}{2(M-1)} \right)$$

$$\times \left[Z\left(\frac{\sin\dfrac{\pi}{M}}{\pi} \right) \int_0^{\pi/2} \frac{\exp\left[-(\log_2 M)Z\left(1 - \cos\dfrac{\pi}{M}\cos\alpha \right) \right]}{1 - \cos\dfrac{\pi}{M}\cos\alpha} d\alpha \right]$$

$$(4.47)$$

where

$$Z \triangleq \frac{\rho E_b}{N_J}. \qquad (4.48)$$

Comparing (4.47) with (4.46), we observe that to determine the worst case partial-band noise jammer, one merely uses a tabulation of $P_s(M)$ versus E_b/N_0 (or Z), such as that found in Table 5-5 of [5], multiplies these error probabilities by Z, and locates the value of Z, say Z_{max} which yields the maximum, say P_{max}, of this product. Then, from (4.47) and (4.48), the worst case jammer strategy corresponds to a partial-band fraction

$$\rho_{wc} = \left(\frac{E_b}{N_J} \right)^{-1} Z_{max} \tag{4.49}$$

and maximum bit error probability

$$P_{b_{max}} = \begin{cases} \left(\dfrac{E_b}{N_J} \right)^{-1} \left[\dfrac{MP_{max}}{2(M-1)} \right]; & \dfrac{E_b}{N_J} \geq Z_{max} \\[4mm] \dfrac{M}{2(M-1)} \left(\dfrac{\sin \dfrac{\pi}{M}}{\pi} \right) \displaystyle\int_0^{\pi/2} \dfrac{\exp\left[-(\log_2 M) \dfrac{E_b}{N_J} \left(1 - \cos \dfrac{\pi}{M} \cos \alpha \right) \right]}{1 - \cos \dfrac{\pi}{M} \cos \alpha} \, d\alpha; \\[6mm] & \dfrac{E_b}{N_J} < Z_{max}. \end{cases} \tag{4.50}$$

Although, for arbitrary M, the single integral of (4.46) cannot be evaluated in closed form, the special case of 2-ary MDPSK (often just abbreviated as DPSK) allows one to obtain the simple result

$$P_s(2) = P_b(2) = \frac{1}{2} \exp\left(-\frac{E_b}{N_0} \right) \tag{4.51}$$

or for FH/DPSK in partial-band noise jamming

$$P_b(2) = \frac{\rho}{2} \exp\left(-\frac{\rho E_b}{N_J} \right). \tag{4.52}$$

Directly differentiating (4.52) with respect to ρ and equating the result to zero yields the worst case performance

$$\rho_{wc} = \left(\frac{E_b}{N_J} \right)^{-1} \tag{4.53}$$

and

$$P_{b_{max}} = \begin{cases} \dfrac{1}{2e\left(\dfrac{E_b}{N_J} \right)}; & \dfrac{E_b}{N_J} \geq 1 \\[4mm] \dfrac{1}{2} \exp\left(-\dfrac{E_b}{N_J} \right); & \dfrac{E_b}{N_J} < 1 \end{cases} \tag{4.54}$$

Figure 4.7 illustrates $ZP_s(M)$ versus Z in dB for $M = 2$, 4, 8, and 16. Table 4.2 provides the corresponding values of Z_{max} and P_{max} from which the asymptotic performance of FH/MDPSK for worst case partial-band noise jamming is computed using (4.49) and (4.50). Comparing Tables 4.1 and 4.2, we observe that for all values of M, FH/MDPSK is more sensitive to worst case partial-band multitone jamming than it is to worst case partial-band noise jamming.

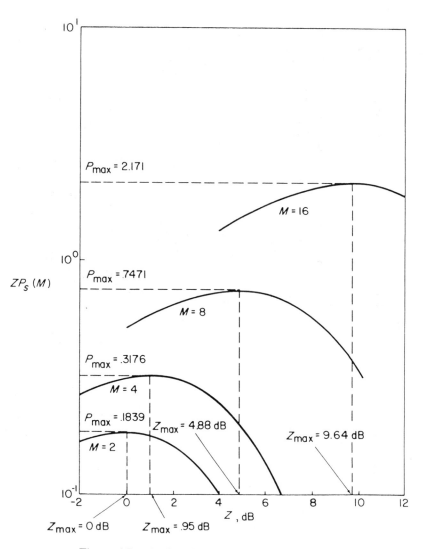

Figure 4.7. A plot of $ZP_s(M)$ versus Z in dB.

Table 4.2
Asymptotic performance of FH/MDPSK
for worst case partial-band noise jamming.

Worst case partial-band fraction $\rho_{wc} = \dfrac{K_\rho}{E_b/N_J}$

Maximum average bit error probability $P_{b_{max}} = \dfrac{K_P}{E_b/N_J}$

M	P_{max}	$K_\rho(Z_{max})$	K_P
2	.1839	1.000	.1839
4	.3176	1.245	.2118
8	.7471	3.076	.4269
16	2.171	9.204	1.1579

4.3 PERFORMANCE OF DQASK IN THE PRESENCE OF ADDITIVE WHITE GAUSSIAN NOISE

Having now discussed the jamming performance of the most classical differentially coherent modulation technique, namely FH/MDPSK, we turn our attention to the SS application of differentially coherent detection to multiple-amplitude-and phase-shift-keying (MAPSK). In particular, we introduce the concept of differentially coherent detection of differentially phase-encoded quadrature amplitude-shift-keyed signals (DQASK) [2] which when augmented with an FH modulation can be used to protect the system from RFI or the intentional jammer [3]. Since we have already seen many times before that the performance of an FH modulation technique in the presence of a partial-band noise jammer is directly obtained from the corresponding performance of the unspread modulation over the AWGN channel, we begin our discussion by first considering the symbol error probability performance of DQASK in such a Gaussian noise environment.

4.3.1 Characterization of the Transmitted Signal

As a brief review of our presentation in the previous chapter, an M-ary (QASK-M) is characterized by transmitting in each symbol interval one of M possible signals of the form

$$s(t) = \sqrt{2}\,\delta[b_n \cos \omega_0 t + a_m \sin \omega_0 t] = \sqrt{2}\,\text{Re}\left\{ s\, e^{j\omega_0 t} \right\} \quad (4.55)$$

where ω_0 is the carrier radian frequency. The total number of signals M is typically the square of an even number K, and the quadrature amplitudes a_m and b_n take on equally likely values m and n with $m, n = \pm 1, \pm 3, \ldots \pm(K-1)$. Here, δ is a parameter which is related to the average power S of the signal set by

$$S = \tfrac{2}{3}(K^2 - 1)\delta^2. \quad (4.56)$$

In order to describe the manner in which a QASK signal may be differentially phase encoded so that differentially coherent phase detection may be performed at the receiver, a polar coordinate representation of the signal set is preferable to the rectangular coordinate representation in (4.55).[3] In particular, if, in the i-th signalling interval $(i - 1)T_s \le t \le iT_s$, it is desired to transmit a signal corresponding to the signal point vector

$$s^{(i)} = \delta \left[b_n^{(i)} - ja_m^{(i)} \right] = A^{(i)} e^{-j\phi^{(i)}}, \qquad (4.57)$$

the differentially phase-encoded QASK signal in this interval would then appear as

$$s^{(i)}(t) = \sqrt{2}\, A^{(i)} \cos\left(\omega_0 t + \theta^{(i)} \right), \qquad (4.58)$$

where

$$\theta^{(i)} = \theta^{(i-1)} - \phi^{(i)}. \qquad (4.59)$$

Note that, while $\phi^{(i)}$ is restricted to take on values

$$\phi = \tan^{-1}\left(\frac{m}{n} \right); \qquad m, n = \pm 1, \pm 3, \dots \pm (K - 1), \qquad (4.60)$$

the actual transmitted phase, $\theta^{(i)}$ in the i-th interval has no such restriction. Consequently, an estimate of $\theta^{(i)}$ alone conveys no information regarding the phase, $\phi^{(i)}$ of the transmitted information symbol. This is a consequence of the differential phase-encoding operation described by (4.59).

Immediately, then, one observes that, in order to estimate $\phi^{(i)}$ in the absence of absolute phase information, what is needed is a receiver which forms estimates of $\theta^{(i)}$ and $\theta^{(i-1)}$ and computes their difference. Furthermore, since the QASK signal set is obviously not constant envelope, the amplitude, $A^{(i)}$, must also be estimated by the receiver in order to complete the decision on $s^{(i)}$. Again, since absolute phase information is assumed to be unavailable at the receiver, the estimate of $A^{(i)}$ is obtained from a non-coherent envelope detector.

In the next section, we describe and derive the performance of a receiver which performs the above two functions, namely, differentially coherent detection of phase and non-coherent detection of envelope (amplitude).

4.3.2 Receiver Characterization and Performance

Figure 4.8 depicts a receiver used to perform differentially coherent detection of differentially phase-encoded QASK. The structure combines the elements of a differentially coherent receiver for a constant envelope modulation such as MPSK with a non-coherent envelope detector. The output from these two receiver components, namely, detected envelope and dif-

---

[3] When we discuss the differentially coherent detection process in the next section, we shall switch back to the rectangular representation to accommodate the rectangular-shaped decision regions.

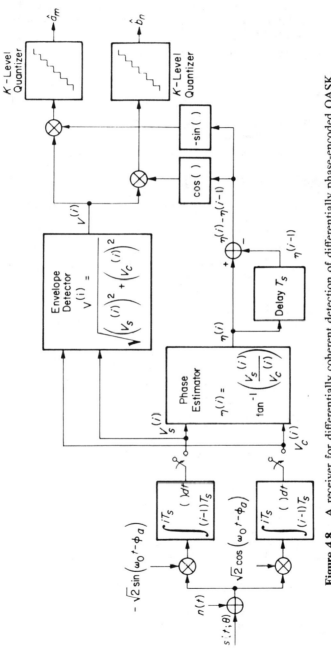

Figure 4.8. A receiver for differentially coherent detection of differentially phase-encoded QASK.

ferential phase are then converted to equivalent in-phase and quadrature signals upon which multilevel decisions are made, as is done in the more conventional coherent QASK receiver.

Appearing at the receiver input in the i-th signalling interval is the transmitted signal of (4.58) to which the channel has added a random phase shift θ and an AWGN which has the usual narrowband representation (repeated here for convenience)

$$n(t) = \sqrt{2}\left[N_c(t)\cos(\omega_0 t + \theta) - N_s(t)\sin(\omega_0 t + \theta)\right], \quad (4.61)$$

where $N_c(t)$ and $N_s(t)$ are statistically independent low-pass Gaussian noise processes with single-sided power spectral density N_0 w/Hz. Thus, the received signal in the i-th signalling interval is of the form

$$y^{(i)}(t) = s[t,\theta] + n(t) = \sqrt{2}\,A^{(i)}\cos\left(\omega_0 t + \theta^{(i)} + \theta\right) + n(t).$$

$$(4.62)$$

The receiver first performs in-phase and quadrature carrier demodulation with a pair of quadrature reference signals of known frequency ω_0 but unknown phase ϕ_a. The results of these demodulations are then passed through integrate-and-dump (I & D) filters whose outputs are given by[4]

$$V_s^{(i)} = \int_{(i-1)T_s}^{iT_s} y^{(i)}(t)\left[-\sqrt{2}\,\sin(\omega_0 t - \phi_a)\right]dt$$

$$= A^{(i)}T_s\sin\left(\theta^{(i)} + \phi_a\right) + n_c\sin\phi_a + n_s\cos\phi_a$$

$$V_c^{(i)} = \int_{(i-1)T_s}^{iT_s} y^{(i)}(t)\left[\sqrt{2}\,\cos(\omega_0 t - \phi_a)\right]dt$$

$$= A^{(i)}T_s\cos\left(\theta^{(i)} + \phi_a\right) + n_c\cos\phi_a - n_s\sin\phi_a, \quad (4.63)$$

where

$$n_c \triangleq \int_{(i-1)T_s}^{iT_s} \cdot N_c(t)\,dt$$

$$n_s \triangleq \int_{(i-1)T_s}^{T_s} N_s(t)\,dt. \quad (4.64)$$

The receiver next generates the equivalent envelope and phase of the I & D outputs, namely,

$$V^{(i)} = \sqrt{\left(V_s^{(i)}\right)^2 + \left(V_c^{(i)}\right)^2}$$

$$\eta^{(i)} = \tan^{-1}\left(\frac{V_s^{(i)}}{V_c^{(i)}}\right). \quad (4.65)$$

---

[4]Without any loss in generality, we shall set $\theta = 0$ for simplicity of notation.

Finally, the differential phase $\eta^{(i)} - \eta^{(i-1)}$ is formed and used to produce the in-phase and quadrature decision variables $V^{(i)} \cos(\eta^{(i)} - \eta^{(i-1)})$ and $-V^{(i)} \sin(\eta^{(i)} - \eta^{(i-1)})$ upon which K-level decisions ($\hat{a}_m$ and $\hat{b}_n$) are made.

At this point, it is convenient to redraw Figure 4.8 in its equivalent form illustrated in Figure 4.9 by recognizing that

$$V^{(i)} \cos(\eta^{(i)} - \eta^{(i-1)}) = \underbrace{V^{(i)} \cos\eta^{(i)} \cos\eta^{(i-1)}}_{V_c(i)} + \underbrace{V^{(i)} \sin\eta^{(i)} \sin\eta^{(i-1)}}_{V_s(i)}$$

$$-V^{(i)} \sin(\eta^{(i)} - \eta^{(i-1)}) = \underbrace{-V^{(i)} \sin\eta^{(i)} \cos\eta^{(i-1)}}_{V_s(i)} + \underbrace{V^{(i)} \cos\eta^{(i)} \sin\eta^{(i-1)}}_{V_c(i)}.$$

$$(4.66)$$

Figure 4.9 has the advantage of resembling a conventional *coherent* QASK receiver [8] with a noisy carrier demodulation reference and thus its error probability performance can be obtained almost by inspection. In particular, from (4.63) and (4.66), we obtain the decision variables

$$U_s^{(i)} = -A^{(i)} T_s \sin(\theta^{(i)} + \phi_a - \eta^{(i-1)}) - n_c \sin(\phi_a - \eta^{(i-1)})$$
$$- n_s \cos(\phi_a - \eta^{(i-1)})$$
$$U_c^{(i)} = A^{(i)} T_s \cos(\theta^{(i)} + \phi_a - \eta^{(i-1)}) + n_c \cos(\phi_a - \eta^{(i-1)})$$
$$- n_s \sin(\phi_a - \eta^{(i-1)}).$$
$$(4.67)$$

Letting

$$\eta_a^{(i-1)} = \eta^{(i-1)} - \theta^{(i-1)} - \phi_a \qquad (4.68)$$

and using (4.59), we can rewrite (4.67) as

$$U_s^{(i)} = A^{(i)} T_s \sin(\phi^{(i)} + \eta_a^{(i-1)}) + n_c \sin(\theta^{(i-1)} + \eta_a^{(i-1)})$$
$$- n_s \cos(\theta^{(i-1)} + \eta_a^{(i-1)})$$
$$= A^{(i)} T_s \sin(\phi^{(i)} + \eta_a^{(i-1)}) - N_1 \cos\eta_a^{(i-1)} + N_2 \sin\eta_a^{(i-1)}$$
$$U_c^{(i)} = A^{(i)} T_s \cos(\phi^{(i)} + \eta_a^{(i-1)}) + n_c \cos(\theta^{(i-1)} + \eta_a^{(i-1)})$$
$$+ n_s \sin(\theta^{(i-1)} + \eta_a^{(i-1)})$$
$$= A^{(i)} T_s \cos(\phi^{(i)} + \eta_a^{(i-1)}) + N_1 \sin\eta_a^{(i-1)} + N_2 \cos\eta_a^{(i-1)} \quad (4.69)$$

where

$$N_1 = n_s \cos\theta^{(i-1)} - n_c \sin\theta^{(i-1)}$$
$$N_2 = n_s \sin\theta^{(i-1)} + n_c \cos\theta^{(i-1)}. \qquad (4.70)$$

Finally, recognizing that (4.69) resembles the decision variables for a coherent QASK receiver whose carrier demodulation reference signals are in error by $\eta_a^{(i-1)}$ radians, we can immediately write down an expression for

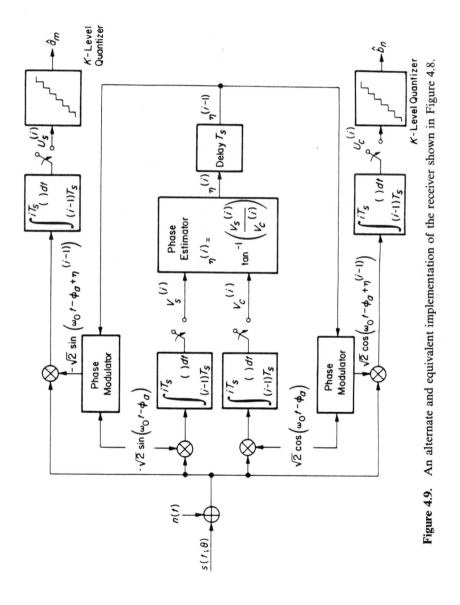

Figure 4.9.   An alternate and equivalent implementation of the receiver shown in Figure 4.8.

the symbol error probability of differentially coherent detected QASK (conditioned on the $(i - 1)$-st symbol SNR), namely,

$$P_s\left(\gamma_s^{(i-1)}\right) = \int_{-\pi}^{\pi} P_s\left(\eta_a^{(i-1)}\right) p_1\left(\eta_a^{(i-1)}; \gamma_s^{(i-1)}\right) d\eta_a^{(i-1)} \qquad (4.71)$$

where $p_1(\eta_a^{(i-1)}; \gamma_s^{(i-1)})$, the probability density function (pdf) of the normalized phase $\eta_a^{(i-1)}$ in the $(i - 1)$-st signalling interval, is given by [2]

$$p_1\left(\eta_a^{(i-1)}; \gamma_s^{(i-1)}\right)$$

$$= \begin{cases} \dfrac{1}{2\pi}\exp\left(-\gamma_s^{(i-1)}\right)\Big\{1 + 2\sqrt{\pi\gamma_s^{(i-1)}}\ \cos\eta_a^{(i-1)}\exp\left(\gamma_s^{(i-1)}\cos^2\eta_a^{(i-1)}\right) \\ \qquad\qquad \times Q\Big[-\sqrt{2\gamma_s^{(i-1)}}\ \cos\eta_a^{(i-1)}\Big]\Big\}; \quad \left|\eta_a^{(i-1)}\right| \le \pi \\ \\ 0; \quad \text{elsewhere} \end{cases}$$

$$(4.72)$$

with the $(i - 1)$-st *transmission interval symbol SNR* $\gamma_s^{(i-1)}$ defined by

$$\gamma_s^{(i-1)} = \frac{\left(A^{(i-1)}\right)^2 T}{N_0}, \qquad (4.73)$$

and $Q(x)$ denoting, as in previous chapters, the Gaussian probability integral. Also from (51) of [8], with ϕ replaced by $\eta_a^{(i-1)}$,

$$P_s\left(\eta_a^{(i-1)}\right) = \frac{4}{K^2}\sum_{j,l} Q\left\{\Delta\left[l + (1 - l)\cos\eta_a^{(i-1)} - j\sin\eta_a^{(i-1)}\right]\right\}$$

$$- \frac{4}{K^2}\sum_{k,l} Q\left\{\Delta\left[k + (1 - k)\cos\eta_a^{(i-1)} + (l - 1)\sin\eta_a^{(i-1)}\right]\right\}$$

$$\times Q\left\{\Delta\left[l + (1 - l)\cos\eta_a^{(i-1)} - (k - 1)\sin\eta_a^{(i-1)}\right]\right\}. \qquad (4.74)$$

In (4.74), the sum over j is for values $\pm 1, \pm 3, \ldots \pm(K - 1)$ while the sums over k and l are for values $0, \pm 2, \pm 4, \ldots, \pm(K - 2)$. Also,

$$\Delta \triangleq \sqrt{\frac{3\gamma_s}{K^2 - 1}} \qquad (4.75)$$

where

$$\gamma_s \triangleq \frac{ST_s}{N_0} \qquad (4.76)$$

is the average symbol SNR of the QASK-K^2 signal set with average power S defined in (4.56).

Finally, the average symbol error probability, P_s, is obtained by averaging (4.71) over the pdf of $\gamma_s^{(i-1)}$. To obtain this pdf, we note that, for a given K, the $(i - 1)$-st symbol signal power $(A^{(i-1)})^2$ ranges over $K(K + 2)/8$ different values. $K/2$ of these values correspond to signal points on the diagonal of any quadrant and occur with probability $4/K^2$. The remaining $K(K - 2)/8$ values correspond to off-diagonal signal points *either* above or below the diagonal of any quadrant and occur with probability $8/K^2$. Thus, $\gamma_s^{(i-1)}$ takes on the discrete set, $\mathcal{R}$, of values $\Delta^2(m^2 + n^2)/2$; $m, n = 1, 3 \ldots, (K - 1)$; $m \leq n$ and the corresponding pdf is then

$$p\left(\gamma_s^{(i-1)}\right) = \begin{cases} \dfrac{4}{K^2}; & \gamma_s^{(i-1)} = m^2\Delta^2; \quad m = 1, 3, \ldots, (K - 1) \\[2mm] \dfrac{8}{K^2}; & \gamma_s^{(i-1)} = \left(\dfrac{m^2 + n^2}{2}\right)\Delta^2; \\[2mm] & m, n = 1, 3, \ldots (K - 1); \qquad m < n \end{cases}$$

(4.77)

where Δ is defined in (4.75). Averaging (4.71) over the pdf of (4.77) gives the desired result

$$P_s = \sum_{\mathcal{R}} P_s\left(\gamma_s^{(i-1)}\right) p\left(\gamma_s^{(i-1)}\right)$$

$$= \int_{-\pi}^{\pi} P_s\left(\eta_a^{(i-1)}\right) \left[\sum_{\mathcal{R}} p_1\left(\eta_a^{(i-1)}; \gamma_s^{(i-1)}\right) p\left(\gamma_s^{(i-1)}\right)\right] d\eta_a^{(i-1)} \quad (4.78)$$

Equivalently, letting

$$p\left(\eta_a^{(i-1)}\right) = \sum_{\mathcal{R}} p_1\left(\eta_a^{(i-1)}; \gamma_s^{(i-1)}\right) p\left(\gamma_s^{(i-1)}\right)$$

$$= \frac{4}{K^2} \sum_{m=1,3,\ldots}^{K-1} p_1\left(\eta_a^{(i-1)}; m^2\Delta^2\right)$$

$$+ \frac{8}{K^2} \sum_{\substack{m, n=1,3\ldots \\ m < n}}^{K-1} p_1\left(\eta_a^{(i-1)}; \left(\frac{m^2 + n^2}{2}\right)\Delta^2\right) \quad (4.79)$$

represent the "effective" pdf of the $(i - 1)$-st symbol phase, then (4.78) becomes the simple result

$$P_s = \int_{-\pi}^{\pi} P_s\left(\eta_a^{(i-1)}\right) p\left(\eta_a^{(i-1)}\right) d\eta_a^{(i-1)}. \quad (4.80)$$

Figure 4.10 is a plot of P_s versus γ_s in decibels as evaluated from (4.80) for $K = 4$ (DQASK-16). Also shown is the corresponding result for coherent detection of QASK which, for $K = 4$, is given by [8]:

$$P_s = 3Q(\Delta)\left[1 - \tfrac{3}{4}Q(\Delta)\right]. \quad (4.81)$$

For comparison, the performance of coherent and differentially coherent detection of MPSK with $M = 16$ (i.e., PSK-16 and DPSK-16) is presented in Figure 4.10 [5]. For small γ_s, the coherent PSK-16 and the DQASK-16 perform almost identically but, for large γ_s, the DQASK-16 approaches the performance of DPSK-16. Also, for large γ_s, coherent QASK-16 is about 4 dB better than coherent PSK-16 showing the more favorable exchange of average power for bandwidth with the QASK-16 than with the PSK-16. While it is true that DQASK-16 suffers a significant performance degrada-

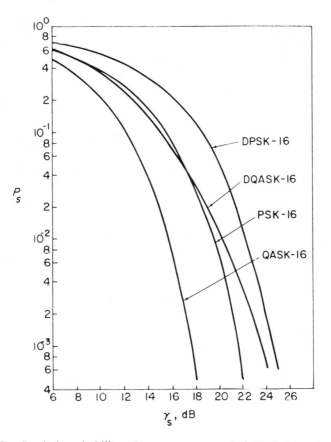

Figure 4.10. Symbol probability of error versus symbol SNR for coherent and differentially coherent detection of QASK-16 and PSK-16.

tion with respect to coherent QASK-16 at large γ_s, we must recall our initial motivation, namely to use DQASK-16 along with frequency hopping (i.e., FH/DQASK) to protect a conventional QASK communication system against jamming. In the next two sections, we present the FH/DQASK performance in the presence of partial-band multitone and noise jamming.

4.4 PERFORMANCE OF FH/DQASK IN THE PRESENCE OF PARTIAL-BAND MULTITONE JAMMING

A frequency-hopped, differentially coherent M-ary QASK modulation (FH/DQASK-M) is characterized by transmitting in the i-th symbol interval $[(i-1)T_s \le t \le iT_s]$ one of M possible signals of the form

$$s^{(i)}(t) = \sqrt{2}\,\delta\left[b_n^{(i)}\cos\left(\omega_h^{(i)}t + \theta^{(i-1)}\right) + a_m^{(i)}\sin\left(\omega_h^{(i)}t + \theta^{(i-1)}\right)\right],$$

$$(4.82)$$

where $\omega_h^{(i)}$ is the particular carrier frequency selected by the hopper for this interval, and $\theta^{(i-1)}$ is again the transmitted phase in the $(i-1)$-st interval. In analyzing the performance of FH/DQASK in the presence of the tone jammer (assuming the i-th transmission interval is jammed)

$$J(t) = \sqrt{2J_0}\,\cos\!\left(\omega_h^{(i)}t + \theta_J\right), \tag{4.83}$$

it is convenient to adopt a vector diagram approach analogous to that taken for MDPSK. As such, the transmitted signal will be represented by a normalized vector with x and y components respectively given by $n = b_n^{(i)}/\delta$ and $m = a_m^{(i)}/\delta$. The jammer is then represented by a normalized vector with phase θ_J and amplitude[5]

$$\beta \triangleq \frac{\sqrt{J_0}}{\delta} = \frac{\sqrt{\dfrac{J}{Q}}}{\delta}. \tag{4.84}$$

Recalling that $M = K^2$, then combining (4.18), (4.19), (4.56), and (4.84),

$$\beta = \sqrt{\frac{K^2 - 1}{3\rho \log_2 K}\left(\frac{N_J}{E_b}\right)}. \tag{4.85}$$

For example, for FH/DQASK-16 ($K = 4$), (4.85) becomes

$$\beta = \sqrt{\frac{5}{2}\left(\frac{N_J}{\rho E_b}\right)}. \tag{4.86}$$

In the remainder of this chapter, we shall deal specifically with FH/DQASK-16 as a matter of convenience. However, whenever results are obtained in their final form, they shall be given in the generalized form suitable to FH/DQASK-K^2, with arbitrary K.

Figure 4.11 is the normalized signal point constellation corresponding to QASK-16. The dashed lines indicate the decision region boundaries appropriate for coherent or differentially coherent detection of the various signal points. Now suppose that we wish to transmit signal point ① with differentially encoded phase in the i-th interval. The signal transmitted in the $(i-1)$-st interval could have been any of the 16 signal points. Thus, the vector representation of Figure 4.12 is adequate for characterizing the signal and jammer in these two intervals. For convenience, we shall always draw the vector representing the signal transmitted in the $(i-1)$-st interval along the positive x-axis. The amplitude A_1 corresponds to the normalized envelope of the signal point transmitted in the $(i-1)$-st interval and, from

---

[5] Note that the normalized vector amplitude β as defined here is not the same as a similar quantity denoted by β in Section 4.1 and defined in (4.17).

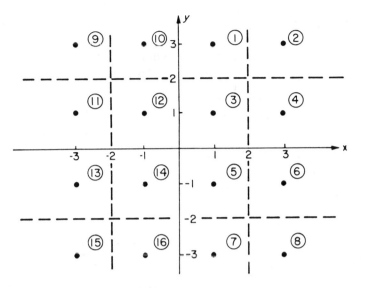

Figure 4.11.   Normalized signal point constellation for QASK-16.

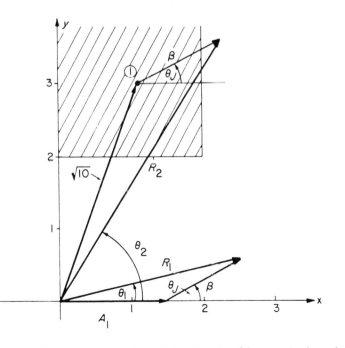

Figure 4.12.   A vector diagram representation of the signal and jammer in the i-th and $(i - 1)$-st transmission intervals (signal point in first quadrant).

Figure 4.11, takes on values

$$A_1 = \begin{cases} \sqrt{2} & \text{with Prob. } 1/4 \\ \sqrt{10} & \text{with Prob. } 1/2 \\ \sqrt{18} & \text{with Prob. } 1/4. \end{cases} \qquad (4.87)$$

From the results of the previous section, we observe that signal point ① will be correctly detected if

$$0 < R_2 \cos(\theta_2 - \theta_1) < 2$$
$$2 < R_2 \sin(\theta_2 - \theta_1) < \infty. \qquad (4.88)$$

The boundaries on the inequalities in (4.88) correspond to the x and y coordinates of the decision region indicated by the shaded area in Figure 4.12.

Expanding the sine and cosine of the difference angle $\theta_2 - \theta_1$ and noting, from Figure 4.12, that

$$R_2 \cos \theta_2 = 1 + \beta \cos \theta_J$$
$$R_2 \sin \theta_2 = 3 + \beta \sin \theta_J$$

$$\cos \theta_1 = \frac{A_1 + \beta \cos \theta_J}{\sqrt{(A_1 + \beta \cos \theta_J)^2 + (\beta \sin \theta_J)^2}} = \frac{A_1 + \beta \cos \theta_J}{\sqrt{A_1^2 + 2\beta A_1 \cos \theta_J + \beta^2}}$$

$$\sin \theta_1 = \frac{\beta \sin \theta_J}{\sqrt{(A_1 + \beta \cos \theta_J)^2 + (\beta \sin \theta_J)^2}} = \frac{\beta \sin \theta_J}{\sqrt{A_1^2 + 2\beta A_1 \cos \theta_J + \beta^2}},$$

$$(4.89)$$

the inequalities of (4.88) can be rewritten as

$$0 < \frac{(1 + \beta \cos \theta_J)(A_1 + \beta \cos \theta_J) + (3 + \beta \sin \theta_J)(\beta \sin \theta_J)}{\sqrt{A_1^2 + 2\beta A_1 \cos \theta_J + \beta^2}} < 2$$

$$2 < \frac{(3 + \beta \sin \theta_J)(A_1 + \beta \cos \theta_J) - (1 + \beta \cos \theta_J)(\beta \sin \theta_J)}{\sqrt{A_1^2 + 2\beta A_1 \cos \theta_J + \beta^2}} < \infty.$$

$$(4.90)$$

Dividing the numerator and denominator of each inequality in (4.90) by A_1

and simplifying results in

$$0 < \frac{1 + \beta \cos \theta_J + \dfrac{\beta}{A_1}(\beta + \cos \theta_J + 3 \sin \theta_J)}{\sqrt{1 + 2\left(\dfrac{\beta}{A_1}\right)\cos \theta_J + \left(\dfrac{\beta}{A_1}\right)^2}} < 2$$

$$2 < \frac{3 + \beta \sin \theta_J + \dfrac{\beta}{A_1}(3 \cos \theta_J - \sin \theta_J)}{\sqrt{1 + 2\left(\dfrac{\beta}{A_1}\right)\cos \theta_J + \left(\dfrac{\beta}{A_1}\right)^2}} < \infty. \tag{4.91}$$

Let $P_{c\text{①}}(\theta_J; \beta, A_1)$ denote the conditional probability of correctly detecting signal point ① for given θ_J, β, and A_1. Then, from (4.91),

$$P_{c\text{①}}(\theta_J; \beta, A_1) = \begin{cases} 1; & \text{for values of } \theta_J \text{ such that} \\ & 0 < X_{1,3}(\theta_J; \beta, A_1) < 2 \text{ and} \\ & 2 < Y_{1,3}(\theta_J; \beta, A_1) < \infty \\ 0; & \text{all other values of } \theta_J \text{ in } (0, 2\pi), \end{cases}$$

$$\tag{4.92}$$

where we have defined the generalized functions

$$X_{i,j}(\theta_J; \beta, A_1) = \frac{i + \beta \cos \theta_J + \dfrac{\beta}{A_1}(\beta + i \cos \theta_J + j \sin \theta_J)}{\sqrt{1 + 2\left(\dfrac{\beta}{A_1}\right)\cos \theta_J + \left(\dfrac{\beta}{A_1}\right)^2}}$$

$$Y_{i,j}(\theta_J; \beta, A_1) = \frac{j + \beta \sin \theta_J + \dfrac{\beta}{A_1}(j \cos \theta_J - i \sin \theta_J)}{\sqrt{1 + 2\left(\dfrac{\beta}{A_1}\right)\cos \theta_J + \left(\dfrac{\beta}{A_1}\right)^2}};$$

$$i, j = \pm 1, \pm 3. \tag{4.93}$$

Further defining

$$G(X) \triangleq \frac{1 - \text{sgn } X}{2} = \begin{cases} 1; & X < 0 \\ 0; & X > 0, \end{cases} \tag{4.94}$$

then, for $a > 0$,

$$1 - G(X - a) = \begin{cases} 0; & -\infty < X < a \\ 1; & a < X < \infty \end{cases}$$

$$G(X - a) - G(X) = \begin{cases} 1; & 0 < X < a \\ 0; & \text{otherwise.} \end{cases} \tag{4.95}$$

In view of (4.95), we may rewrite (4.92) as

$$P_{c\circled{1}}(\theta_J; \beta, A_1) = \left[G\left(X_{1,3}(\theta_J; \beta, A_1) - 2\right) - G\left(X_{1,3}(\theta_J; \beta, A_1)\right)\right]$$
$$\times \left[1 - G\left(Y_{1,3}(\theta_J; \beta, A_1) - 2\right)\right]. \tag{4.96}$$

As our next example, consider the problem of correctly detecting signal point ⑩ of Figure 4.11. The vector diagram describing this situation is given in Figure 4.13. Noting again that the shaded area corresponds to the correct decision region, analogous to (4.88), we then have

$$-2 < R_2\cos(\theta_2 - \theta_1) < 0$$
$$2 < R_2\sin(\theta_2 - \theta_1) < \infty. \tag{4.97}$$

Once again expanding the sine and cosine of the difference angle $\theta_2 - \theta_1$ and making use of relations similar to (4.89), we obtain the equivalent inequalities

$$-2 < X_{-1,3}(\theta_J; \beta, A_1) < 0$$
$$2 < Y_{-1,3}(\theta_J; \beta, A_1) < \infty, \tag{4.98}$$

where $X_{i,j}(\theta_J, \beta, A_1)$ and $Y_{i,j}(\theta_J; \beta, A_1)$ are given by (4.93). Noting from (4.94) that, for $a > 0$,

$$G(X) - G(X + a) = \begin{cases} 1; & -a < X < 0 \\ 0; & \text{otherwise} \end{cases} \tag{4.99}$$

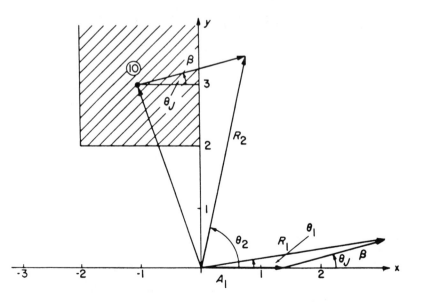

Figure 4.13. A vector diagram of the signal and jammer in the i-th and $(i - 1)$-st transmission intervals (signal point in second quadrant).

then letting $P_{c\text{⑩}}(\theta_J; \beta, A_1)$ denote the conditional probability of correctly detecting signal point ⑩, we have, from (4.98) together with (4.95) and (4.99), that

$$P_{c\text{⑩}}(\theta_J; \beta, A_1) = [G(X_{-1,3}(\theta_J; \beta, A_1)) - G(X_{-1,3}(\theta_J; \beta, A_1) + 2)]$$

$$\times [1 - G(Y_{-1,3}(\theta_J; \beta, A_1) - 2)]. \tag{4.100}$$

At this point, one can write down the remainder of the conditional probabilities of correct decision by inspection. Without going into great detail, the results are given as follows:[6]

$$P_{c\text{②}}(\theta_J; \beta, A_1) = [1 - G(X_{3,3} - 2)][1 - G(Y_{3,3} - 2)]$$

$$P_{c\text{③}}(\theta_J; \beta, A_1) = [G(X_{1,1} - 2) - G(X_{1,1})][G(Y_{1,1} - 2) - G(Y_{1,1})]$$

$$P_{c\text{④}}(\theta_J; \beta, A_1) = [1 - G(X_{3,1} - 2)][G(Y_{3,1} - 2) - G(Y_{3,1})]$$

$$P_{c\text{⑤}}(\theta_J; \beta, A_1) = [G(X_{1,-1} - 2) - G(X_{1,-1})]$$

$$\times [G(Y_{1,-1}) - G(Y_{1,-1} + 2)]$$

$$P_{c\text{⑥}}(\theta_J; \beta, A_1) = [1 - G(X_{3,-1} - 2)][G(Y_{3,-1}) - G(Y_{3,-1} + 2)]$$

$$P_{c\text{⑦}}(\theta_J; \beta, A_1) = [G(X_{1,-3} - 2) - G(X_{1,-3})]G(X_{1,-3} + 2)$$

$$P_{c\text{⑧}}(\theta_J; \beta, A_1) = [1 - G(X_{3,-3} - 2)]G(Y_{3,-3} + 2)$$

$$P_{c\text{⑨}}(\theta_J; \beta, A_1) = G(X_{-3,3} + 2)[1 - G(Y_{-3,3} - 2)]$$

$$P_{c\text{⑪}}(\theta_J; \beta, A_1) = G(X_{-3,1} + 2)[G(Y_{-3,1} - 2) - G(Y_{-3,1})]$$

$$P_{c\text{⑫}}(\theta_J; \beta, A_1) = [G(X_{-1,1}) - G(X_{-1,1} + 2)]$$

$$\times [G(Y_{-1,1} - 2) - G(Y_{-1,1})]$$

$$P_{c\text{⑬}}(\theta_J; \beta, A_1) = G(X_{-3,-1} + 2)[G(Y_{-3,-1}) - G(Y_{-3,-1} + 2)]$$

$$P_{c\text{⑭}}(\theta_J; \beta, A_1) = [G(X_{-1,-1}) - G(X_{-1,-1} + 2)]$$

$$\times [G(Y_{-1,-1}) - G(Y_{-1,-1} + 2)]$$

$$P_{c\text{⑮}}(\theta_J; \beta, A_1) = G(X_{-3,-3} + 2)G(Y_{-3,-3} + 2)$$

$$P_{c\text{⑯}}(\theta_J; \beta, A_1) = [G(X_{-1,-3}) - G(X_{-1,-3} + 2)]G(Y_{-1,-3} + 2). $$

$$\tag{4.101}$$

Since, as previously mentioned, the jammer phase θ_J is uniformly distributed in the interval $(0, 2\pi)$, the average symbol error probability (condi-

---

[6]For simplicity of notation, we delete the dependence of $X_{i,j}$ and $Y_{i,j}$ on θ_J, β, and A_1.

tioned on A_1) is then given by

$$P_s(\beta, A_1) = 1 - \frac{1}{2\pi} \int_0^{2\pi}\left[1 - \frac{1}{16}\sum_{k=1}^{16} P_{c\text{\textcircled{k}}}(\theta_J; \beta, A_1)\right] d\theta_J. \quad (4.102)$$

Finally, making use of (4.86) and (4.87) and the fact that only the fraction ρ of the total number of hop intervals are jammed, then the average unconditional symbol error probability is given by

$$P_s = \rho\left\{\frac{1}{4}P_s\left(\sqrt{\frac{5}{2}\left(\frac{N_J}{\rho E_b}\right)},\sqrt{2}\right) + \frac{1}{2}P_s\left(\sqrt{\frac{5}{2}\left(\frac{N_J}{\rho E_b}\right)},\sqrt{10}\right)\right.$$
$$\left. + \frac{1}{4}P_s\left(\sqrt{\frac{5}{2}\left(\frac{N_J}{\rho E_b}\right)},\sqrt{18}\right)\right\}. \quad (4.103)$$

Before leaving this subject, we note that the result of substituting (4.96), (4.100), and (4.101) in (4.102) can be put into a compact form. In particular, by replacing each $G(X)$ term in (4.96), (4.100), and (4.101) with its equivalent form $1 - G(-X)$ and summing all terms, we obtain the following result:

$$P_s(\beta, A_1) = \frac{1}{2\pi}\int_0^{2\pi} P_s(\theta_J; \beta, A_1)\, d\theta_J, \quad (4.104)$$

where

$$P_s(\theta_J; \beta, A_1)$$

$$\triangleq 1 - \frac{1}{16}\sum_{k=1}^{16} P_{c\text{\textcircled{k}}}(\theta_J; \beta, A_1)$$

$$= \frac{1}{16}\sum_{m=\pm 1}\sum_{j=\pm 1, \pm 3}\sum_{l=0, \pm 2}\left[G\left(mX_{m(1-l),\, j} + l\right) + G\left(mY_{j,\, m(1-l)} + l\right)\right]$$

$$- \frac{1}{16}\sum_{m=\pm 1}\sum_{n=\pm 1}\sum_{l=0, \pm 2}\sum_{k=0, \pm 2} G\left(mX_{m(1-l),\, n(1-k)} + l\right)$$

$$\times G\left(nY_{m(1-l),\, n(1-k)} + k\right). \quad (4.105)$$

For the more general case of FH/DQASK-K^2 with arbitrary K, the summation on j is for values $j = \pm 1, \pm 3, \ldots, \pm(K-1)$, while the summations on l and k are for values $l, k = 0, \pm 2, \pm 4, \ldots, \pm(K-2)$.

It is of interest to evaluate the limit of P_s of (4.103) as E_b/N_J approaches zero ($\beta \to \infty$). From (4.93), we first note that

$$\lim_{\beta \to \infty} X_{i,\, j}(\theta_J; \beta, A_1) = \infty$$

$$\lim_{\beta \to \infty} Y_{i,\, j}(\theta_J; \beta, A_1) = A_1\sin\theta_J + j\cos\theta_J - i\sin\theta_J. \quad (4.106)$$

Then, substituting (4.106) into (4.105) and simplifying gives

$$\lim_{\beta \to \infty} P_s(\theta_J; \beta, A_1)$$

$$= \frac{1}{16} \left\{ 12 + \sum_{m=\pm 1} \sum_{l=0, \pm 2} G\big(m[A_1 \sin \theta_J + m(1-l)\cos \theta_J - 3\sin \theta_J] + l\big) \right\},$$

$$(4.107)$$

which when averaged over θ_J results in

$$\lim_{\beta \to \infty} P_s(\beta, A_1) = \frac{1}{2\pi} \int_0^{2\pi} \lim_{\beta \to \infty} P_s(\theta_J; \beta, A_1) \, d\theta_J$$

$$= \frac{15}{16} \text{ independent of } A_1. \qquad (4.108)$$

Thus, applying (4.108) to (4.103) gives the final desired result, namely,

$$\lim_{\beta \to \infty} P_s = \lim_{E_b/N_J \to 0} P_s = \rho\left(\frac{15}{16}\right). \qquad (4.109)$$

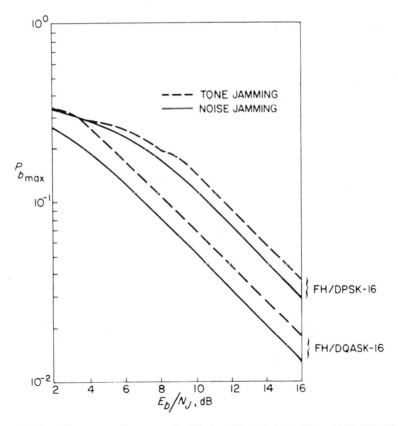

Figure 4.14.   Worst case P_b versus E_b/N_J for FH/DQASK-16 and FH/DPSK-16.

Finally, using the same relation between average symbol and bit error probabilities as for FH/MDPSK, namely (4.15), then the worst case jamming strategy for FH/DQASK-16 can be determined to be

$$
\rho_{wc} = \begin{cases} \dfrac{2.4}{E_b/N_J}; & E_b/N_J \geq 2.4 \\[2mm] 1; & E_b/N_J < 2.4 \end{cases} \tag{4.110}
$$

and

$$
P_{b_{max}} = \begin{cases} \dfrac{0.67}{E_b/N_J}; & E_b/N_J \geq 2.4 \\[2mm] \dfrac{8}{15} P_s \Big|_{\rho=1}; & E_b/N_J < 2.4 \end{cases} \tag{4.111}
$$

where $P_s\big|_{\rho=1}$ is given by (4.103) with $\rho = 1$. Figure 4.14 illustrates this worst case jammer bit error probability performance.

4.5 PERFORMANCE OF FH/DQASK IN THE PRESENCE OF PARTIAL-BAND NOISE JAMMING

Since we are ignoring the background AWGN, the results for the performance in the presence of a partial-band noise jammer are then directly obtained from the Gaussian results in Section 4.3 by replacing N_0 with N_J/ρ. In particular, the average symbol error probability performance is given by (4.80), multiplied by ρ, namely

$$
P_s = \rho \int_{-\pi}^{\pi} P_s\big(\eta_a^{(i-1)}\big) p\big(\eta_a^{(i-1)}\big) \, d\eta_a^{(i-1)}, \tag{4.112}
$$

where the dependence of the integrand on ρ is entirely contained within the parameter Δ of (4.75) which now becomes

$$
\Delta = \sqrt{\frac{6\rho \log_2 K}{K^2 - 1} \left(\frac{E_b}{N_J} \right)}. \tag{4.113}
$$

Once again using (4.15) to obtain P_b from P_s, we obtain the worst case jammer performance given by

$$
\rho_{wc} = \begin{cases} \dfrac{5.0}{E_b/N_J}; & E_b/N_J \geq 5.0 \\[2mm] 1; & E_b/N_J < 5.0 \end{cases} \tag{4.114}
$$

and

$$
P_{b_{max}} = \begin{cases} \dfrac{0.51}{E_b/N_J}; & E_b/N_J \geq 5.0 \\[2mm] \dfrac{8}{15} P_s \Big|_{\rho=1}; & E_b/N_J < 5.0 \end{cases} \tag{4.115}
$$

where now $P_s|_{\rho-1}$ corresponds to (4.112) with $\rho = 1$. The behavior of $P_{b_{max}}$ of (4.115) as a function of E_b/N_J in dB is also illustrated in Figure 4.14.

Finally, for the purpose of comparison, the results of (4.43) and (4.50), corresponding to the worst case tone and noise jammer performances of FH/DPSK-16, have been superimposed on Figure 4.14. We note that, asymptotically, FH/DQASK-16 outperforms FH/DPSK-16 by 3.283 dB for worst case tone jamming and 3.55 dB for worst case noise jamming.

4.6 REFERENCES

[1] S. W. Houston, "Modulation techniques for communication, Part I: Tone and noise jamming performance of spread spectrum M-ary FSK and 2, 4-ary DPSK waveforms," *NAECON'75 Record*, pp. 51–58.

[2] M. K. Simon, G. K. Huth, and A. Polydoros, "Differentially coherent detection of QASK for frequency-hopping systems; Part I: Performance in the presence of a Gaussian noise background," *IEEE Trans. Commun.*, COM-30, no. 1, pp. 158–164, January 1982.

[3] M. K. Simon, "Differentially coherent detection of QASK for frequency-hopping systems; Part II: Performance in the presence of jamming," *IEEE Trans. Commun.*, COM-30, no. 1, pp. 164–172, January 1982.

[4] M. K. Simon, "The Performance of M-ary DPSK/FH in the Presence of Partial-Band Multitone Jamming," *IEEE Trans. Commun.*, Special Issue on Spread Spectrum Communications, COM-30, no. 5, pp. 953–958, May 1982.

[5] W. C. Lindsey and M. K. Simon, *Telecommunication Systems Engineering*, Englewood Cliffs, NJ: Prentice-Hall, 1973, Chapter 5.

[6] J. T. Fleck and E. A. Trabka, "Error Probabilities of Multiple-State Differentially Coherent Phase Shift Keyed Systems in the Presence of White, Gaussian Noise," Detect Memo No. 2A in Investigation of Digital Data Communication Systems, Report No. UA-1420-S-1, J. G. Lawton, ed., Cornell Aeronautical Laboratory, Inc., Buffalo, NY, January 1961. Available as ASTIA Document No. AD 256 584.

[7] J. H. Roberts, S. O. Rice, and R. F. Pawula, "Distribution of the phase angle between two vectors perturbed by Gaussian noise," *IEEE Trans. Commun.*, COM-30, no. 8, pp. 1828–1841, August 1982.

[8] M. K. Simon and J. G. Smith, "Carrier synchronization and detection of QASK signal sets," *IEEE Trans. Commun.*, COM-22, no. 2, pp. 98–106, February 1974.

VOLUME I
INDEX

VOLUME II
INDEX